AF576040

Wine Aromas

Wine Aromas

Editors

Bin Tian
Jicheng Zhan

Basel • Beijing • Wuhan • Barcelona • Belgrade • Novi Sad • Cluj • Manchester

Editors

Bin Tian
Department of Wine, Food and Molecular Biosciences
Lincoln University
Lincoln
New Zealand

Jicheng Zhan
College of Food Science and Engineering
China Agricultural University
Beijing
China

Editorial Office
MDPI
St. Alban-Anlage 66
4052 Basel, Switzerland

This is a reprint of articles from the Special Issue published online in the open access journal *Fermentation* (ISSN 2311-5637) (available at: https://www.mdpi.com/journal/fermentation/special_issues/wine_aromas).

For citation purposes, cite each article independently as indicated on the article page online and as indicated below:

Lastname, A.A.; Lastname, B.B. Article Title. *Journal Name* **Year**, *Volume Number*, Page Range.

ISBN 978-3-0365-9559-7 (Hbk)
ISBN 978-3-0365-9558-0 (PDF)
doi.org/10.3390/books978-3-0365-9558-0

© 2023 by the authors. Articles in this book are Open Access and distributed under the Creative Commons Attribution (CC BY) license. The book as a whole is distributed by MDPI under the terms and conditions of the Creative Commons Attribution-NonCommercial-NoDerivs (CC BY-NC-ND) license.

Contents

MDPI

Article

Characterisation of Tannin and Aroma Profiles of Pinot Noir Wines Made with or without Grape Pomace

Pradeep M. Wimalasiri [1], Jicheng Zhan [2] and Bin Tian [1,*]

[1] Faculty of Agriculture and Life Sciences, Lincoln University, Lincoln 7647, New Zealand
[2] Beijing Key Laboratory of Viticulture and Enology, College of Food Science and Nutritional Engineering, China Agricultural University, Beijing 100083, China
* Correspondence: bin.tian@lincoln.ac.nz; Tel.: +64-3-423-0645

Abstract: The contribution of grape pomace on tannin concentration, tannin composition and aroma profile of Pinot noir wine was studied using different fermentation media to make up four treatments: GJ-P, grape juice plus pomace; MJ-P, model juice plus pomace; GJ, grape juice; MJ, model juice. The MJ-P treatment showed significantly lower amounts of tannins, mean degree of polymerisation (mDP), similar amounts of anthocyanin, and a similar secondary aroma profile compared to the GJ-P treatment. Grape pomace addition significantly increased the tannin concentration in wines. This study was also revealed the importance of phenolics present in grape juice in tannin polymerisation and final tannin concentration in wines. Grape pomace addition significantly reduced some important aroma compounds such as acetate esters (except ethyl acetate), most of the volatile fatty acids, a few ethyl esters and β-damascenone but increased some primary aromas in wines due to the presence of their aroma precursors in skins. Hence, these results indicate that grape pomace may bind or delay the release of some aroma compounds and/or lose these compounds during cap management in GJ-P and MJ-P treatments compared to the respective juice treatments.

Keywords: aroma profile; grape pomace; model juice; Pinot noir; tannin

Citation: Wimalasiri, P.M.; Zhan, J.; Tian, B. Characterisation of Tannin and Aroma Profiles of Pinot Noir Wines Made with or without Grape Pomace. *Fermentation* **2022**, *8*, 718. https://doi.org/10.3390/fermentation8120718

Academic Editor: Manuel Malfeito Ferreira

Received: 31 October 2022
Accepted: 6 December 2022
Published: 8 December 2022

Publisher's Note: MDPI stays neutral with regard to jurisdictional claims in published maps and institutional affiliations.

Copyright: © 2022 by the authors. Licensee MDPI, Basel, Switzerland. This article is an open access article distributed under the terms and conditions of the Creative Commons Attribution (CC BY) license (https://creativecommons.org/licenses/by/4.0/).

1. Introduction

Tannins, anthocyanin, and the aroma profile of red wines collectively determine the final wine quality. Wine phenolics mainly originated from grape skins, seeds, and stems if used in the winemaking process, of which extraction could be influenced by various winemaking techniques [1–3]. Volatile aroma compounds in wine are usually divided into three groups based on their origin: primary aroma compounds are derived from grapes; secondary aroma compounds are mainly originated from the metabolism of yeast and bacteria; and tertiary aroma compounds are formed from barrel maturation or bottle ageing [4,5].

In winemaking, tannin origin, composition, and extractability determine the final phenolic and colour attributes of the wine [6–8]. Ultimately, wine colour, flavour and mouthfeel are largely determined by the relative proportions of tannin extracted from different grape tissues in the winemaking process. Making wine from Pinot noir grapes has long been a challenge due to its delicate flavour, light colour, and sometimes poor ageing potential of the wine [2]. According to previous studies, Pinot noir grapes have a higher seed tannin concentration compared to skin tannin levels, which results in a lower skin-to-seed tannin ratio [2,9,10]. This tannin distribution in Pinot noir berries may contribute to lower tannin concentration in wine and increase seed-originated tannin proportion in the wine [11,12]. Previous studies suggested that the perception of astringency in wine is related to berry tannin concentration, composition, and molecular size of the tannins. Moreover, a greater degree of polymerization and a greater percentage of galloylation (G%), are reported to provoke the sensation of astringency in wine, and lower molecular weight tannins could contribute to bitterness in wines [6,13,14]. Hence, the extraction of optimum proportions of tannins from different tissues is important in wine quality. In the present study, wines prepared with grape juice and model juice with and without grape pomace help to understand the effect of grape juice and pomace on final wine tannin concentration and composition.

The factors affecting the formation of wine aromas are complex. In general, wine aroma is greatly influenced by the grape variety, relative proportions of grape skins, vintage, viticultural practices, maceration and winemaking practices, regional variation, etc. [15–19]. During fermentation, yeast converts grape sugars into alcohol and fermentation drives complex chemical reactions that affect the aroma and flavour of the wine. Primary aromas such as monoterpenes, C_{13}-norisoprenoids, methoxypyrazines, some volatile phenols such as eugenol, C_6 compounds and thiols are mainly found in grape skin, in free volatile form or bound into non-volatile precursors such as glycoside. Later, glycosidically bounded aroma compounds are released during fermentation or ageing and contribute to primary aroma and wine quality [15,20]. Most of the previous studies have also confirmed that there is a positive correlation between skin contact time and the concentration of primary aromas observed in final wine [21]. In contrast, fermentation-derived aroma compounds make up the largest percentage of the total aroma composition of wine comprising volatile fatty acids, higher alcohols, ethyl esters and acetate esters but most of them do not significantly contribute to the wine aroma due to their higher aroma thresholds in wine [21]. However, the dividing line between primary aromas and secondary aromas is not clear-cut in terms of their origin. For example, yeast also could indirectly contribute to the primary aromas by converting odourless grape-derived precursors into important aroma compounds through enzymatic hydrolysis in wine [22].

There is very limited studies investigating the effect of grape pomace and juice on Pinot noir wine composition. In this study, Pinot noir grape pomace and juice were separated, and model juice was used to substitute the grape juice, in order to have a better understanding on the contribution of individual grape components (grape juice and grape pomace) to the tannin and aroma composition of resultant Pinot noir wine.

2. Materials and Methods

2.1. Chemicals

All chemicals used for model juice preparation, phenolic analysis and HPLC were purchased from Sigma-Aldrich (Auckland, New Zealand). Purchased aroma standards used in GC-MS include ethyl decanoate and *cis*-3-hexen-1-ol (Fluka Chemicals, Ronkonkoma, NY, USA), deuterated benzaldehyde-d6 and hexanoic-d11-acid (Isotec, St. Louis, MO, USA). The deuterated octanoic-d2-acid, isoamyl-d3-acetate, ethyl-d5-hexanoate, hexyl-d3-acetate, ethyl-d5-octanoate, d5-ethyl butyrate, and d5-ethyl decanoate were previously synthesized and stored at Lincoln University, New Zealand [17].

2.2. Microvinification

Pinot noir grapes were harvested in 2020 from the Akarua vineyard in Central Otago, New Zealand. In this study, four treatments of wines were prepared through fermentation of: (1) grape juice and pomace (GJ-P), (2) model juice and grape pomace (MJ-P), (3) grape juice (GJ), and (4) model juice (MJ). Winemaking in each treatment was triplicated in 2 L plastic containers.

Hand-destemmed grapes (1.6 kg) were used in each replicate of the GJ-P treatment. Another 1.6 kg of hand-destemmed grapes was pressed to obtain 1 L of grape juice that was used for each replicate of the GJ treatment, and the pressed grape pomace was combined with 1 L of the model juice to prepare each replicate of the MJ-P treatment. Each replicate of the MJ treatment was prepared using 1 L of model juice without pomace.

Model juice included three different solutions, which were sterilized separately at 121 °C for 20 min and then combined aseptically [23]: solution A (110 g/L D-glucose, 110 g/L D-fructose, 10 mg/L ergosterol, and 1 mL/L Tween 80), solution B (6 g/L L-(+)-

tartaric acid, 3 g/L L-(−)-malic acid, and 0.5 g/L citric acid), and solution C (1.7 g/L yeast nitrogen base with amino acids, 2 g/L casamino acids, 0.2 g/L $CaCl_2$, 0.8 g/L arginine-HCl, 1 g/L L-(2)-proline, and 0.1 g/L L-(2)-tryptophan). The final pH was adjusted to 3.21 using 2 M sodium hydroxide to match the grape juice.

The EC1118 yeast (Lallemand, Montreal, QC, Canada) was inoculated at 0.25 g/L for all winemaking treatments. The fermentation temperature was maintained in the range of 25 to 30 °C. The weights of ferments were recorded to monitor the progress of fermentation. For the fermentations with addition of pomace, the cap was plunged twice a day until the end of fermentation. Free-run wines were then collected and added with 50 mg/L of SO_2. After two days of cold settling at 4 °C, wine samples were racked off the lees and bottled.

2.3. *Oenological Parameters*

Total soluble solids (TSS), residual sugar, alcohol content, pH, and titratable acidity were measured using standard protocols [24]. Briefly, the total soluble solid (TSS) in the juice was analysed using an Atago 101 digital handheld refractometer with an automatic temperature compensation function (Atago Co., Ltd., Tokyo, Japan). The Rebelein method was used to determine the residual sugar concentration in wines. Alcohol concentration was measured by an ebulliometer (Dujardin Salleron, France). Titratable acidity and pH were determined using a Suntex SP-701 pH meter. Titratable acidity (TA) was measured by titrating 10 mL of juice/wine with a standardised 0.1 N sodium hydroxide (NaOH) solution to a pH endpoint of 8.2. Yeast assimilable nitrogen (YAN) was determined using a commercial enzyme test kit (Vintessential Laboratories, Dromana, Victoria, Australia).

2.4. *Spectrophotometric Analysis*

Wine samples were centrifuged at 3000× *g* for 5 min (Model Heraeus Multifuge X1R, Thermo Fisher Scientific, Lower Saxony, Germany) prior to analysis. Spectrophotometric measurements were carried out on a Shimadzu 1800 UV-Visible spectrophotometer (Shimadzu, Tokyo, Japan).

2.4.1. Methyl Cellulose Precipitable (MCP) Tannin

Tannin in wine samples was determined using the 1 mL methyl cellulose precipitation (MCP) method [25]. The absorbance of the sample and the respective control at 280 nm was measured using the spectrophotometer. The absorbance difference between treatment and control was used to calculate the concentration of MCP tannins against an epicatechin calibration curve.

2.4.2. Total Phenolics

Total phenolic content was performed using the microscale Folin–Ciocalteu assay [26]. The sample and standard solutions were measured at 765 nm using a UV–Vis spectrophotometer.

2.5. *Analysis of Flavan-3-Ols Monomers by HPLC*

Monomeric phenolic fractions in wine samples were separated using the optimised solid-phase extraction (SPE) method [27]. In brief, an Oasis HLB cartridge (3 mL, 60 mg, 30 µm) (Waters, Rydalmere, NSW, Australia) was conditioned with 2 mL of methanol followed by 2 mL of water. Wine samples were centrifuged at 3000× *g* for 5 min before loading into the SPE cartridge. The wine sample (1 mL) was then loaded under gravity and completely dried with a gentle stream of nitrogen. The monomeric phenolic fraction was eluted by adding 40 mL of 95% acetonitrile and 5% 0.01 M hydrochloric acid. The elute was vacuum evaporated completely at 36 °C and then dissolved in 1 mL of 10% ethanol/0.1% formic acid.

HPLC analysis was carried out according to the previously published method [28]. In brief, the monomeric phenolic fraction (10 µL) was loaded at 1 mL/min and the elution of monomeric phenolics was monitored by absorbance at 280, 320, 360 and 520 nm in the diode array detector (DAD) and the corresponding excitation at 280 nm and emission at 320 nm in the fluorescence detector (FLD). Phenolic compounds were identified by comparing their retention times and the spectra with those of standards. Quantification of phenolic compounds was carried out by area measurements at 280 nm, 320 nm and FLD separately. Quantitative assays were achieved using external calibration curves for all standard phenolics by the dissolution of the standard solution accordingly.

2.6. Phloroglucinolysis

Wine tannin was characterised by acid catalysis in the presence of excess phloroglucinol followed by reversed-phase HPLC [29]. Phloroglucinolysis was carried out with modifications, which provided information on subunit composition, conversion yield, and mean degree of polymerization (mDP). Wine samples (4 mL) were rotary evaporated to 0.2–0.3 mL at 40 °C to remove ethanol, and re-dissolved in 4 mL of deionised water. After that, the sample was loaded onto a pre-conditioned 0.36 g C18 SEP-PAK cartridge (WAT051910, Global Science, Auckland, New Zealand) with 5 mL of methanol, 7.5 mL ethyl acetate and 7.5 mL of deionised water. The loaded C18 cartridge was washed with 7.5 mL of deionised water and allowed to dry with nitrogen gas at a 3 L/min flow rate for 60 min. After that, 5 mL of ethyl acetate was passed through the cartridge to wash off the monomeric phenols and the required polymeric fraction was eluted with 5 mL of methanol. This methanolic solution was rotary evaporated at 40 °C and reconstituted with methanol to a final volume of 1 mL. Then, 0.5 mL of this prepared methanolic solution was reacted with an equal volume of phloroglucinol reagent (A solution of 50 g/L phloroglucinol in methanol, containing 0.1 M hydrochloric acid, and 10 g/L ascorbic acid) in a screw cap tube in a water bath at 50 °C for 20 min. To stop the reaction, 5 mL of 40 mM sodium acetate solution was added into the screw cap tube and vortexed the mixture for dissolving. This solution was filtered through a 0.45 μm pore size 13 mm diameter PTFE filter into the HPLC vial (the first few drops discarded).

Reversed-phase HPLC analysis was carried out using an Agilent 1100 series HPLC (Waldbronn, Germany) equipped with a quaternary pump, diode array detector (DAD) and fluorescence detector (FLD) was used to identify and quantify proanthocyanidin cleavage products. A sample fraction of 40 μL was injected and separated using an ACE C18-PFP 150 × 4.6 mm, particle size 3 μm column (Advanced Chromatography Technologies, Aberdeen, Scotland) that was thermostat at 25 °C. Two mobile phases were used: 2% v/v aqueous acetic acid (A), and 2% v/v acetic acid in methanol (B). The flow rate was set at 0.8 mL/min, with the mobile phase set initially as a linear pump gradient from 5 to 10% B in 5 min, a gradient from 10 to 40% B in 25 min, and a gradient from 40 to 100% B in 1 min and held for 5 min, then reduced to 5% B in 1 min and held for 5 min, before equilibrating the column to the initial running conditions. The detection and quantification of proanthocyanidin cleavage products were conducted at 280 in the DAD. The corresponding excitation at 280 nm and emission at 320 nm in the FLD were used for assisting in the identification and confirmation.

To estimate the molarity of proanthocyanidin cleavage products (terminal and extension subunits), the areas obtained from HPLC analysis were divided by their respective molar response factors (relative to (+)-catechin) [29] to convert into catechin equivalent molar response areas. Then, the catechin equivalent molar response areas were used to calculate the molarity of proanthocyanidin cleavage products using a catechin calibration curve. After that, the relative molar composition of terminal subunits (flavan-3-ol monomers) and extension subunits (phloroglucinol adducts) were calculated separately. The mDP was calculated as the sum of all subunits (terminal and extension subunits in moles) divided by the sum of all terminal subunits (in moles). The per cent conversion yield was determined as the proportion of epicatechin equivalent total tannin concentration determined by phloroglucinolysis divided by the epicatechin equivalent total tannin results from the MCP tannin assay.

2.7. Volatile Aroma Compounds

Volatile aroma compounds were analysed according to the methods described previously [17,30]. Three different methods were used to determine three groups of aroma compounds: esters and alcohols, volatile fatty acids, and low-concentration compounds.

The wine sample (0.9 mL) was pipetted together with 8.06 mL of pH 3.5 acidified water, followed by 40 µL of deuterated internal standard solution and 4.5 g of sodium chloride into a 20 mL SPME vial. The samples were incubated and agitated for 10 min at 60 °C. The SPME fibre (Stableflex DVB/CAR/PDMS, Sigma-Aldrich, St. Louis, MO, USA) was conditioned for 60 min at 60 °C. Desorption of the fibre occurred in the injection port at 270 °C for 5 min. The GC–MS was equipped with dual columns in series: a Rtx-Wax column (30.0 m × 0.25 mm ID × 0.5 µm film thickness, polyethene glycol, Restek, Bellefonte, PA, USA) and a Rxi–1 MS column (15 m × 0.25 mm ID × 0.5 µm film thickness, 100% dimethyl polysiloxane, Restek). Helium was used as the carrier gas with a linear velocity of 33.5 cm/s. Splitless injection was used for the initial 3 min and then switched to a 20:1 split ratio. The GC oven temperature was held at 35 °C for 3 min, heated to 250 °C at 4 °C/min and then held for 10 min. The interface and MS source were set to 250 °C and 200 °C, respectively. The MS source was operated in electron impact (EI) mode with ionization energy of 70 eV. Analysis of the chromatograms was performed on GCMS Solution software, version 2.5 (Shimadzu, Auckland, New Zealand).

For volatile fatty acids analysis, the HS-SPME extraction and GC operational conditions were changed. The helium gas flow was set at a constant linear velocity of 46.8 cm/s. The GC oven temperature was held at 50 °C for 3 min, then increased to 240 °C at 10 °C/min, further increased to 250 °C at 30 °C/min and held for 5 min. The column configuration and other operating conditions remained the same as for esters and alcohol analysis.

For low-concentration compounds analysis, the method is similar to that for esters and alcohols, but the acquisition mode was changed to the selected ion monitoring (SIM) to increase the sensitivity, and the GC oven temperature ramp was modified to 35 °C for 3 min, increased to 105 °C at 3 °C/min and held for 10 min, increased to 140 °C at 2 °C/min and held for 10 min, and finally increased to 250 °C at 4 °C/min and held for 10 min.

2.8. Odour Activity Values

The odour activity value (OAV) of each aroma compound was calculated as the ratio between the concentration of an individual aroma compound and the perception threshold of the corresponding compound found in the literature. All aroma compounds with OAV above 0.1 were considered significant aroma compounds in this study. Important aroma compounds were categorised into seven aroma series (fruity, floral, spicy, chemical, fatty, woody, and green) based on their odour description [28]. Woody, green, and spicy aroma series were combined to make herbaceous aroma series in this work due to the lower OAVs detected in wines.

2.9. Statistical Analysis

Data were presented as mean ± SD of three replicates. Data obtained from HPLC, GC-MS, and OAVs were analysed by one-way analysis of variance (ANOVA) with a post hoc analysis using the Tukey's honestly significant difference (HSD) test at a significance level of 0.05 (Minitab Inc., State College, PA, USA, version 18.1). Tannin composition data from phloroglucinolysis was analysed using a two-sample *t*-test at 0.05 level of significance using Minitab 18. All aroma compounds analysed in this study were subjected to principal component analysis (PCA). R statistical software (R Core Team 2019, Vienna, Austria) and mixOmics package from Bioconductor (https://bioconductor.org/ accessed on 21 September 2022) was used for statistical analysis.

3. Results and Discussion

3.1. Oenological Testing

At harvest, the total soluble solids (TSS) in grapes were measured at 23.4°Brix, and TSS in model juice was determined at 22.2°Brix. YAN in grape juice and model juice was measured at 236 mg/L and 362 mg/L, respectively. The pH and TA in juice and wine, alcohol content, residual sugar, total phenolics, and tannins in wine are shown in Table 1. The GJ-P wine showed significantly higher pH, total phenolics and tannin concentration compared to all other treatments. The pH and TA of the wines ranged from 3.40 to 3.16 pH and 10.91–8.54 g/L TA. Comparatively higher pH in GJ-P is likely due to the extraction of potassium from grape pomace during fermentation. However, when comparing MJ-P and GJ treatments they did not show a significant difference in pH between them because the model juice pH had to be adjusted to match grape juice pH by adding 2 M sodium hydroxide before inoculating and it ultimately caused significantly lower TA in juice and wine. The highest TA was observed in GJ treatment as expected because of the absence of grape pomace. The alcohol content ranged from 10.9 to 13.4% and reflected the differences in juice soluble solids and final residual sugar amounts in the wines.

Table 1. General oenological parameters.

	GJ-P	MJ-P	GJ	MJ
Juice				
pH	3.20 ± 0.02 a	3.22 ± 0.01 b	3.13 ± 0.06 c	3.21 ± 0.00 b
TA (g/L)	9.57 ± 0.18 b	6.38 ± 0.12 b	9.70 ± 0.22 a	6.10 ± 0.00 c
Wine				
pH	3.40 ± 0.03 a	3.22 ± 0.02 b	3.16 ± 0.02 c	3.23 ± 0.02 b
TA (g/L)	9.62 ± 0.19 b	8.34 ± 0.08 c	10.91 ± 0.58 a	8.54 ± 0.00 c
Alcohol (%)	13.4 ± 0.2 a	10.9 ± 0.5 b	14.2 ± 0.2 a	11.0 ± 0.4 b
Residual sugar (g/L)	2.78 ± 0.42 b	2.22 ± 0.08 b	1.58 ± 0.16 c	3.70 ± 0.13 a
Total phenolics (mg/L)	2705 ± 69 a	2152 ± 34 b	274 ± 13 c	127 ± 3 d
Tannin (mg/L)	1225 ± 119 a	821 ± 22 b	ND	ND

ND-Not Detected; different lowercase letters in rows indicate a significant difference among treatments ($p < 0.05$). GJ-P, grape juice plus pomace; MJ-P, model juice plus pomace; GJ, grape juice; MJ, model juice.

Comparing to MJ-P treatment, GJ-P treatment showed significantly higher MCP tannin and total phenolic concentrations in the wine. This may be due to two reasons. First, the phenolics present in grape juice, which could be involved in the polymerisation of tannins by interacting with phenolics extracted from skins [31]. Second, the higher concertation of alcohol may have increased the extraction of tannins from grape skin and seeds [32,33]. As expected, MJ and GJ treatments showed a lot lower total phenolic concertation and zero MCP tannin concertation compared to the wines fermented with grape pomace, as vast majority of the phenolics in the wine are derived from the skin and seed tissues [2,34]. The total phenolics observed in MJ treatment were mainly because the Folin–Ciocalteu method is an antioxidant assay, which measures the reductive capacity of antioxidants in the sample, such as amino acids, sugars, etc. [35].

3.2. Monomeric Phenolics Composition

Concentrations of monomeric phenolics determined by HPLC were shown in Table 2. The flavan-3-ols, predominantly catechin and epicatechin represented 52–54% of total monomeric phenolics quantified by HPLC in GJ-P and MJ-P treatments. Significantly higher monomeric flavan-3-ol concentrations were observed in the GJ-P treatment and reflected the same trend as that of tannin and total phenolics concentrations observed in the wines.

Table 2. Monomeric phenolic composition of the wines.

Phenolics (mg/L)	GJ-P	MJ-P	GJ	MJ
Flavan-3-ols				
Catechin	356 ± 16 a	306 ± 24 b	2.40 ± 0.37 c	ND
Epicatechin	95.7 ± 6.8 a	81.8 ± 5.3 b	ND	ND
Procyanidin B1	24.2 ± 1.3 a	17.3 ± 1.8 b	ND	ND
Procyanidin B2	21.9 ± 1.4 a	13.4 ± 1.4 b	ND	ND
Anthocyanins				
Delphenidin-3-*O*-glucoside	26.7 ± 2.1 a	31.2 ± 2.2 a	ND	ND
Cyanidin-3-*O*-glucoside	1.04 ± 0.11 a	1.21 ± 0.18 a	ND	ND
Malvidin-3-*O*-glucoside	217 ± 4 a	196 ± 22 a	10.7 ± 0.8 b	ND
Peonidin-3-*O*-glucoside	21.8 ± 1.2 a	21.8 ± 3.5 a	0.86 ± 0.17 b	ND
Flavonols				
Kaempferol	0.72 ± 0.54 ab	1.24 ± 0.19 a	0.14 ± 0.07 b	ND
Quercetin	5.45 ± 0.78 a	5.58 ± 0.58 a	0.45 ± 0.20 b	ND
Rutin	0.50 ± 0.02 a	0.78 ± 0.33 a	ND	ND
Hydroxybenzoic acids				
Gallic acid	13.8 ± 1.1 a	12.3 ± 0.6 a	0.35 ± 0.05 b	0.18 ± 0.06 b
Protocatechuic acid	1.52 ± 0.04 a	0.97 ± 0.07 b	0.67 ± 0.07 c	ND
Syringic acid	3.84 ± 1.30 a	2.78 ± 0.94 ab	0.95 ± 0.12 bc	ND
Vanilic acid	3.82 ± 0.07 a	1.63 ± 0.66 b	1.27 ± 0.11 a	3.70 ± 0.17 b
Hydroxycinnamic acids				
Caffeic acid	0.00 ± 0.00 b	ND	0.09 ± 0.01 a	ND
Caftaric acid	2.60 ± 0.13 b	3.20 ± 0.09 a	0.85 ± 0.06 c	ND
Cinnamic acid	0.10 ± 0.00 b	0.13 ± 0.02 a	ND	ND
Ferulic acid	0.05 ± 0.09 b	0.29 ± 0.04 a	0.03 ± 0.05 b	ND
p-Coumaric acid	0.00 ± 0.00 b	0.00 ± 0.00 b	0.28 ± 0.01 a	ND
Stilbenes				
Resveratrol	0.06 ± 0.05 b	0.15 ± 0.02 a	ND	ND

ND-Not Detected; different lowercase letters in rows indicate a significant difference among treatments ($p < 0.05$). GJ-P, grape juice plus pomace; MJ-P, model juice plus pomace; GJ, grape juice; MJ, model juice.

The main anthocyanin form found in Pinot noir wine is malvidin-3-*O*-glucoside [36]. Malvidin-3-O-glucoside, delphenidin-3-*O*-glucoside, cyanidin-3-*O*-glucoside, and peonidin-3-*O*-glucoside were quantified within the concentration range previously reported in Pinot noir wines [28,37,38]. There was no significant difference in the concentrations of monomeric anthocyanins between GJ-P and MJ-P treatment. This confirms that the model juice could extract anthocyanin from the grape skins effectively as grape juice and stabilize it in the wine matrix. A lower concentration of malvidin-3-*O*-glucoside was detected in GJ treatment confirming the possible leaching of anthocyanin during pressing grapes.

Flavonols in wines are mainly originated from the grape skins [39] and cluster sunlight exposure appears to be the primary factor determining flavonol levels in grapes [40]. Hence, this could be the reason for observing no significant differences in flavonol concentrations between GJ-P and MJ-P treatments because both treatments contained a similar amount of skin. Lower concentrations of kaempferol and quercetin in GJ treatment may be due to leaching from skins because they are not present in the pulp and are mainly located in grape skins. A recent study on Vranac grapes also confirmed that quercetin and kaempferol are not present in grape pulp [41].

Hydroxybenzoic acids in the GJ-P treatment, including gallic acid, protocatechuic acid, syringic acid, and vanillic acid were quantified within the concentration range previously reported in Pinot noir wine [28,42]. According to previous studies [41,43], gallic acid is mainly located in grape seeds, while syringic acid, protocatechuic acid, and vanilic acid are present in grape skins, seeds, and pulp in variable amounts. Hence, the presence of hydroxybenzoic acids (except gallic acid) in both MJ-P and GJ treatments confirms the presence of these compounds in both skins and pulp of Pinot noir berries because these compounds could quickly release from pulp to grape juice during pressing. Interestingly, the sum of hydroxybenzoic acids in MJ-P and GJ treatments is also similar to the concentrations observed in the GJ-P wine. The main hydroxycinnamic acid found in the wines was caftaric acid and it agrees with previous studies [28,44]. Hydroxycinnamic acids are mainly found in grape pulp, and also present in skin, seeds and stems in variable amounts.

Moreover, their concentration varies a lot between different tissues and different grape varieties [43,45,46]. The lower concentrations of caftaric acid, ferulic acid, and p-coumaric acid in GJ treatment confirm the presence of these compounds in grape pulp in very low concentrations in Pinot noir berries and/or these compounds are oxidised by endogenous tyrosinase when grapes are crushed [46].

3.3. *Tannin Composition and Mean Degree of Polymerisation (mDP)*

Analysis of tannin composition was carried out in GJ-P and MJ-P treatments using phloroglucinolysis (Table 3). GJ-P treatment showed significantly higher epicatechin gallate terminal and extension subunit proportions, and significantly lower epicatechin terminal and extension subunit proportions compared to MJ-P treatment. This could be mainly associated with higher seed tannin extraction during fermentation in GJ-P treatment due to the presence of higher alcohol concentration compared to MJ-P treatment [47].

Table 3. Tannin mean degree of polymerisation (mDP) and composition in wines.

	GJ-P	MJ-P
Concentrations		
Total Terminal (nmol/L)	270 ± 73 a	330 ± 81 a
Total Extension (nmol/L)	1368 ± 153 a	999 ± 113 b
Terminal subunit composition (%)		
C	59.1 ± 0.2 a	59.8 ± 1.2 a
EC	32.1 ± 0.3 b	33.4 ± 0.3 a
ECG	8.8 ± 0.5 a	6.9 ± 0.9 b
Extension subunit composition (%)		
P-C	10.4 ± 0.6 a	11.7 ± 0.7 a
P-EC	73.0 ± 0.3 b	73.8 ± 0.1 a
P-ECG	7.3 ± 0.1 a	5.7 ± 0.6 b
P-EGC	9.3 ± 0.7 a	8.9 ± 0.2 a
Tannin characteristics		
mDP	6.2 ± 0.6 a	4.1 ± 0.4 b
Yield	38.7 ± 4.5 a	47.1 ± 7.6 a

Different lowercase letters in rows indicate a significant difference among treatments ($p < 0.05$). GJ-P, grape juice plus pomace; MJ-P, model juice plus pomace; GJ, grape juice; MJ, model juice.

GJ-P treatment showed a significantly higher mean degree of polymerisation than MJ-P treatment, agreeing with a previous study [2]. It was mainly due to the significantly higher total extension subunit concertation recorded in GJ-P treatment because the phenolic acids in grape juice and anthocyanins leached into grape juice may contribute to the tannin polymerisation reactions in GJ-P treatment. Many previous studies have shown that anthocyanin is associated with polymerisation reactions with tannins present in the must during winemaking [31,48,49].

3.4. *Aroma Compound Analysis*

Forty-six aroma compounds were identified and quantified in the wines, including nineteen esters, seven volatile fatty acids, eight higher alcohols, one aldehyde, four volatile phenols, three C13-norisoprenoids and four monoterpenes (Table 4). All aroma compounds showed significant differences between the treatments.

Table 4. Aroma compound concentrations and odour activity values.

Aroma Compounds *	Aroma Threshold	Aroma Series	Concentration of Aroma Compounds				Odour Activity Values			
			GJ-P	MJ-P	GJ	MJ	GJ-P	MJ-P	GJ	MJ
Acetate Esters										
Ethyl acetate (mg/L)	7.50	3, 1	112.8 ± 8.5 a	57.3 ± 4.0 c	86.9 ± 5.5 b	34.4 ± 0.2 d	15.07	7.64	11.59	4.59
Isobutyl acetate	1605	1	46.9 ± 3.9 bc	34.0 ± 3.6 c	263.9 ± 16.3 a	60.3 ± 0.3 b	-	-	0.16	-
2-Methylbutyl acetate (mg/L)	0.313	1	0.255 ± 0.012 b	0.207 ± 0.018 b	5.009 ± 0.145 a	0.254 ± 0.006 b	0.81	0.66	16.01	0.81
Isoamyl acetate (mg/L)	0.030	1	0.208 ± 0.009 b	0.141 ± 0.008 b	2.535 ± 0.169 a	0.171 ± 0.003 b	6.93	4.70	84.67	5.70
Hexyl acetate	700	1	6.73 ± 0.34 b	5.79 ± 0.40 b	55.24 ± 7.26 a	4.57 ± 0.48 b	-	-	-	-
Octyl acetate	50,000	1, 2	4.97 ± 0.34 b	4.82 ± 0.15 b	6.24 ± 0.31 a	0.96 ± 0.05 c	-	-	-	-
Volatile fatty acids										
Acetic acid (mg/L)	200	3	139 ± 7 c	174 ± 4 b	172 ± 10 b	640 ± 3 a	0.70	0.87	0.86	3.20
Butyric acid (mg/L)	0.173	4	1.136 ± 0.025 a	0.960 ± 0.017 b	1.100 ± 0.026 a	0.466 ± 0.003 c	6.59	5.55	6.36	2.69
Isobutyric acid (mg/L)	2.30	4	2.696 ± 0.091 b	2.363 ± 0.020 c	3.514 ± 0.127 a	1.571 ± 0.010 d	1.17	1.03	1.53	0.68
2-Methylbutyric acid (mg/L)	3.00	4	0.665 ± 0.087 a	0.363 ± 0.016 b	0.777 ± 0.027 a	0.207 ± 0.003 c	0.22	0.12	0.26	-
Isovaleric acid (mg/L)	0.0334	4	0.638 ± 0.032 a	0.461 ± 0.017 c	0.588 ± 0.006 b	0.178 ± 0.002 d	19.10	13.80	17.60	5.33
Hexanoic acid (mg/L)	0.420	4	2.45 ± 0.07 b	2.08 ± 0.05 c	2.63 ± 0.02 a	0.87 ± 0.02 d	5.83	4.95	6.26	2.08
Octanoic acid (mg/L)	0.500	4	1.86 ± 0.07 b	1.49 ± 0.22 c	3.10 ± 0.04 a	1.25 ± 0.04 c	3.72	2.98	6.20	2.50
Alcohols										
Isoamyl alcohol (mg/L)	30.0	3	177 ± 2 a	117 ± 5 c	145 ± 6 b	58 ± 1 d	5.90	3.90	4.83	1.92
cis-3-Hexen-1-ol	1000	4, 5	50.1 ± 0.5 a	28.3 ± 1.1 a	41.3 ± 16.6 a	1.0 ± 0.1 b	-	-	-	-
trans-3-Hexen-1-ol	1000	5	22.1 ± 0.5 a	18.9 ± 0.3 a	20.6 ± 5.4 a	ND	-	-	-	-
trans-2-Hexen-1-ol	1000	5	9.52 ± 3.22 a	6.53 ± 0.34 ab	8.08 ± 0.63 a	2.63 ± 0.07 b	-	-	-	-
Hexanol (mg/L)	1.10	5	1.180 ± 0.083 a	1.131 ± 0.067 a	0.657 ± 0.072 b	0.015 ± 0.001 c	1.07	1.03	0.60	-
1-Heptanol	200	4	53.9 ± 6.4 a	55.7 ± 11.0 a	7.8 ± 0.4 b	7.8 ± 0.4 b	0.27	0.28	-	-
Phenylethyl alcohol (mg/L)	14.0	2	45.0 ± 5.0 a	40.4 ± 0.2 ab	38.2 ± 0.4 b	19.0 ± 0.1 c	3.21	2.89	2.73	1.36
1-Octanol	800	2	35.8 ± 3.3 a	31.3 ± 2.8 a	18.0 ± 0.3 b	14.2 ± 2.2 b	-	-	-	-
Aldehydes										
Benzaldehyde	2000		12.4 ± 2.9 a	10.7 ± 1.2 ab	11.3 ± 1.0 ab	6.9 ± 0.9 b	-	-	-	-
Ethyl Esters										
ethyl isobutyrate	15.0	1	27.3 ± 2.1 b	21.7 ± 0.4 c	41.2 ± 3.2 a	17.7 ± 0.3 c	1.82	1.45	2.75	1.18
Ethyl butyrate	400	1	422 ± 27 a	226 ± 17 b	406 ± 21 a	90 ± 3 c	1.06	0.57	1.02	0.23
Ethyl lactate (mg/L)	150	1, 4	8.10 ± 0.34 b	9.23 ± 0.67 a	9.10 ± 0.79 a	5.99 ± 0.08 b	-	-	-	-
Ethyl 2-methylbutyrate	18.0	1	3.59 ± 0.47 b	2.43 ± 0.28 bc	5.78 ± 0.95 a	1.29 ± 0.05 c	0.20	0.14	0.32	-
Ethyl pentanoate	5.00	1	1.41 ± 0.21 a	1.38 ± 0.12 a	1.48 ± 0.06 a	0.95 ± 0.07 b	0.28	0.28	0.30	0.19
Ethyl hexanoate	14.0	1, 2	806 ± 25 b	496 ± 25 c	919 ± 24 a	228 ± 7 d	57.57	35.43	65.64	16.29
Ethyl heptanoate	2.20	1	1.72 ± 0.18 b	2.46 ± 0.27 a	0.92 ± 0.01 c	0.97 ± 0.08 c	0.78	1.12	0.42	0.44
2-Phenylethyl acetate	250	2, 5	12.6 ± 1.3 b	12.1 ± 1.1 b	244.0 ± 9.7 a	22.7 ± 1.3 b	-	-	0.98	-
Ethyl octanoate (mg/L)	0.580	1, 2	1.239 ± 0.064 a	0.866 ± 0.037 b	1.256 ± 0.006 a	0.500 ± 0.008 c	2.14	1.49	2.17	0.86
Diethyl succinate	1,200,000	1, 2	240.9 ± 28.4 a	248.5 ± 16.9 a	152.0 ± 13.6 b	13.8 ± 0.3 c	-	-	-	-
Ethyl cinnamate	1.10	5	0.970 ± 0.131 a	0.747 ± 0.151 a	0.846 ± 0.102 a	0.141 ± 0.024 b	0.88	0.68	0.77	0.13
Ethyl hydrocinnamate	1.60	1, 2	0.705 ± 0.069 a	0.688 ± 0.114 a	0.550 ± 0.004 a	0.197 ± 0.052 b	0.44	0.43	0.34	0.12
Ethyl decanoate (mg/L)	0.200	1, 3, 4	1.06 ± 0.04 b	0.85 ± 0.01 c	1.34 ± 0.08 a	0.26 ± 0.01 d	5.30	4.26	6.70	1.32
Volatile phenols										
Phenol	5900	3, 5	3.87 ± 0.12 a	3.01 ± 0.19 b	2.49 ± 0.13 c	2.57 ± 0.14 c	-	-	-	-
Guaiacol	9.50	3, 5	6.26 ± 0.13 a	5.30 ± 0.47 b	5.27 ± 0.16 b	4.10 ± 0.01 c	0.66	0.56	0.55	0.43
4-Ethylguaiacol	33.0	5	0.123 ± 0.028 a	0.097 ± 0.034 ab	0.126 ± 0.007 a	0.043 ± 0.003 b	-	-	-	-
Eugenol	6.00	5	3.02 ± 0.48 a	1.99 ± 0.10 b	1.81 ± 0.28 b	0.94 ± 0.02 c	0.50	0.33	0.30	0.16
C13-norisoprenoids										
β-Damascenone	7.00	1, 2	6.57 ± 0.16 b	2.22 ± 0.21 c	14.11 ± 2.22 a	0.42 ± 0.01 c	0.94	0.32	2.01	-
α-Ionone	2.60	1	0.117 ± 0.008 b	0.137 ± 0.009 a	0.050 ± 0.005 c	0.047 ± 0.000 c	-	-	-	-
β-Ionone	5.00	1, 2	0.711 ± 0.021 a	0.663 ± 0.032 a	0.396 ± 0.016 b	0.351 ± 0.031 b	0.14	0.13	-	-
Monoterpenes										
Geraniol	30.0	1, 2	8.59 ± 0.11 a	4.73 ± 0.16 c	6.10 ± 0.12 b	ND	0.29	0.16	0.20	-
Linalool	25.2	2, 1	17.7 ± 0.5 a	10.7 ± 0.3 c	15.4 ± 0.6 b	2.4 ± 0.1 d	0.70	0.42	0.61	-
Citronellol	100	2	19.2 ± 0.8 a	13.5 ± 0.7 b	12.0 ± 0.5 b	4.9 ± 0.1 c	0.19	0.14	0.12	-
Nerol	300	2	5.16 ± 0.40 a	3.84 ± 0.36 b	2.92 ± 0.21 c	ND	-	-	-	-

* Concentrations are expressed in μg/L unless otherwise noted; ND: Not Detected; different lowercase letters in rows indicate a significant difference among treatments ($p < 0.05$). GJ-P, grape juice plus pomace; MJ-P, model juice plus pomace; GJ, grape juice; MJ, model juice. Aroma Series: 1-fruity; 2-floral; 3-chemical; 4-fatty; 5-herbaceous.

When comparing GJ-P and MJ-P treatments, about half of the analysed aroma compounds did not show a significant difference between GJ-P and MJ-P treatments, which includes aldehydes, acetate esters except ethyl acetate, higher alcohols except isoamyl alcohol, some ethyl esters including ethyl-2-methylbutyrate, ethyl pentanoate, 2-phenylethyl acetate, diethyl succinate, ethyl cinnamate, and ethyl hydrocinnamate, 4-ethylguaiacol in volatile phenols, and β-Ionone in C13-norisoprenoids. Among these compounds, ethyl cinnamate, ethyl hydrocinnamate, and benzaldehyde were identified as important aroma compounds contributing to varietal characteristics in Pinot Noir wine [17,50]. These results indicate that grape skins largely determine the aroma profile of Pinot noir wine because MJ-P treatment showed a similar aroma profile to GJ-P treatment even in the absence of grape juice. Furthermore, most of these compounds were secondary aroma compounds, mainly originated from the metabolism of yeast, and enzyme activity during fermentation [4]. Secondary aromas are the most abundant aroma fraction in Pinot noir wine and form the aroma base of all wines [18,28].

Compared to MJ-P treatment, twenty-one compounds out of total forty-six aroma compounds analysed in this work showed significantly higher concentrations in GJ-P treatment including, monoterpenes, most of the volatile phenols, C13-norisoprenoids, volatile fatty acids except acetic acid, few ethyl esters and isoamyl alcohol. Primary aroma compounds (monoterpenes, eugenol and C13-norisoprenoids) are found in both grape skins and juice [15] hence, higher concentrations are expected in GJ-P treatment compared to MJ-P treatment. In addition, the aroma precursors originating from grape juice may also contribute to result in significantly higher concentrations of these compounds in GJ-P treatment. Previous studies have shown that precursors of monoterpenes, C13 -norisoprenoids, and volatile phenols such as eugenol are derived in the berries at earlier stages of berry development, and they have been identified as monoglucosides and diglycosides forms in both grape skin and juice [20,51]. Ethyl decanoate, ethyl octanoate, and ethyl butanoate were identified as important wine aroma compounds contributing to the fruity aroma characteristics in New Zealand Pinot noir wine [17]. Apart from the primary aromas, a higher concentration of isoamyl alcohol in GJ-P treatment may be associated with the lower YAN levels recorded in GJ-P treatment because a previous study has demonstrated that lower amounts of YAN (200–300 mg/L) led to a higher concentrations of isoamyl alcohol, 2-methyl propanol, 2-methyl butanol, and 2-phenyl ethanol in wine [52]. In addition, higher concentration of butyric acid, isobutyric acid, their corresponding ethyl esters, and branched chain fatty acids including 2-methylbutyric acid and isovaleric acid in GJ-P treatment also agrees with the lower YAN levels in wine but also has been shown that yeast strain has a profound effect on the concentration of medium chain fatty acids and their corresponding ethyl esters in wines [52,53].

When comparing GJ-P and GJ treatments, sixteen aroma compounds including acetate esters (except ethyl acetate), most of the volatile fatty acids, a few ethyl esters and β-damascenone showed significantly lower concentrations in GJ-P treatment. A previous study on Muscat blanc grapes has also shown that fermentation with skins greatly reduced ho-trienol, β-damascenone, fatty acids and esters in wine [21]. According to a previous study [5], it is possible that skins in some way inhibited the formation of these compounds by providing either competitive substrates or enzyme inhibitors or adsorbing them on their surface. It is also possible that cap management during fermentation in GJ-P and MJ-P treatments had a significant effect on aroma compound evaporation. A previous study has also shown that solid parts of the grapes inhibit the biosynthesis of volatile fatty acids in yeast cells and result in lower concentrations in the wines [54]. However, a recent study on Pinot noir grapes fermented with and without grape skins has shown fairly similar results to this study but wines showed no significant difference in most of the volatile fatty acids analysed [55]. This may be associated with the vintage differences, and higher cap management frequency used in winemaking.

Moreover, eighteen aroma compounds including all monoterpenes, C13-norisoprenoids (except β-damascenone), volatile phenols (except 4-ethylguaiacol), most of the higher alcohols, few ethyl esters, and ethyl acetate showed significantly higher concentrations in GJ-P treatment compared to GJ treatment. Most of the primary aromas, such as monoterpenes, C13-norisoprenoids, and volatile phenols (eugenol) are largely found in grape skins in free volatile form or bound into non-volatile precursors [15,55] and hence higher concentrations are expected in GJ-P treatment. However, the higher concentration of higher alcohols in GJ-P treatment may be linked to the presence of more amino acids in the fermentation media because GJ-P treatment contained grape pomace during the fermentation. Previous studies have shown that the total production of higher alcohols increases as the concentrations of corresponding amino acids increase in grape must during fermentation [4,53,56]. Interestingly twelve aroma compounds including butyric acid, 2-methylbutyric acid, *cis*-3-hexen-1-ol, *trans*-3-hexen-1-ol, *trans*-2-hexen-1-ol, benzaldehyde, ethyl butyrate, ethyl pentanoate, ethyl octanoate, ethyl cinnamate, ethyl hydrocinnamate, and eugenol showed no significant difference between GJ-P and GJ treatments and it confirms that their concentrations are not affected by the grape solids in the media.

MJ treatment had the lowest concentration in most of the aroma compounds analysed in this study, which is mainly due to the absence of primary aroma compounds and aroma precursors which comes from grapes in MJ treatment. Hence, primary aroma compounds including C13-norisoprenoids, monoterpenes, C_6 alcohols (hexanol, *cis*-3-hexen-1-ol, *trans*-3-hexen-1-ol, and *trans*-2-hexen-1-ol) were not detected or detected below the limit of quantification (LOQ) in MJ treatment. MJ treatment showed a higher concentration of acetic acid (640 mg/L) compared to other treatments, which is still below the sensory threshold in red wine and well below the legal limits of acetic acid in wine [57,58]. Higher concentration of acetic acid in MJ treatment may be due to the high level of nitrogen source added in model juice, and previous studies have shown that higher YAN level in must resulted in a higher acetic acid concentration and volatile acidity in wines [52,59]. The higher acetic acid concentration in MJ treatment may also be associated with the absence of tannins and lower alcohol levels. In addition, no cap management could be another reason for relatively high acetic acid in MJ as more acetic acid is preserved in wine.

Medium-chain fatty acids (C_6–C_{12}), hexanoic acid and octanoic acid were detected in MJ treatment because they are mainly produced by yeast during sugar metabolism in fermentation [56]. Corresponding ethyl esters of medium-chain fatty acids were also found in MJ treatment such as ethyl hexanoate, ethyl heptanoate, ethyl octanoate, ethyl decanoate because medium-chain fatty acids are the initial substrate for the final formation of ethyl esters in wines [56,60]. The presence of higher alcohols in MJ treatment indicates that the amino acids in model juice have been converted to higher alcohols through the Ehrlich pathway [4]. For example, isoamyl alcohol and phenylethyl alcohol in MJ treatment is derived from leucine, and phenylalanine amino acids in the model juice, respectively. These higher alcohols are the initial substrate for the final formation of acetate esters in wine. For example, isobutyl acetate in MJ treatment is derived from the isoamyl alcohol (derived from leucine).

3.4.1. Odour Activity Value (OAV) Analysis

Odour activity values (OAVs) are commonly used to assess the contribution of volatile compounds to the overall wine aroma in many studies [61–64]. According to previous studies, aroma compounds with an OAV above 0.1 is considered important contributors to the overall wine aroma due to the synergistic effect of certain aroma compounds in wine [61,63]. Hence, all aroma compounds with OAV above 0.1 were considered significant aroma compounds in this study.

All the aroma series showed significant differences between treatments and the aroma series intensity patterns showed that the aroma profile of the wines mainly consisted of fruity, floral, fatty, chemical, and herbaceous aromas (Figure 1). GJ treatment showed a significantly higher total odour activity value (ΣOAV) in all aroma series except the chemical aroma series. It was mainly due to the significantly higher concentrations of 2-methylbutyl acetate, isoamyl acetate, isobutyric acid, hexanoic acid, octanoic acid, ethyl isobutyrate, and ethyl decanoate.

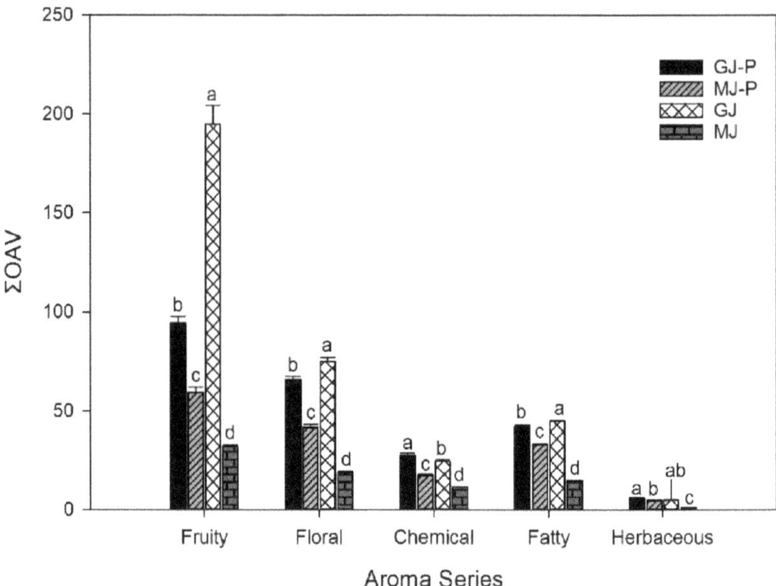

Figure 1. Comparison of aroma series based on ΣOAV between four treatments. Different letters in the same aroma series means a significant difference ($p < 0.05$). GJ-P, grape juice plus pomace; MJ-P, model juice plus pomace; GJ, grape juice; MJ, model juice.

The fruity aroma series was the major aroma series having the highest intensity in all the treatments, which agrees with previous observations in Pinot noir wines [28,61]. The major contributors to the fruity aroma series include two acetate esters (ethyl acetate, and isoamyl acetate) and two ethyl esters (ethyl hexanoate and ethyl decanoate) in all the wines and accounted for 85% to 90% of the ΣOAV of fruity aroma series. Previous studies also confirm that isoamyl acetate, ethyl hexanoate and ethyl decanoate are important wine aroma compounds contributing to the fruity aroma characteristics in Pinot noir wine [28,61]. When comparing GJ and GJ-P treatments, the higher ΣOAV of fruity aroma series in GJ treatment was mainly due to the higher concentrations of isoamyl acetate and 2-methylbutyl acetate. Their concentrations in the GJ treatment were twelve times and twenty times higher than that in the GJ-P treatment. This is in agreement with a previous study [55], reporting that the concentration of isoamyl acetate in the wine fermented without grape skins was eleven times higher than in the wine made with grape skins. In general, a higher concentration of acetate esters observed in GJ compared to GJ-P treatment could be due to grape skins binding or delaying the release of some volatile compounds in wine [5], and no cap management in GJ treatment during fermentation that could better preserve esters formed in wines.

Floral and fatty aroma series were also two other major aroma series in this work. For the floral aroma series, the ΣOAV was dominated by three compounds (ethyl hexanoate, phenylethyl alcohol, and ethyl octanoate) and the fatty series was dominated by six compounds (isovaleric acid, butyric acid, hexanoic acid, ethyl decanoate, octanoic acid, and isobutyric acid). However, β-damascenone was also a major contributor to the fruity aroma series in GJ treatment, while isobutyric acid was not a major contributor to the fatty aroma series in MJ treatment. Compared to the GJ-P treatment, significantly higher concentrations of ethyl hexanoate and β-damascenone in the GJ treatment led to significantly higher ΣOAV of floral aroma series in the resultant wine. A higher concentration of β-damascenone in GJ treatment indicates grape juice is a good source of β-damascenone, which can be better preserved in the wine by avoiding cap management.

3.4.2. Principal Component Analysis

The score plot and correlation circle plot obtained from principal component analysis (PCA) is shown in Figure 2. The ellipses in the score plot represent 95% confidence intervals for the means of the treatments. Ninety-one per cent of the differences in the concentration of aroma compounds between treatments can be explained by the first two principal components (PCs).

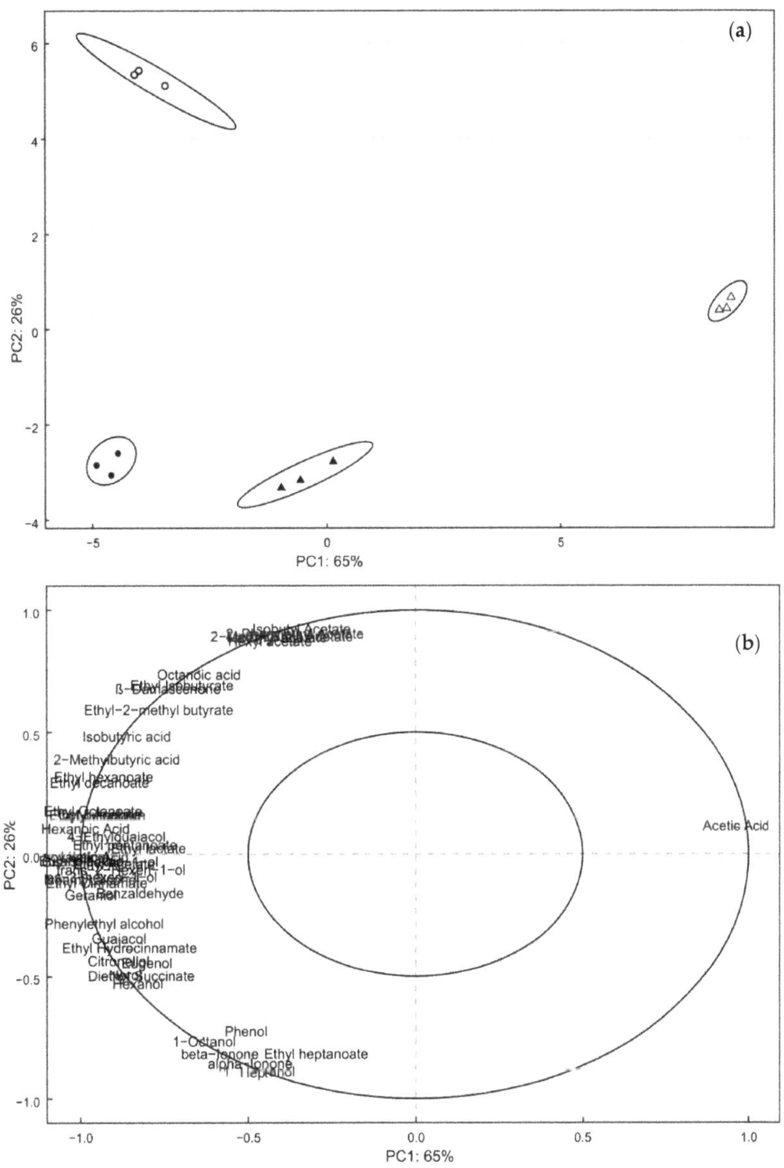

Figure 2. (**a**) Score-plot and (**b**) correlation circle obtained from principal component analysis of aroma compounds determined in wines, GJ-P, grape juice plus pomace (•); MJ-P, model juice plus pomace (▲); GJ, grape juice (○); MJ, model juice (△).

Triplicate ferments from each treatment were tightly clustered together on the score plot and it indicates good reproducibility of the treatments. The PC1 explained 65% of the differences and separated all four treatments. Loading vectors of PC1 and PC2 are given in Table S1. The MJ treatment was grouped away from all other treatments toward the left on the plot, showing a distinctive aroma profile. It was largely characterised by the higher concentration of acetic acid in MJ treatment. While PC2 explained 26% of the differences and wines prepared with grape pomace (GJ-P and MJ-P) were grouped separately from wines made with juice (GJ and MJ). Wines made with grape pomace were characterised mostly by primary aromas, and higher alcohols, including 1-heptanol, α-ionone, ethyl heptanoate, β-ionone, 1-octanol, phenol, hexanol, diethyl succinate, nerol, eugenol, respectively. Because it is generally accepted that most of the primary aroma compounds such as monoterpenes, C13-norisoprenoids, C6 compounds and volatile phenols (eugenol) are largely found in grape skins [15,55]. The higher concentrations of higher alcohols indicate the presence of a higher concentration of amino acids in grape pomace added treatments, which can act as aroma precursors to form higher alcohols through the Ehrlich pathway during fermentation as discussed above [4]. In contrast, GJ treatment was mainly characterised by the higher concentrations of acetate esters (except ethyl acetate) compared to grape pomace-added treatments. This may be associated with the effect of skins in GJ-P and MJ-P treatments, which may adsorb aroma compounds into skins or provide competitive substrates or enzyme inhibitors to reduce the formation of acetate esters in wines [5], and lower fermentation temperatures and aeration in GJ treatment during fermentation could enhance acetate ester concentration in wines (Figure S1) [65,66].

4. Conclusions

Tannin extraction is important in red wine production, and this study investigated the impact of phenolics in grape juice and polyphenols in grape pomace on tannin concentration and tannin composition. Grape skins and seeds are the most important source of tannins extracted into wine during fermentation. This study revealed the important role of phenolic compounds in grape juice in the polymerisation and composition of tannins in the wines. Inclusion of grape pomace in the fermentation enhanced the concentrations of primary aromas including monoterpenes, and C13-norisoprenoids, which are mainly responsible for floral and fruity notes in wines. The presence of corresponding aroma precursors are well documented in grape skins. Fermentation derived aroma compounds, e.g., esters, volatile fatty acids, are better preserved in the fermentation without pomace addition. These observations reflect the differences in aroma profiles between red winemaking and white winemaking.

Supplementary Materials: The following supporting information can be downloaded at: https://www.mdpi.com/article/10.3390/fermentation8120718/s1, Table S1: Loading vectors of first principal component (PC1) and second principal component (PC2), Figure S1: Fermentation curves: (a) total mass reduction of the ferments; (b) must temperature fluctuation during fermentation.

Author Contributions: Conceptualization, P.M.W., J.Z. and B.T.; methodology, P.M.W. and B.T.; software, P.M.W.; formal analysis, P.M.W.; investigation, B.T. and P.M.W.; resources, B.T.; writing—original draft preparation, P.M.W.; writing—review and editing, J.Z. and B.T.; supervision, B.T.; project administration, B.T.; funding acquisition, B.T. All authors have read and agreed to the published version of the manuscript.

Funding: This research received no external funding.

Institutional Review Board Statement: Not applicable.

Informed Consent Statement: Not applicable.

Data Availability Statement: All data can be found in the manuscript and the Supplementary Materials.

Acknowledgments: Authors gratefully acknowledge Jason Breitmeyer for his assistance with the GC-MS analysis, Jenny Zhao for her assistance with the HPLC analysis, and Richard Hider for his assistance with other chemical analyses.

Conflicts of Interest: The authors declare no conflict of interest.

References

1. Souquet, J.-M.; Labarbe, B.; Le Guernevé, C.; Cheynier, V.; Moutounet, M. Phenolic composition of grape stems. *J. Agric. Food Chem.* **2000**, *48*, 1076–1080. [CrossRef] [PubMed]
2. Sparrow, A.M.; Dambergs, R.G.; Bindon, K.A.; Smith, P.A.; Close, D.C. Interactions of Grape Skin, Seed, and Pulp on Tannin and Anthocyanin Extraction in Pinot noir Wines. *Am. J. Enol. Vitic.* **2015**, *66*, 472–481. [CrossRef]
3. Tian, B.; Harrison, R.; Morton, J.; Jaspers, M. Influence of skin contact and different extractants on extraction of proteins and phenolic substances in Sauvignon Blanc grape skin. *Aust. J. Grape Wine Res.* **2020**, *26*, 180–186. [CrossRef]
4. Belda, I.; Ruiz, J.; Esteban-Fernández, A.; Navascués, E.; Marquina, D.; Santos, A.; Moreno-Arribas, M.V. Microbial contribution to wine aroma and its intended use for wine quality improvement. *Molecules* **2017**, *22*, 189. [CrossRef]
5. Callejón, R.M.; Margulies, B.; Hirson, G.D.; Ebeler, S.E. Dynamic Changes in Volatile Compounds during Fermentation of Cabernet Sauvignon Grapes with and without Skins. *Am. J. Enol. Vitic.* **2012**, *63*, 301–312. [CrossRef]
6. Vidal, S.; Francis, L.; Guyot, S.; Marnet, N.; Kwiatkowski, M.; Gawel, R.; Cheynier, V.; Waters, E.J. The mouth-feel properties of grape and apple proanthocyanidins in a wine-like medium. *J. Sci. Food Agric.* **2003**, *83*, 564–573. [CrossRef]
7. Cortell, J.M.; Halbleib, M.; Gallagher, A.V.; Righetti, T.L.; Kennedy, J.A. Influence of Vine Vigor on Grape (*Vitis vinifera* L. Cv. Pinot Noir) and Wine Proanthocyanidins. *J. Agric. Food Chem.* **2005**, *53*, 5798–5808. [CrossRef]
8. Koyama, K.; Goto-Yamamoto, N.; Hashizume, K. Influence of maceration temperature in red wine vinification on extraction of phenolics from berry skins and seeds of grape (*Vitis vinifera*). *Biosci. Biotechnol. Biochem.* **2007**, *71*, 958–965. [CrossRef]
9. Carew, A.L.; Kerslake, F.L.; Bindon, K.A.; Smith, P.A.; Close, D.C.; Dambergs, R.G. Viticultural and Controlled Phenolic Release Treatments Affect Phenolic Concentration and Tannin Composition in Pinot noir Wine. *Am. J. Enol. Vitic.* **2020**, *71*, 256–265. [CrossRef]
10. Harbertson, J.F.; Kennedy, J.A.; Adams, D.O. Tannin in Skins and Seeds of Cabernet Sauvignon, Syrah, and Pinot noir Berries during Ripening. *Am. J. Enol. Vitic.* **2002**, *53*, 54. [CrossRef]
11. Dambergs, R.; Sparrow, A.; Carew, A.; Scrimgeour, N.; Wilkes, E.; Godden, P.; Herderich, M.; Johnson, D. Quality in a cool climate–maceration techniques in Pinot Noir production. *J. Wine Vitic.* **2012**, *18*, 18–26.
12. Patricia, A.M.; Kennedy, J.A. Compositional investigation of phenolic polymers isolated from *Vitis vinifera* L. Cv. Pinot Noir during fermentation. *J. Agric. Food Chem.* **2007**, *55*, 5670. [CrossRef]
13. Gawel, R. Red wine astringency: A review. *Aust. J. Grape Wine Res.* **1998**, *4*, 74–95. [CrossRef]
14. Wollmann, N.; Hofmann, T. Compositional and Sensory Characterization of Red Wine Polymers. *J. Agric. Food Chem.* **2013**, *61*, 2045–2061. [CrossRef]
15. Slegers, A.; Angers, P.; Ouellet, É.; Truchon, T.; Pedneault, K. Volatile compounds from grape skin, juice and wine from five interspecific hybridgrape cultivars grown in Québec (Canada) for wine production. *Molecules* **2015**, *20*, 10980–11016. [CrossRef] [PubMed]
16. Olejar, K.J.; Vasconcelos, M.C.; King, P.D.; Smart, R.E.; Ball, K.; Field, S.K. Herbicide reduction through the use of weedmat undervine treatment and the lack of impact on the aromatic profile and volatile composition of Malbec wines. *Food Chem.* **2021**, *343*, 128474. [CrossRef]
17. Tomasino, E.; Harrison, R.; Breitmeyer, J.; Sedcole, R.; Sherlock, R.; Frost, A. Aroma composition of 2-year-old New Zealand Pinot Noir wine and its relationship to sensory characteristics using canonical correlation analysis and addition/omission tests. *Aust. J. Grape Wine Res.* **2015**, *21*, 376–388. [CrossRef]
18. Longo, R.; Carew, A.; Sawyer, S.; Kemp, B.; Kerslake, F. A review on the aroma composition of *Vitis vinifera* L. Pinot noir wines: Origins and influencing factors. *Crit. Rev. Food Sci. Nutr.* **2021**, *61*, 1589–1604. [CrossRef] [PubMed]
19. Borren, E.; Tian, B. The important contribution of non-Saccharomyces yeasts to the aroma complexity of wine: A review. *Foods* **2021**, *10*, 13. [CrossRef]
20. Michlmayr, H.; Nauer, S.; Brandes, W.; Schümann, C.; Kulbe, K.D.; Andrés, M.; Eder, R. Release of wine monoterpenes from natural precursors by glycosidases from *Oenococcus oeni*. *Food Chem.* **2012**, *135*, 80–87. [CrossRef]
21. Lukić, I.; Lotti, C.; Vrhovsek, U. Evolution of free and bound volatile aroma compounds and phenols during fermentation of Muscat blanc grape juice with and without skins. *Food Chem.* **2017**, *232*, 25–35. [CrossRef] [PubMed]
22. Maicas, S.; Mateo, J.J. Hydrolysis of terpenyl glycosides in grape juice and other fruit juices: A review. *Appl. Microbiol. Biotechnol.* **2005**, *67*, 322–335. [CrossRef] [PubMed]
23. Ciani, M.; Ferraro, L. Enhanced Glycerol Content in Wines Made with Immobilized Candida stellata Cells. *Appl. Environ. Microbiol.* **1996**, *62*, 128–132. [CrossRef]
24. Iland, P.G. *Chemical Analysis of Grapes and Wine: Techniques and Concepts*, 2nd ed.; Campbelltown, S., Ed.; Patrick Iland Wine Promotions: Athelstone, SA, Australia, 2013.

25. Mercurio, M.D.; Dambergs, R.G.; Herderich, M.J.; Smith, P.A. High throughput analysis of red wine and grape phenolics-adaptation and validation of methyl cellulose precipitable tannin assay and modified Somers color assay to a rapid 96 well plate format. *J. Agric. Food Chem.* **2007**, *55*, 4651. [CrossRef] [PubMed]
26. Waterhouse, A.L. Determination of Total Phenolics. In *Current Protocols in Food Analytical Chemistry*; Wrolstad, R.E., Ed.; John Wiley & Sons, Inc.: New York, NY, USA, 2002; pp. I1.1.1–I1.1.8.
27. Jeffery, D.W.; Mercurio, M.D.; Herderich, M.J.; Hayasaka, Y.; Smith, P.A. Rapid isolation of red wine polymeric polyphenols by solid-phase extraction. *J. Agric. Food Chem.* **2008**, *56*, 2571. [CrossRef]
28. Wimalasiri, P.M.; Olejar, K.J.; Harrison, R.; Hider, R.; Tian, B. Whole bunch fermentation and the use of grape stems: Effect on phenolic and volatile aroma composition of *Vitis vinifera* cv. Pinot Noir wine. *Aust. J. Grape Wine Res.* **2021**, *28*, 395–406. [CrossRef]
29. Kennedy, J.A.; Jones, G.P. Analysis of Proanthocyanidin Cleavage Products Following Acid-Catalysis in the Presence of Excess Phloroglucinol. *J. Agric. Food Chem.* **2001**, *49*, 1740–1746. [CrossRef]
30. Olejar, K.J.; Breitmeyer, J.; Wimalasiri, P.M.; Tian, B.; Field, S.K. Detection of sub-aroma threshold concentrations of wine methoxypyrazines by multidimensional GCMS. *Analytica* **2021**, *2*, 1–13. [CrossRef]
31. Teng, B.; Hayasaka, Y.; Smith, P.A.; Bindon, K.A. Effect of Grape Seed and Skin Tannin Molecular Mass and Composition on the Rate of Reaction with Anthocyanin and Subsequent Formation of Polymeric Pigments in the Presence of Acetaldehyde. *J. Agric. Food Chem.* **2019**, *67*, 8938–8949. [CrossRef]
32. Hernández-Jiménez, A.; Kennedy, J.A.; Bautista-Ortín, A.B.; Gómez-Plaza, E. Effect of ethanol on grape seed proanthocyanidin extraction. *Am. J. Enol. Vitic.* **2012**, *63*, 57–61. [CrossRef]
33. Canals, R.; Llaudy, M.; Valls, J.; Canals, J.; Zamora, F. Influence of ethanol concentration on the extraction of color and phenolic compounds from the skin and seeds of Tempranillo grapes at different stages of ripening. *J. Agric. Food Chem.* **2005**, *53*, 4019–4025. [CrossRef] [PubMed]
34. Cerpa-Calderón, F.K.; Kennedy, J.A. Berry Integrity and Extraction of Skin and Seed Proanthocyanidins during Red Wine Fermentation. *J. Agric. Food Chem.* **2008**, *56*, 9006–9014. [CrossRef] [PubMed]
35. Everette, J.D.; Bryant, Q.M.; Green, A.M.; Abbey, Y.A.; Wangila, G.W.; Walker, R.B. Thorough study of reactivity of various compound classes toward the Folin–Ciocalteu reagent. *J. Agric. Food Chem.* **2010**, *58*, 8139–8144. [CrossRef]
36. Dimitrovska, M.; Bocevska, M.; Dimitrovski, D.; Murkovic, M. Anthocyanin composition of Vranec, Cabernet Sauvignon, Merlot and Pinot Noir grapes as indicator of their varietal differentiation. *Eur. Food Res. Technol.* **2011**, *232*, 591–600. [CrossRef]
37. Casassa, L.F.; Sari, S.E.; Bolcato, E.A.; Diaz-Sambueza, M.A.; Catania, A.A.; Fanzone, M.L.; Raco, F.; Barda, N. Chemical and Sensory Effects of Cold Soak, Whole Cluster Fermentation, and Stem Additions in Pinot noir Wines. *Am. J. Enol. Vitic.* **2019**, *70*, 19–33. [CrossRef]
38. Cortell, J.M.; Halbleib, M.; Gallagher, A.V.; Righetti, T.L.; Kennedy, J.A. Influence of vine vigor on grape (*Vitis vinifera* L. Cv. Pinot Noir) anthocyanins. 2. Anthocyanins and pigmented polymers in wine. *J. Agric. Food Chem.* **2007**, *55*, 6585. [CrossRef]
39. Waterhouse, A.L.; Sacks, G.L.; Jeffery, D.W. *Understanding Wine Chemistry*; John Wiley & Sons, Ltd.: Chichester, UK, 2016. [CrossRef]
40. Price, S.F.; Breen, P.J.; Valladao, M.; Watson, B.T. Cluster sun exposure and quercetin in Pinot noir grapes and wine. *Am. J. Enol. Vitic.* **1995**, *46*, 187–194.
41. Šuković, D.; Knežević, B.; Gašić, U.; Sredojević, M.; Ćirić, I.; Todić, S.; Mutić, J.; Tešić, Ž. Phenolic profiles of leaves, grapes and wine of grapevine variety vranac (*Vitis vinifera* L.) from Montenegro. *Foods* **2020**, *9*, 138. [CrossRef]
42. Van Leeuw, R.; Kevers, C.; Pincemail, J.; Defraigne, J.O.; Dommes, J. Antioxidant capacity and phenolic composition of red wines from various grape varieties: Specificity of Pinot Noir. *J. Food Compos. Anal.* **2014**, *36*, 40–50. [CrossRef]
43. Doshi, P.; Adsule, P.; Banerjee, K.; Oulkar, D. Phenolic compounds, antioxidant activity and insulinotropic effect of extracts prepared from grape (*Vitis vinifera* L.) byproducts. *J. Food Sci. Technol.* **2015**, *52*, 181–190. [CrossRef]
44. Gomes, T.M.; Toaldo, I.M.; da Silva Haas, I.C.; Burin, V.M.; Caliari, V.; Luna, A.S.; de Gois, J.S.; Bordignon-Luiz, M.T. Differential contribution of grape peel, pulp, and seed to bioaccessibility of micronutrients and major polyphenolic compounds of red and white grapes through simulated human digestion. *J. Funct. Foods* **2019**, *52*, 699–708. [CrossRef]
45. Garrido, J.; Borges, F. Wine and grape polyphenols—A chemical perspective. *Food Res. Int.* **2013**, *54*, 1844–1858. [CrossRef]
46. Vendramin, V.; Viel, A.; Vincenzi, S. Caftaric Acid Isolation from Unripe Grape: A "Green" Alternative for Hydroxycinnamic Acids Recovery. *Molecules* **2021**, *26*, 1148. [CrossRef] [PubMed]
47. Rousserie, P.; Lacampagne, S.; Vanbrabant, S.; Rabot, A.; Geny-Denis, L. Influence of berry ripeness on seed tannins extraction in wine. *Food Chem.* **2020**, *315*, 126307. [CrossRef]
48. Cheynier, V.; Dueñas-Paton, M.; Salas, E.; Maury, C.; Souquet, J.M.; Sarni-Manchado, P.; Fulcrand, H. Structure and Properties of Wine Pigments and Tannins. *Am. J. Enol. Vitic.* **2006**, *57*, 298–305. [CrossRef]
49. Kumar, L.; Tian, B.; Harrison, R. Interactions of *Vitis vinifera* L. cv. Pinot Noir grape anthocyanins with seed proanthocyanidins and their effect on wine color and phenolic composition. *LWT-Food Sci. Technol.* **2022**, *162*, 113428. [CrossRef]
50. Moio, L.; Etievant, P.X. Ethyl anthranilate, ethyl cinnamate, 2,3-dihydrocinnamate, and methyl anthranilate: Four important odorants identified in Pinot noir wines of Burgundy. *Am. J. Enol. Vitic.* **1995**, *46*, 392–398.
51. Gunata, Z.; Bitteur, S.; Brillouet, J.-M.; Bayonove, C.; Cordonnier, R. Sequential enzymic hydrolysis of potentially aromatic glycosides from grape. *Carbohydr. Res.* **1988**, *184*, 139–149. [CrossRef]

52. Vilanova, M.; Ugliano, M.; Varela, C.; Siebert, T.; Pretorius, I.; Henschke, P. Assimilable nitrogen utilisation and production of volatile and non-volatile compounds in chemically defined medium by *Saccharomyces cerevisiae* wine yeasts. *Appl. Microbiol. Biotechnol.* **2007**, *77*, 145–157. [CrossRef]
53. HERNANDEZ-ORTE, P.; Bely, M.; Cacho, J.; Ferreira, V. Impact of ammonium additions on volatile acidity, ethanol, and aromatic compound production by different *Saccharomyces cerevisiae* strains during fermentation in controlled synthetic media. *Aust. J. Grape Wine Res.* **2006**, *12*, 150–160. [CrossRef]
54. Valero, E.; Millán, C.; Ortega, J.M. Influence of pre-fermentative treatment on the fatty acid content of *Saccharomyces cerevisiae* (M330-9) during alcoholic fermentation of grape must. *J. Biosci. Bioeng.* **2001**, *91*, 117–122. [CrossRef] [PubMed]
55. Qiao, Y.; Hawkins, D.; Parish-Virtue, K.; Fedrizzi, B.; Knight, S.J.; Deed, R.C. Contribution of Grape Skins and Yeast Choice on the Aroma Profiles of Wines Produced from Pinot Noir and Synthetic Grape Musts. *Fermentation* **2021**, *7*, 168. [CrossRef]
56. Carpena, M.; Fraga-Corral, M.; Otero, P.; Nogueira, R.A.; Garcia-Oliveira, P.; Prieto, M.A.; Simal-Gandara, J. Secondary aroma: Influence of wine microorganisms in their aroma profile. *Foods* **2020**, *10*, 51. [CrossRef] [PubMed]
57. Cliff, M.A.; Pickering, G.J. Determination of odour detection thresholds for acetic acid and ethyl acetate in ice wine. *J. Wine Res.* **2006**, *17*, 45–52. [CrossRef]
58. Corison, C.; Ough, C.; Berg, H.; Nelson, K. Must acetic acid and ethyl acetate as mold and rot indicators in grapes. *Am. J. Enol. Vitic.* **1979**, *30*, 130–134.
59. BELL, S.J.; Henschke, P.A. Implications of nitrogen nutrition for grapes, fermentation and wine. *Aust. J. Grape Wine Res.* **2005**, *11*, 242–295. [CrossRef]
60. Saerens, S.; Delvaux, F.; Verstrepen, K.; Van Dijck, P.; Thevelein, J.; Delvaux, F. Parameters affecting ethyl ester production by *Saccharomyces cerevisiae* during fermentation. *Appl. Environ. Microbiol.* **2008**, *74*, 454–461. [CrossRef]
61. Cai, J.; Zhu, B.-Q.; Wang, Y.-H.; Lu, L.; Lan, Y.-B.; Reeves, M.J.; Duan, C.-Q. Influence of pre-fermentation cold maceration treatment on aroma compounds of Cabernet Sauvignon wines fermented in different industrial scale fermenters. *Food Chem.* **2014**, *154*, 217–229. [CrossRef]
62. Ferreira, V.; López, R.; Cacho, J.F. Quantitative determination of the odorants of young red wines from different grape varieties. *J. Sci. Food Agric.* **2000**, *80*, 1659–1667. [CrossRef]
63. Zea, L.; Moyano, L.; Moreno, J.; Medina, M. Aroma series as fingerprints for biological ageing in fino sherry-type wines. *J. Sci. Food Agric.* **2007**, *87*, 2319–2326. [CrossRef]
64. Zhu, L.-X.; Zhang, M.-M.; Shi, Y.; Duan, C.-Q. Evolution of the aromatic profile of traditional Msalais wine during industrial production. *Int. J. Food Prop.* **2019**, *22*, 911–924. [CrossRef]
65. Killian, E.; Ough, C. Fermentation esters—Formation and retention as affected by fermentation temperature. *Am. J. Enol. Vitic.* **1979**, *30*, 301–305.
66. Varela, C.; Cuijvers, K.; Van Den Heuvel, S.; Rullo, M.; Solomon, M.; Borneman, A.; Schmidt, S. Effect of aeration on yeast community structure and volatile composition in uninoculated chardonnay wines. *Fermentation* **2021**, *7*, 97. [CrossRef]

Article

Effect of Pre-Fermentative Bentonite Addition on Pinot Noir Wine Colour, Tannin, and Aroma Profile

Pradeep M. Wimalasiri [1], Tanya Rutan [2] and Bin Tian [1,*]

[1] Faculty of Agriculture and Life Sciences, Lincoln University, Lincoln 7647, New Zealand
[2] Bragato Research Institute, Blenheim 7201, New Zealand
* Correspondence: bin.tian@lincoln.ac.nz

Abstract: Pinot noir is a grape variety with thin grape skin, which means the extraction of colour and polyphenols is more challenging than other red grape varieties. The aim of this study was to investigate the impact of protein removal by adding bentonite prior to fermentation on Pinot noir wine composition. Four treatments were conducted, including the control without bentonite addition and Pinot noir wines produced with the addition of three different types of bentonites before cold soaking. The juice and wine samples were analysed for pathogenesis-related proteins, tannin, wine colour parameters, and aroma composition. The results showed that bentonite addition at 0.5 g/L had little impact on tannin and aroma compounds but more impact on wine colour, especially significantly higher level of SO_2 resistant pigments observed in Na bentonite addition treatment. This study indicates the potential use of bentonite to modulate the Pinot noir juice composition that may facilitate the extraction of colour components from grape into juice, which plays an important role in colour stabilization in finished wine.

Keywords: anthocyanins; bentonite; cold soaking; colour; pathogenesis-related proteins; Pinot noir; tannin; wine aroma

Citation: Wimalasiri, P.M.; Rutan, T.; Tian, B. Effect of Pre-Fermentative Bentonite Addition on Pinot Noir Wine Colour, Tannin, and Aroma Profile. *Fermentation* **2022**, *8*, 639. https://doi.org/10.3390/fermentation8110639

Academic Editors: Chih Yao Hou and Amparo Gamero

Received: 16 October 2022
Accepted: 10 November 2022
Published: 14 November 2022

Publisher's Note: MDPI stays neutral with regard to jurisdictional claims in published maps and institutional affiliations.

Copyright: © 2022 by the authors. Licensee MDPI, Basel, Switzerland. This article is an open access article distributed under the terms and conditions of the Creative Commons Attribution (CC BY) license (https://creativecommons.org/licenses/by/4.0/).

1. Introduction

The extraction of anthocyanins and tannins is critical for red wine production as they are associated with important sensory attributes, e.g., wine colour, taste, and mouthfeel [1]. Pinot noir is a grape variety with thin grape skin, so the winemaking techniques that can enhance the extraction of colours and polyphenols are useful tools to produce quality Pinot noir wine.

Previous studies [2,3] have reported that tannins in wine can interact with proteins, which influences the perception of astringency [4]. Tannins and proteins can bind with each other through the hydrogen bonding and hydrophobic interactions [5]. From a winemaking point of view, the tannin–protein interaction may also influence the extraction of those components from grape into juice and wine. As the major soluble proteins in white grape juice and wine, pathogenesis-related (PR) proteins have been well studied as they could cause protein haze formation in white wine [6–8]. PR proteins have been reported to be present in both the grape skin and pulp of but not in the seed in Sauvignon blanc [9] and their concentrations were gradually increased in grape skin and pulp during ripening [10], but the diversity of PR proteins decreased during grape maturation [11]. The level of PR proteins in grapes can also be influenced by UV radiation and fungal infection [12]. To remove PR proteins in white wine in order to prevent protein haze formation [13], bentonite is commonly used as a fining agent. Bentonite is a negatively charged clay belonging to the group of montmorillonites (hydrated aluminium silicates), which is able to swell when combined with water and produce a gel-type suspension [14,15]. Bentonite interacts with positively charged wine proteins to form flocculation and precipitate [16]. The exchangeable cations mainly Na^+, Ca^{2+}, and Mg^{2+} localized on the external surface of clay particles and in-between the layers are important for balancing the net negative charge of the clay [17]. The three major types of bentonite used in the wine industry are sodium (Na) bentonite, calcium (Ca) bentonite, and sodium and calcium combined (NaCa) bentonite, with Na bentonite having the greatest swelling capacity and Ca bentonite having the most compact lees [18]. In red wine production, fining is also commonly carried out but for different purposes, e.g., adjusting wine colour using charcoal and lowering tannins using egg albumin. Bentonite has been used in finished red wines to achieve colloidal stability and reduce astringency [14,19,20].

In red wine, proteins have also been reported and characterized in Cabernet Sauvignon [21] and a recent study suggested the PR proteins can limit the retention of added condensed tannins during red wine fermentation [22]. Thus, the removal of or reduction in proteins in red grapes may facilitate the extraction of phenolic substances from grape into juice and wine during fermentation. The typical bentonite addition rates for protein stabilization in white wines are between 0 and 1 g/L and the protein content in red grape juice is normally lower than in white grape juice, so in this study three common types of bentonites are added at 0.5 g/L, with the aim to investigate the impact of early protein removal by adding bentonite on Pinot noir wine composition.

2. Materials and Methods

2.1. Grapes and Wine Samples

Pinot noir grapes (clone/rootstock: AM 10-5/Swartzman) were harvested in 2021 from the Pegasus Bay vineyard located in north Canterbury. Four treatments, including control, were carried out in this study. In each treatment, the wines were produced in triplicate and, in each replicate, 700 g of destemmed and crushed grapes were placed in a 1 L plastic bucket. Three types of bentonite (Enartis Pacific, Napier, New Zealand), including sodium (Na), calcium (Ca), and sodium and calcium combined (NaCa) bentonites, were added at a rate of 0.5 g/L into grape must prior to the cold maceration (5 days). After five days cold maceration, the grape must was warmed up to room temperature at 20 °C and then inoculated with EC1118 yeast. Alcoholic fermentation was carried out in a temperature-controlled room at 28 °C and cap management was conducted once a day. The fermentation was considered finished when the residual sugar is less than 2 g/L and all the ferments have gone through three days of post-fermentation maceration. The free-run wines from different treatments were collected at the end of alcoholic fermentation for chemical analysis. No malolactic fermentation was carried out in this study as the aim of this study was to investigate the immediate effect of bentonite addition on Pinot noir wine composition at the end of alcoholic fermentation.

2.2. Oenological Parameters

The free-run wines collected at the end of alcoholic fermentation were analysed for pH, titratable acidity (TA), and alcohol content according to the methods described previously [23].

2.3. Methylcellulose Precipitable Tannins

The total tannins were measured in the free-run wines collected at the end of the alcoholic fermentation using the methylcellulose precipitation method [24].

2.4. Analysis of Pathogenesis-Related Proteins by HPLC

Two major pathogenesis-related (PR) proteins, thaumatin-like proteins (TLPs) and chitinases, were measured in grape juice at crushing, after 5 days' cold maceration, and in resultant wines according to the method published previously [25]. In brief, the juice/wine samples (50 µL) were loaded at 1 mL/min and the elution of proteins was monitored by absorbance at 210, 220, 260, 280, and 320 nm. The identification of thaumatin-like proteins (TLPs) and chitinases was assigned from the 210 nm chromatogram by comparison of the peak retention times to the purified TLPs and chitinases. The protein was quantified through a comparison with the peak area of thaumatin from *Thaumatococcus daniellii* (Sigma–Aldrich) and thus the protein concentration was expressed as thaumatin equivalents.

2.5. Colour Parameters in Resultant Wines

The wine samples were analysed for colour parameters using the modified Somer's assay [24]. All wine samples were diluted by 10 times in the model wine (0.5% w/v tartaric acid in 12% v/v ethanol adjusted to pH 3.4 with 5 M NaOH) and treated with 0.375% w/v sodium metabisulphite, 0.1% v/v acetaldehyde, and 1 M HCl, respectively. The absorbance of the mixtures was measured at 280 nm, 420 nm, and 520 nm, respectively, after incubation using UV-Visible spectrophotometer (Model UV-1800, Shimadzu Corporation, Kyoto, Japan) and colour parameters were calculated using the formulas described in the method.

2.6. Aroma Profiling by SPME GC-MS

The volatile aroma compounds in the wine samples were analysed according to the previously published method [26]. Two different methods were used to quantify two groups of aroma compounds: esters and alcohols and low concentration compounds. In brief, the esters and alcohols were analysed using a headspace solid-phase micro-extraction (HS-SPME)-GC/MS method. The wine samples (0.9 mL) were pipetted together with 8.06 mL of pH 3.5 acidified water, followed by 40 μL of deuterated internal standard solution, and 4.5 g of sodium chloride into a 20 mL SPME vial. The samples were incubated and agitated for 10 min at 60 °C. The SPME fibre was conditioned for 60 min at 60 °C before being desorbed in the injection port at 270 °C for 5 min. The GC/MS was equipped with dual columns. The MS source was operated in the electron impact (EI) mode with an ionization energy of 70 eV. The analysis of the chromatograms was performed on GC/MS Solution software, version 2.5 (Shimadzu, Auckland, NZ). The compounds of low concentration were analysed by changing the acquisition mode to the selected ion monitoring (SIM) to increase the sensitivity.

2.7. Statistical Analysis

All data were presented as means and standard deviations of three replicates, which were analysed using one-way analysis of variance (ANOVA), followed by a post hoc analysis (Tukey's test $p \leq 0.05$) using Minitab 18 (Minitab Inc., State College, PA, USA).

3. Results

3.1. Fermentation Kinetics and Oenological Parameters

The grapes were harvested with total soluble solids (TSS) at 22° Brix and the titratable acidity (TA) and pH at harvest were measured at 9.51 g/L and 3.14, respectively. The addition of bentonite prior to the cold soaking did not show a significant impact on fermentation dynamics. The fermentation progress was measured as ferment cumulative weight loss and all the fermentations in different treatments were finished after eight days (Figure 1). The pH, TA, and alcohol content in resultant wines were determined at the ranges of 3.62–3.67, 6.71–6.88 g/L, and 12.09–12.31% abv, respectively (Table 1). No significant difference in pH, TA, and alcohol content was observed in the resultant wines between treatments. There was also no significant difference in the tannin concentration between treatments, with the concentration ranging from 890 mg/L to 929 mg/L.

Table 1. Oenological parameters and tannins measured in resultant wines.

	Control	Na	Ca	NaCa
pH	3.65 ± 0.11 a	3.63 ± 0.03 a	3.62 ± 0.02 a	3.67 ± 0.01 a
TA (g/L)	6.88 ± 0.08 a	6.73 ± 0.15 a	6.88 ± 0.04 a	6.71 ± 0.04 a
Ethanol (%)	12.11 ± 0.22 a	12.14 ± 0.26 a	12.09 ± 0.32 a	12.31 ± 0.22 a
Tannin (mg/L)	890 ± 98 a	929 ± 61 a	929 ± 95 a	918 ± 17 a

Different letters in the same row indicate significant differences statistically between treatments.

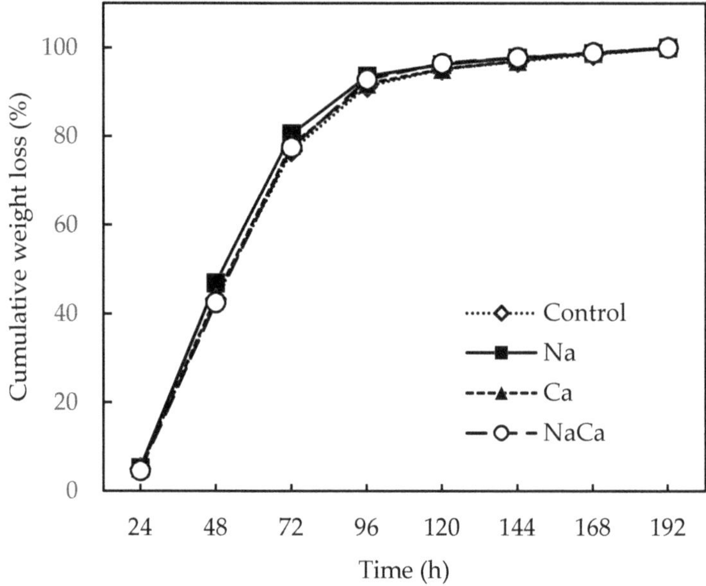

Figure 1. Fermentation progress of Pinot noir wine with or without bentonite addition, measured as ferment cumulative weight loss.

3.2. Pathogenesis-Related Proteins in Juice

The free run Pinot noir grape juice was analysed for both TLPs and chitinases, with concentrations determined at 70.4 mg/L and 58.7 mg/L, respectively (Table 2). After five days cold soaking, the concentrations of both TLPs and chitinases decreased to 37.5 mg/L and 38.7 mg/L, respectively.

Table 2. Analysis of pathogenesis-related proteins in juice with and without bentonite addition.

PR Proteins (mg/L)	Juice at Crushing	Juice after Cold Soaking			
		Control	Na	Ca	NaCa
TLPs	70.4 ± 0.6 a	37.5 ± 4.0 b	16.3 ± 6.1 c	25.5 ± 4.5 bc	28.2 ± 5.9 bc
Chitinases	58.7 ± 2.7 a	38.7 ± 4.8 b	28.1 ± 10.0 b	33.1 ± 5.6 b	33.9 ± 6.8 b

Different letters in the same row indicate significant differences statistically between treatments.

In the treatments with a bentonite addition, the concentration of TLPs and chitinases was decreased even further, with the Na bentonite-added treatment showing the lowest level of TLPs and chitinases at 16.3 mg/L and 28.1 mg/L, respectively. Compared to the control, the concentrations of TLPs and chitinases were not significantly different in treatments added with a Ca bentonite or NaCa bentonite. After fermentation, no PR proteins were observed in wines from any treatments (data not shown).

3.3. Wine Colour and Phenolics

The colour parameters measured by the modified Somer's assay are shown in Table 3. At the end of fermentation, all colour parameters except chemical age I, hue, and total phenolics showed significant differences between the treatments. However, Na bentonite addition seems to have a greater effect on wine colour compared to the other types of bentonites used in this study. The Na bentonite addition showed a significantly higher chemical age II, colour density, SO_2-corrected colour density, and SO_2 resistant pigments, compared to the control treatment. In contrast, the addition of Ca or NaCa bentonite showed no significant difference in chemical age I, colour density, SO_2 corrected colour density, hue, and SO_2 resistant pigments in wines compared to the control treatment.

Table 3. Colour parameters of resultant wines measured by modified Somer's assay.

Parameters	Control	Na	Ca	NaCa
Chemical age I	0.44 ± 0.02 a	0.47 ± 0.01 a	0.47 ± 0.02 a	0.45 ± 0.01 a
Chemical age II	0.13 ± 0.01 a	0.16 ± 0.01 b	0.17 ± 0.01 b	0.15 ± 0.01 ab
Degree of ionization of anthocyanins (%)	21.11 ± 1.38 a	25.56 ± 1.92 b	26.83 ± 1.90 b	23.18 ± 0.97 ab
Total anthocyanin (mg/L)	170.89 ± 8.06 a	150.89 ± 7.82 b	139.44 ± 7.00 b	153.89 ± 2.84 ab
Colour density	6.37 ± 0.21 a	7.32 ± 0.28 b	7.02 ± 0.50 ab	6.48 ± 0.35 ab
SO_2 corrected colour density	6.45 ± 0.17 a	7.26 ± 0.30 b	6.99 ± 0.36 ab	6.52 ± 0.34 ab
Hue	0.96 ± 0.03 a	1.04 ± 0.03 a	1.01 ± 0.06 a	0.98 ± 0.03 a
SO_2 resistant pigments	1.45 ± 0.05 a	1.67 ± 0.09 b	1.62 ± 0.10 ab	1.48 ± 0.10 ab
Total phenolics	32.88 ± 0.68 a	33.50 ± 0.75 a	30.00 ± 3.00 a	31.42 ± 0.40 a

Different letters in the same row indicate significant differences statistically between treatments.

The total anthocyanin concentration was reduced by 10–18% in all the bentonite-treated wines compared to the control treatment, although the reduction degree varied in each wine. For example, the NaCa bentonite addition did not show a significant difference in the total anthocyanin concentration, while the Na and Ca bentonite addition significantly reduced the total anthocyanin level in wines compared to the control treatment. The degree of ionization of anthocyanin (DIA) in wines ranged from 21.11% to 25.74%, with a significantly higher DIA observed in the Na and Ca bentonite treatments compared to the control treatment. A significantly higher colour density and SO_2 corrected colour density values were observed in Na-bentonite-added treatment. Furthermore, the Na bentonite addition also showed a significantly higher SO_2 resistant pigments compared to the control treatment. However, the Na bentonite-added treatment did not show significantly higher total phenolic concentration but still showed averagely higher total phenolic results compared to the control treatment.

3.4. Wine Aroma Composition Affected by Bentonite Addition

There were 36 aroma compounds, including esters, higher alcohols, terpenes, and volatile phenols, analysed in the resultant Pinot noir wines, and the concentrations of these aroma compounds are shown in Table 4.

There was no significant difference in the concentrations of aroma compounds between treatments except for three aroma compounds, ethyl cinnamate, hexyl acetate, and *cis*-3-hexenol, which showed significantly lower concentrations in the treatments added with bentonite compared to the control. The significantly higher concentrations of ethyl cinnamate, hexyl acetate and *cis*-3-hexenol in the control Pinot noir wines were determined at 1.15 μg/L, 4.3 μg/L, and 51.08 μg/L, respectively. The wines from Na bentonite treatment showed significantly lower concentrations of ethyl cinnamate and hexyl acetate at 0.61 μg/L and 3.48 μg/L, respectively. The wines from the Ca bentonite treatment showed significantly lower concentrations of ethyl cinnamate and *cis*-3-hexenol at 0.55 μg/L and 36.54 μg/L, respectively. The wines from the NaCa bentonite treatment showed significantly lower concentrations of ethyl cinnamate, hexyl acetate, and *cis*-3-hexenol at 0.71 μg/L, 3.54 μg/L, and 38.33 μg/L, respectively.

In general, the addition of bentonite at the level of 0.5 g/L prior to the cold soaking seems to have no impact on the aroma profile in resultant Pinot noir wines except for the aforementioned three aroma compounds.

Table 4. Quantification of aroma compounds in resultant Pinot noir wines.

Aroma Compounds *	Control	Na	Ca	NaCa
Esters				
Ethyl 2-Methyl Butyrate	5.75 ± 0.41 a	5.41 ± 0.57 a	5.07 ± 0.49 a	5.46 ± 0.58 a
Ethyl Hydrocinnamate	0.78 ± 0.17 a	0.59 ± 0.03 a	0.57 ± 0.05 a	0.60 ± 0.06 a
Ethyl Cinnamate	1.15 ± 0.30 a	0.61 ± 0.06 b	0.55 ± 0.08 b	0.71 ± 0.03 b
Ethyl Acetate (mg/L)	58.03 ± 5.80 a	53.46 ± 7.39 a	50.59 ± 2.88 a	57.39 ± 4.93 a
Ethyl Isobutyrate	29.86 ± 2.60 a	26.62 ± 0.74 a	26.08 ± 1.87 a	26.78 ± 2.25 a
Ethyl Butanoate	292.78 ± 18.94 a	285.52 ± 28.12 a	255.54 ± 13.50 a	290.97 ± 29.70 a
Ethyl Isovalerate	5.92 ± 1.27 a	5.48 ± 0.18 a	5.29 ± 0.70 a	6.12 ± 0.79 a
Ethyl Pentanoate	1.77 ± 0.14 a	1.67 ± 0.10 a	1.64 ± 0.10 a	1.76 ± 0.13 a
Ethyl Hexanoate	1020.45 ± 16.90 a	1057.21 ± 66.13 a	1004.36 ± 51.83 a	998.97 ± 60.26 a
Ethyl Lactate	3413.55 ± 161.66 a	3304.15 ± 233.72 a	3258.34 ± 44.85 a	3432.74 ± 123.63 a
Ethyl Heptanoate	8.18 ± 0.29 a	7.66 ± 1.11 a	8.19 ± 0.35 a	7.70 ± 0.86 a
Ethyl Octanoate	2369.14 ± 34.52 a	2659.39 ± 285.58 a	2446.15 ± 131.14 a	2298.64 ± 175.76 a
Isoamyl Acetate	188.70 ± 7.77 a	167.30 ± 7.93 a	156.20 ± 22.90 a	172.45 ± 3.08 a
Isobutyl Acetate	49.99 ± 2.58 a	45.93 ± 2.52 a	37.65 ± 9.11 a	46.00 ± 2.77 a
Octyl Acetate	10.00 ± 0.56 a	10.73 ± 0.45 a	9.15 ± 2.31 a	9.35 ± 0.91 a
2-Phenylethyl Acetate	17.88 ± 1.03 a	16.63 ± 1.50 a	15.21 ± 1.84 a	16.12 ± 0.71 a
Hexyl Acetate	4.30 ± 0.23 a	3.48 ± 0.07 b	4.31 ± 0.49 a	3.54 ± 0.20 b
Higher alcohols				
Trans-2-Hexenol	4.19 ± 0.47 a	4.63 ± 0.48 a	4.07 ± 0.24 a	3.94 ± 0.79 a
1-Octanol	42.49 ± 3.03 a	38.87 ± 5.98 a	34.35 ± 8.90 a	40.42 ± 6.70 a
Isoamyl Alcohol (mg/L)	201.74 ± 6.79 a	199.58 ± 6.91 a	194.38 ± 10.16 a	201.97 ± 2.98 a
Hexanol	2670.13 ± 124.75 a	2445.67 ± 124.61 a	2440.31 ± 77.77 a	2429.43 ± 121.52 a
Trans-3-Hexenol	51.54 ± 3.63 a	49.32 ± 0.71 a	47.94 ± 2.59 a	49.45 ± 2.46 a
Cis-3-Hexenol	51.08 ± 6.76 a	39.56 ± 6.15 ab	36.54 ± 1.98 b	38.33 ± 2.48 b
1-Heptanol	80.03 ± 4.61 a	72.40 ± 4.13 a	72.21 ± 3.32 a	76.55 ± 3.01 a
Phenyethyl Alcohol (mg/L)	59.41 ± 3.70 a	57.66 ± 4.00 a	57.28 ± 1.07 a	58.40 ± 1.65 a
Terpenes				
Linalool	36.65 ± 2.31 a	36.21 ± 1.21 a	35.20 ± 3.22 a	37.91 ± 0.97 a
Citronellol	24.72 ± 0.83 a	24.06 ± 0.18 a	23.89 ± 0.96 a	24.93 ± 0.94 a
Nerol	8.58 ± 0.54 a	8.61 ± 0.60 a	8.13 ± 0.69 a	9.11 ± 0.39 a
β-Damascenone	12.44 ± 0.45 a	11.82 ± 0.46 a	12.30 ± 1.12 a	12.50 ± 0.25 a
Geraniol	9.75 ± 0.39 a	9.73 ± 0.20 a	9.38 ± 0.40 a	10.06 ± 0.47 a
α-Ionone	ND	ND	ND	ND
β-Ionone	0.63 ± 0.06 a	0.64 ± 0.02 a	0.65 ± 0.03 a	0.61 ± 0.00 a
Volatile phenols				
Guaiacol	5.45 ± 0.61 a	4.02 ± 0.68 a	3.81 ± 1.27 a	5.40 ± 1.18 a
Phenol	3.80 ± 0.27 a	3.65 ± 0.12 a	3.98 ± 0.26 a	4.21 ± 0.37 a
4-Ethyl Guaiacol	0.10 ± 0.02 a	0.09 ± 0.01 a	0.10 ± 0.02 a	0.10 ± 0.00 a
Eugenol	3.79 ± 0.16 a	4.06 ± 0.49 a	4.49 ± 0.23 a	4.34 ± 0.22 a

* concentration of aroma compounds is expressed as μg/L, except of three compounds specified in the table that are expressed as mg/L; ND: not detected; different letters in the same row indicate significant differences statistically between treatments.

4. Discussion

As expected, the addition of bentonite prior to the cold soaking had no impact on general oenological parameters. After cold soaking, the concentration of the PR proteins was significantly reduced in all treatments, including the control. The reduction in PR proteins during cold soaking might be due to their interactions with phenolic compounds [27] and a consequent complex may precipitate during the cold soaking process. Treatments added with bentonite showed a further reduction in PR proteins, especially a significant reduction in TLPs observed in Na bentonite treatment. Bentonite can bind PR proteins and remove them, which has been well studied in white wine [28]. The greater reduction in PR proteins by adding Na bentonite could be associated with its higher swelling ability and cation exchange capacity compared to Ca and NaCa bentonites [29].

The reduction in anthocyanins due to the bentonite addition observed in this study is in agreement with previous studies [14,20,30]. The decrease in total anthocyanins could be associated with the binding ability of bentonite with anthocyanins. The bentonite indirectly binds phenolic compounds that have complexed with proteins and they can also bind anthocyanins, with a resulting loss in wine colour [31]. However, a previous study [32] reported that the addition of Na bentonite at rate of 15 g/100 g of grapes during pre-fermentation maceration increased the malvidin concentration in wines due to the removal of suspended solids and yeast lees in the fermenting juice, which consequently prevents the adsorption of anthocyanins into solids. The contradictory results might be due to the much higher dosage rate of bentonite.

The degree of ionization of anthocyanin (DIA) in wines ranged from 21.11% to 25.74%, which is in the range previously reported in young red wines [33]. A significantly higher DIA were observed in Na and Ca treatments compared to the control treatment. However, bentonite has proven to be an important agent for the removal of colouring matters and is constituted of ionized anthocyanins (flavylium cations), tannins, polysaccharides, and proteins when added in a 0.2–0.5 g/L dose rate in wines [19]. The increased degree of ionization of anthocyanin in bentonite-treated wines may be due to the lower anthocyanin levels observed in Na and Ca treatments, which has been observed in our previous study that wines with lower anthocyanin (e.g., older wines) showed a higher degree of ionization of anthocyanins (data not shown).

The higher colour density and SO_2-corrected colour density values were observed in Na bentonite treatments, which agrees with a previous study on Monastrell wines [31]. The increase in colour density may be attributed to the self-association and co-pigmentation of anthocyanins and the higher concentration of ionized anthocyanins found in wine. Interestingly, some studies [14,30,34] suggested that the addition of bentonite (0.5 g/L) at the end of fermentation could reduce the colour density in wines by 1–7%. This suggests that the wine colour could be influenced by the timing of bentonite addition and vary between grape varieties.

The Na bentonite addition also showed significantly higher SO_2 resistant pigments compared to the control treatment. This might associate with the reduced concentrations of PR proteins observed in Na bentonite treatment. A previous study [22] reported that PR proteins could limit the retention of added tannins and, thus, reducing the PR proteins is likely to aid tannin retention, which may favour the formation of polymeric pigments by the interaction with anthocyanins and increase the colour density in wine.

Although 36 aroma compounds have been analysed in resultant Pinot noir wines, only three aroma compounds—ethyl cinnamate, hexyl acetate, and *cis*-3-hexenol—showed significant differences between the treatments. Ethyl cinnamate could contribute to sweet floral notes to the wine, but it has a relatively short life in wine as it can slowly hydrolyse to form alcohol [35], so its reduction in bentonite-added treatments may have little impact on Pinot noir in long term. The concentration of hexyl acetate, contributing red berry aromas to the wine, was significantly reduced in the Na and NaCa bentonite treatments, but its concentrations in all the treatments were almost 100 times lower than its perception threshold at 400 μg/L [36]. The *cis*-3-hexenol has been reported to be associated with grassy herbaceous notes in wine [37]. The lower level of *cis*-3-hexenol in Pinot noir wines produced from bentonite treatments may reduce the green notes that are normally negatively associated with Pinot noir wine quality [38].

5. Conclusions

This study reveals that the bentonite addition prior to fermentation had a significant impact on the Pinot noir wine colour but little impact on the aroma profile at the end of fermentation. One aroma compound, *cis*-3-hexenol, showed significantly lower levels in bentonite-added treatments, which may positively contribute to Pinot noir wine quality by reducing the negative green notes. The significantly higher level of colour density and SO_2 resistant pigments observed in Na bentonite treatment suggests more colour stability in resultant wine, which is beneficial for varieties similar to Pinot noir that are challenging in colour extraction. In this study, PR proteins still remained after cold soaking, indicating the dosage rate at 0.5 g/L is not sufficient to remove all the PR proteins in the must. Further studies could investigate if a higher dosage rate of bentonite addition could further increase the formation of un-bleachable pigments and even increase the extraction of tannins in Pinot noir wine, without having a negative impact on the aroma profile.

Author Contributions: Conceptualization, B.T. and T.R.; methodology, B.T.; investigation, P.M.W.; writing—original draft preparation, P.M.W. and B.T.; writing—review and editing, P.M.W., T.R. and B.T.; supervision, B.T.; project administration, B.T.; funding acquisition, T.R. All authors have read and agreed to the published version of the manuscript.

Funding: This research was funded by New Zealand Winegrowers Research Centre, operating as Bragato Research Institute.

Institutional Review Board Statement: Not applicable.

Informed Consent Statement: Not applicable.

Data Availability Statement: Not applicable.

Acknowledgments: Authors are grateful for assistance received from Jenny Zhao for HPLC analysis and Jason Breitmeyer for GC-MS analysis.

Conflicts of Interest: The authors declare no conflict of interest.

References

1. Harrison, R. Practical interventions that influence the sensory attributes of red wines related to the phenolic composition of grapes: A review. *Int. J. Food Sci. Technol.* **2018**, *53*, 3–18. [CrossRef]
2. González-Muñoz, B.; Garrido-Vargas, F.; Pavez, C.; Osorio, F.; Chen, J.; Bordeu, E.; O'Brien, J.A.; Brossard, N. Wine astringency: More than just tannin–protein interactions. *J. Sci. Food Agric.* **2022**, *102*, 1771–1781. [CrossRef]
3. Brandão, E.; Silva, M.S.; García-Estévez, I.; Williams, P.; Mateus, N.; Doco, T.; de Freitas, V.; Soares, S. The role of wine polysaccharides on salivary protein-tannin interaction: A molecular approach. *Carbohydr. Polym.* **2017**, *177*, 77–85. [CrossRef]
4. McRae, J.M.; Kennedy, J.A. Wine and grape tannin interactions with salivary proteins and their impact on astringency: A review of current research. *Molecules* **2011**, *16*, 2348–2364. [CrossRef]
5. McManus, J.P.; Davis, K.G.; Beart, J.E.; Gaffney, S.H.; Lilley, T.H.; Haslam, E. Polyphenol interactions. Part 1. Introduction; some observations on the reversible complexation of polyphenols with proteins and polysaccharides. *J. Chem. Soc. Perkin Trans.* **1985**, *2*, 1429–1438. [CrossRef]
6. Le Bourse, D.; Conreux, A.; Villaume, S.; Lameiras, P.; Nuzillard, J.M.; Jeandet, P. Quantification of chitinase and thaumatin-like proteins in grape juices and wines. *Anal. Bioanal. Chem.* **2011**, *401*, 1541–1549. [CrossRef]
7. Sauvage, F.X.; Bach, B.; Moutounet, M.; Vernhet, A. Proteins in white wines: Thermo-sensitivity and differential adsorbtion by bentonite. *Food Chem.* **2010**, *118*, 26–34. [CrossRef]
8. Marangon, M.; Van Sluyter, S.C.; Haynes, P.A.; Waters, E.J. Grape and wine proteins: Their fractionation by hydrophobic interaction chromatography and identification by chromatographic and proteomic analysis. *J. Agric. Food Chem.* **2009**, *57*, 4415–4425. [CrossRef]
9. Tian, B.; Harrison, R.; Morton, J.; Deb-Choudhury, S. Proteomic analysis of Sauvignon blanc grape skin, pulp and seed and relative quantification of pathogenesis-related proteins. *PLoS ONE* **2015**, *10*, e0130132. [CrossRef]
10. Tian, B.; Harrison, R.; Morton, J.; Jaspers, M. Changes in pathogenesis-related proteins and phenolics in *Vitis vinifera* L. cv. 'Sauvignon Blanc' grape skin and pulp during ripening. *Sci. Hortic.* **2019**, *243*, 78–83. [CrossRef]
11. Monteiro, S.; Piçarra-Pereira, M.A.; Loureiro, V.B.; Teixeira, A.R.; Ferreira, R.B. The diversity of pathogenesis-related proteins decreases during grape maturation. *Phytochemistry* **2007**, *68*, 416–425. [CrossRef]
12. Tian, B.; Harrison, R.; Jaspers, M.; Morton, J. Influence of ultraviolet exclusion and of powdery mildew infection on Sauvignon Blanc grape composition and on extraction of pathogenesis-related proteins into juice. *Aust. J. Grape Wine Res.* **2015**, *21*, 417–424. [CrossRef]
13. Batista, L.; Monteiro, S.; Loureiro, V.B.; Teixeira, A.R.; Ferreira, R.B. The complexity of protein haze formation in wines. *Food Chem.* **2009**, *112*, 169–177. [CrossRef]
14. Dordoni, R.; Galasi, R.; Colangelo, D.; De Faveri, D.M.; Lambri, M. Effects of fining with different bentonite labels and doses on colloidal stability and colour of a Valpolicella red wine. *Int. J. Food Sci. Technol.* **2015**, *50*, 2246–2254. [CrossRef]
15. Lambri, M.; Dordoni, R.; Silva, A.; De Faveri, D.M. Comparing the impact of bentonite addition for both must clarification and wine fining on the chemical profile of wine from Chambave Muscat grapes. *Int. J. Food Sci. Technol.* **2012**, *47*, 1–12. [CrossRef]
16. Lambri, M.; Dordoni, R.; Silva, A.; De Faveri, D.M. Effect of bentonite fining on odor-active compounds in two different white wine styles. *Am. J. Enol. Vitic.* **2010**, *61*, 225–233.

17. Au, P.-I.; Leong, Y.-K. Rheological and zeta potential behaviour of kaolin and bentonite composite slurries. *Colloids Surf. A Physicochem. Eng. Asp.* **2013**, *436*, 530–541. [CrossRef]
18. Gougeon, R.D.; Soulard, M.; Miehé-Brendlé, J.; Chézeau, J.-M.; Le Dred, R.; Jeandet, P.; Marchal, R. Analysis of two bentonites of enological interest before and after commercial activation by solid Na_2CO_3. *J. Agric. Food Chem.* **2003**, *51*, 4096–4100. [CrossRef]
19. Ribéreau-Gayon, P.; Glories, Y.; Maujean, A.; Dubourdieu, D. *Handbook of Enology, Volume 2: The Chemistry of Wine-Stabilization and Treatments*; John Wiley & Sons: Hoboken, NJ, USA, 2021; pp. 281–530.
20. Stankovic, S.; Jovic, S.; Zivkovic, J.; Pavlovic, R. Influence of age on red wine colour during fining with bentonite and gelatin. *Int. J. Food Prop.* **2012**, *15*, 326–335. [CrossRef]
21. Mainente, F.; Zoccatelli, G.; Lorenzini, M.; Cecconi, D.; Vincenzi, S.; Rizzi, C.; Simonato, B. Red wine proteins: Two dimensional (2-D) electrophoresis and mass spectrometry analysis. *Food Chem.* **2014**, *164*, 413–417. [CrossRef]
22. Springer, L.F.; Sherwood, R.W.; Sacks, G.L. Pathogenesis-related proteins limit the retention of condensed tannin additions to red wines. *J. Agric. Food Chem.* **2016**, *64*, 1309–1317. [CrossRef]
23. Iland, P.; Bruer, N.; Edwards, G.; Weeks, S.; Wilkes, E. *Chemical Analysis of Grapes and Wine: Techniques and Concepts*; Patrick Iland Wine Promotions: Campbelltown, SA, Australia, 2004.
24. Mercurio, M.D.; Dambergs, R.G.; Herderich, M.J.; Smith, P.A. High throughput analysis of red wine and grape phenolics adaptation and validation of methyl cellulose precipitable tannin assay and modified somers color assay to a rapid 96 well plate format. *J. Agric. Food Chem.* **2007**, *55*, 4651–4657. [CrossRef]
25. Tian, B.; Harrison, R.; Morton, J.; Jaspers, M. Influence of skin contact and different extractants on extraction of proteins and phenolic substances in Sauvignon Blanc grape skin. *Aust. J. Grape Wine Res.* **2020**, *26*, 180–186. [CrossRef]
26. Wimalasiri, P.M.; Olejar, K.J.; Harrison, R.; Hider, R.; Tian, B. Whole bunch fermentation and the use of grape stems: Effect on phenolic and volatile aroma composition of Vitis vinifera cv. Pinot Noir wine. *Aust. J. Grape Wine Res.* **2022**, *28*, 395–406. [CrossRef]
27. Ozdal, T.; Capanoglu, E.; Altay, F. A review on protein–phenolic interactions and associated changes. *Food Res. Int.* **2013**, *51*, 954–970. [CrossRef]
28. Ferreira, R.B.; Picarra-Pereira, M.A.; Monteiro, S.; Loureiro, V.B.; Teixeira, A.R. The wine proteins. *Trends Food Sci. Technol.* **2002**, *12*, 230–239. [CrossRef]
29. Wijeyesekera, D.; Loh, E.; Diman, S.; John, A.; Zainorabidin, A.; Ciupala, M. Sustainability study of the application of geosynthetic clay liners in hostile and aggressive environments. *Int. J. Sustain. Dev.* **2012**, *5*, 81–96.
30. González-Neves, G.; Favre, G.; Gil, G. Effect of fining on the colour and pigment composition of young red wines. *Food Chem.* **2014**, *157*, 385–392. [CrossRef]
31. Donovan, J.L.; McCauley, J.C.; Nieto, N.T.; Waterhouse, A.L. *Effects of Small-Scale Fining on the Phenolic Composition and Antioxidant Activity of Merlot Wine*; ACS Publications: Washington, DC, USA, 1998.
32. Gómez-Plaza, E.; Gil-Muñoz, R.; López-Roca, J.; De La Hera-Orts, M.; Martínez-Cultíllas, A. Effect of the addition of bentonite and polyvinylpolypyrrolidone on the colour and long-term stability of red wines. *J. Wine Res.* **2000**, *11*, 223–231. [CrossRef]
33. Basalekou, M.; Pappas, C.; Kotseridis, Y.; Tarantilis, P.A.; Kontaxakis, E.; Kallithraka, S. Red Wine Age Estimation by the Alteration of Its Color Parameters: Fourier Transform Infrared Spectroscopy as a Tool to Monitor Wine Maturation Time. *J. Anal. Methods Chem.* **2017**, *2017*, 5767613–5767619. [CrossRef]
34. Ghanem, C.; Taillandier, P.; Rizk, M.; Rizk, Z.; Nehme, N.; Souchard, J.-P.; El Rayess, Y. Analysis of the impact of fining agents types, oenological tannins and mannoproteins and their concentrations on the phenolic composition of red wine. *LWT-Food Sci. Technol.* **2017**, *83*, 101–109. [CrossRef]
35. Ferreira, V.; López, R.; Cacho, J.F. Quantitative determination of the odorants of young red wines from different grape varieties. *J. Sci. Food Agric.* **2000**, *80*, 1659–1667. [CrossRef]
36. Benkwitz, F.; Tominaga, T.; Kilmartin, P.A.; Lund, C.; Wohlers, M.; Nicolau, L. Identifying the chemical composition related to the distinct aroma characteristics of New Zealand Sauvignon blanc wines. *Am. J. Enol. Vitic.* **2012**, *63*, 62–72. [CrossRef]
37. Herbst-Johnstone, M.; Araujo, L.; Allen, T.; Logan, G.; Nicolau, L.; Kilmartin, P. Effects of mechanical harvesting on 'Sauvignon blanc' aroma. In Proceedings of the I International Workshop on Vineyard Mechanization and Grape and Wine Quality 978, Piacenza, Italy, 27–29 June 2012; pp. 179–186.
38. Parr, W.V.; Grose, C.; Hedderley, D.; Maraboli, M.M.; Masters, O.; Araujo, L.D.; Valentin, D. Perception of quality and complexity in wine and their links to varietal typicality: An investigation involving Pinot noir wine and professional tasters. *Food Res. Int.* **2020**, *137*, 109423. [CrossRef]

 *fermentation*

Article

Enhancing Ethanol Tolerance via the Mutational Breeding of *Pichia terricola* H5 to Improve the Flavor Profiles of Wine

Jie Gao [1,†], Xiuli He [1,2,†], Weidong Huang [1], Yilin You [1,*] and Jicheng Zhan [1,*]

1 Beijing Key Laboratory of Viticulture and Enology, College of Food Science and Nutritional Engineering, China Agricultural University, Tsinghua East Road 17, Haidian District, Beijing 100083, China; bs20213060545@cau.edu.cn (J.G.); 18813019165@163.com (X.H.); huanggwd@263.net (W.H.)

2 Key Laboratory of Beijing Yanjing Beer Brewing Technology, Beijing Yan Jing Brewery Company Ltd., Beijing 101300, China

* Correspondence: yilinyou@cau.edu.cn (Y.Y.); zhanjicheng@cau.edu.cn (J.Z.)

† These authors contributed equally to this work.

Abstract: Although using non-*Saccharomyces* yeasts during alcoholic fermentation can improve the wine aroma, most of them are not ethanol tolerant; therefore, in 2017, this study screened 85 non-*Saccharomyces* yeasts isolated and identified from 24 vineyards in seven Chinese wine-producing regions, obtaining *Pichia terricola* strain H5, which displayed 8% ethanol tolerance. Strain H5 was subjected to ultraviolet (UV) irradiation and diethyl sulfate (DES) mutagenesis treatment to obtain mutant strains with different fermentation characteristics from the parental H5. Compared with strain H5, the UV-irradiated strains, UV5 and UV8, showed significantly higher ethanol tolerance and fermentation capacity. Modified aroma profiles were also evident in the fermentation samples exposed to the mutants. Increased ethyl caprate, ethyl caprylate, and ethyl dodecanoate content were apparent in the UV5 samples, providing the wine with a distinctly floral, fruity, and spicy profile. Fermentation with strain UV8 produced a high ethyl acetate concentration, causing the wine to present a highly unpleasant odor. To a certain extent, UV irradiation improved the ethanol tolerance and fermentation ability of strain H5, changing the wine aroma profile. This study provides a theoretical basis for the industrial application of non-*Saccharomyces* yeasts that can improve wine flavor.

Keywords: non-*Saccharomyces* yeasts; ethanol tolerance; ultraviolet irradiation; diethyl sulfate mutagenesis; fermentation

Citation: Gao, J.; He, X.; Huang, W.; You, Y.; Zhan, J. Enhancing Ethanol Tolerance via the Mutational Breeding of *Pichia terricola* H5 to Improve the Flavor Profiles of Wine. *Fermentation* **2022**, *8*, 149. https://doi.org/10.3390/fermentation8040149

Academic Editor: Agustín Aranda

Received: 22 February 2022
Accepted: 26 March 2022
Published: 28 March 2022

Publisher's Note: MDPI stays neutral with regard to jurisdictional claims in published maps and institutional affiliations.

Copyright: © 2022 by the authors. Licensee MDPI, Basel, Switzerland. This article is an open access article distributed under the terms and conditions of the Creative Commons Attribution (CC BY) license (https://creativecommons.org/licenses/by/4.0/).

1. Introduction

Wine making involves microbial interaction. Although bacteria, fungi, and yeasts participate during the fermentation process, yeasts play a leading role [1]. *Saccharomyces cerevisiae* is mainly responsible for alcoholic fermentation and is the dominant strain during the wine-making process; however, in addition to *Saccharomyces cerevisiae*, many non-*Saccharomyces* yeasts are also involved [2]. Non-*Saccharomyces* yeasts refer to a large group of yeasts other than *Saccharomyces cerevisiae*, which exist during all the stages of wine making after the grapes are harvested, including *Metschnikowia pulcherrima*, *Starmerella stellata*, *Torulaspora delbrueckii*, *Debaryomyces hansenii*, *Pichia fermentans*, *Hanseniaspora uvarum*, *Pichia terricola*, etc. [3–5]. They mainly improve the aroma of wine via metabolism and autolysis, and participate in the formation of complex, fresh flavor substances [6]. These yeasts improve the taste of wine, providing it with fruity, floral, and nutty aroma characteristics, which positively impacts the wine quality [7,8]. Previous studies have shown an increase in the acetate, phenethyl acetate, and 2-phenylethanol content in wine produced via the combined fermentation of *Hanseniaspora vineae* and *Saccharomyces cerevisiae*, while the ethyl ester proportion decreases, enhancing the fruity flavor and complexity of the wine [8]. Whitener et al. [9] studied *Kazachstania gamospora* fermentation, revealing that it produced more floral-scented esters (such as phenethyl propionate) than commercial

Saccharomyces cerevisiae. Mixed fermentation with *Pichia kluyveri* and *Saccharomyces cerevisiae* can increase the concentration of volatile mercaptans, especially hexyl 3-mercaptoacetate, which is recommended for the production of Sauvignon Blanc, Chardonnay, and Riesling wines [10].

Non-*Saccharomyces* yeasts represent the dominant microbiota during the early stage of wine making [11]; however, many factors affect the growth and metabolism of non-*Saccharomyces* yeasts during the complex fermentation process, such as ethanol, SO_2, temperature, sugar, and acid. Of these, ethanol stress is a vital influencing factor. The ethanol concentration continues to rise as fermentation progresses, leading to a continuous decline in the number of non-*Saccharomyces* yeasts during the middle and late stages, with some even dying [12], reducing their positive effect on the final wine quality. Different methods have been adopted for selection and cultivation to improve the yeast tolerance to high ethanol stress conditions. Using chemical mutagens and ultraviolet (UV) irradiation for random mutagenetic breeding presents advantages, such as a short mutagenetic cycle and high mutagenetic efficiency, providing practical strategies to modify the physiological yeast characteristics [13,14]. UV mutagenesis can induce cytosine to thymine nucleotide base substitution and DNA photoproducts produced by UV radiation, such as cyclobutane pyrimidine dimers (CPDS) and (6-4) pyrimidine–pyrimidone photoproducts (6-4 PPs), to block the DNA replication process [15,16]. Diethyl sulfate (DES) can alkylate some oxygen sites, including the ethylation of the sixth guanine oxygen atom. The mutagenicity of DES manifests in the occurrence of mutations via direct mispairing [17,18].

This study subjected strain H5 to UV irradiation and DES mutagenesis to obtain mutants with higher ethanol tolerance. Applying these mutants to fermentation revealed that their fermentation characteristics differed from the parental H5. Mutant UV5 added a distinctly floral, fruity, and spicy aroma to the sample, while strain UV8 produced a high ethyl acetate concentration, causing a highly unpleasant odor.

2. Materials and Methods

2.1. *Yeast Strains and Growth Conditions*

In 2017, the 85 non-*Saccharomyces* yeasts used in this experiment were identified and isolated from 24 vineyards in seven Chinese wine-producing regions. The yeast information is shown in Table S1.

The strains were stored at −80 °C in cryovials containing 20% (v/v) glycerol. The strains were cultured and propagated in YPD liquid medium (2% glucose, 2% peptone, and 1% yeast extract (all w/v)). During the study, the strains were cultured on the YPD agar medium and stored at 4 °C.

2.2. *Screening the Non-Saccharomyces Yeasts for Ethanol Tolerance*

2.2.1. High-Throughput 2,3,5-Triphenyltetrazolium Chloride (TTC) Medium Method for First Stage Selection

The method described in the previous study [19] was used with minor changes. Yeast cell suspensions were diluted in gradients and plated on the TTC lower medium (YPD agar medium at pH 5.5), while three parallel inoculations were performed for each yeast. The plates were incubated at 28 °C for 48 h and stained by a lay of TTC upper medium (0.5% glucose, 1.5% agar, and 0.05% TTC). The red colonies were selected after incubation at 28 °C in the dark for 3 h. TTC, a redox substance that can react with yeast metabolites, can be reduced to red by hydrogen. It can be used to judge the dehydrogenase activity in yeast, that is, the alcohol production capacity of yeast.

2.2.2. Detecting Non-*Saccharomyces* Yeast Activity under Ethanol Stress

The loopful volume of the cell suspensions (OD600 = 1.2) were streaked on YPD solid plates at ethanol volume fractions of 4%, 6%, 8%, 9%, 10%, 11%, and 12% (v/v). Three parallels were set for each treatment and incubated at 28 °C for 72 h. The colony growth on the plate was monitored, with the untreated plate as a control.

2.2.3. The Influence of Ethanol Addition on Yeast Cell Growth

Here, 2% of the yeast cell suspensions, which were diluted to OD600 of 1.2, were inoculated into the YPD liquid medium at ethanol concentrations of 0%, 4%, 6%, 8%, and 10% (*v*/*v*), and cultured at 28 °C in a rotary shaker at 180 rpm. Each treat was set in three parallels, and the untreated medium was used as a control. The absorbance of the yeast cultures was measured at 600 nm at 3-h intervals (at 6-h intervals in stationary phase) for 30 h to determine their growth rates under ethanol stress.

2.3. *Mutation Procedure of Non-Saccharomyces Yeast by DES and UV*

To obtain strains with higher ethanol tolerance, the cell suspension of strain H5 was subjected to DES mutagenesis and UV irradiation, using a method described by Yi et al. [20] with minor modifications. The treatment conditions with a lethality rate of about 80% were set as the mutation conditions [21]. During UV irradiation, the cell suspension (OD600 = 1.2) was spread on the YPD agar plate by the gradient dilution coating plate method, and irradiated with UV (20 W) at a distance of 50 cm for 0, 10, 20, 30, 40, 50, 60, 70, 80, 90, 100, 110, and 120 s. After irradiation, the plates were immediately shielded from light and incubated at 28 °C in the dark for 48 h. During DES treatment, 10 mL of the H5 bacterial suspension cultured to the mid-log phase (OD600 = 1.2) was centrifuged at 805× *g* for 10 min. The precipitate was washed twice with phosphate buffer (pH 7.2), and then redissolved in 10 mL of phosphate buffer. The cell suspension was treated with 10 mL of 1% DES in phosphate buffer (pH 7.2) for 0, 2, 4, 6, 8, 10, 12, 14, 16, 18, and 20 min, after which the procedure was terminated with 10 mL of 25% sodium thiosulfate. After the gradient dilution of the treated cells with sterile saline, they were spread on a YPD agar plate and incubated at 28 °C for 48 h.

The lethality was computed as follows:

Lethality (%) = (Number of strains before mutation − Number of strains after mutation)/(Number of strains before mutation) × 100.

2.4. *Identification of Yeast*

The strains requiring identification were inoculated in YPD liquid medium and incubated for 12 h at 28 °C while exposed to constant shaking at 160 rpm. The yeast cells were harvested after the centrifugation of 2 mL of the cell suspension, and the ribosomal DNA genomes of yeast were extracted using a TIANamp Yeast DNA Kit (by TIANGEN Biotech (Beijing) Co., Ltd., Beijing, China). ITS1 (5′-TCCGTAGGTGAACCTGCGG-3′) and ITS4 (5′-TCCTCCGCTTATTGATATGC-3′) [22] primers were used for PCR amplification of the 5.8S-ITS rDNA sequence [23]. The PCR cycling conditions were as follows: 95 °C pre-degeneration for 5 min; 95 °C degeneration for 1 min, 52 °C annealing for 2 min, 72 °C extension for 2 min 20 s, repeated 35 times; complementary extension at 72 °C for 10 min. The reaction system consisted of 50 μL 2×TSINGKE® MasterMix 25 μL, 10 μ MF-primer 1 μL, 10 μM R-primer 1 μL, template DNA 1 μL, and DD water 22 μL. The PCR products were detected via 1% agarose gel electrophoresis and sent to Beijing Tsingke Biological Technology for sequencing. The gene sequences of the strains were searched on the NCBI website (https://blast.ncbi.nlm.nih.gov/Blast.cgi) (21 February 2022), and compared with similar sequences using MEGABLAST to preliminarily determine the strain species based on the similarity of homologous sequences. The strains were then compared with the ITS sequences of the most similar type strains found in the MEGABLAST search by pairwise Blast alignment.

2.5. *The Practical Fermentation of Selected Strains*

Cabernet Sauvignon grape juice (2018, Changli Hebei) was used for single-strain fermentation. Grape juice stored at −20 °C was thawed at 4 °C. After centrifugation at 13,919× *g* for 15 min at 4 °C, the grape juice was sterilized using 0.65 μm and then 0.45 μm filter membranes. Separate 500 mL fermentation bottles were filled with 200 mL of sterile grape juice. The strains were pre-cultured in YPD medium at 28 °C and 120 rpm for 24 h and then statically incubated in sterile grape juice at 28 °C for 72 h to obtain 10^7 CFU/mL cells. The pre-cultured strains were inoculated with 1% inoculum at 0 d of fermentation. With commercial *Saccharomyces cerevisiae* S1 as the control group, all strains were subjected to single-strain fermentation, and each fermentation group was triplicated. Each group was subjected to static fermentation at 20 °C, and samples were taken at 0, 2, 4, 7, 10, 13, and 16 d of fermentation to determine the number of yeast colonies, °Brix, and pH value, as well as the residual sugar, glycerin, ethanol, acid, and volatile aroma compound content. The remaining samples were stored at −20 °C for high-performance liquid chromatography (HPLC) and headspace-solid phase microextraction/gas chromatography–mass spectrometry flame-ionization detection (HS-SPME/GC-MS-FID).

2.6. Determining the Physicochemical Indexes via HPLC

A method described by Sun et al. [24] was used to determine the content of glucose, fructose, glycerol, and ethanol. The chromatographic conditions included an Aminex HPX-87H (300 mm × 7.8 mm) analytical column at a temperature of 55 °C; the differential detector was Waters-2414 (Waters, Dublin, Ireland); an internal temperature of 40 °C; respective sample injection volumes of 10 µL, a mobile phase of 0.005 mol/L H_2SO_4, a flow rate of 0.5 mL/min, and isocratic elution. Qualitative analysis was based on retention time. Quantitative calculation occurred according to peak areas.

The oxalic acid, citric acid, tartaric acid, malic acid, succinic acid, lactic acid, and acetic acid content were determined according to a method described by Chen et al. [25], with minor modifications. The chromatographic conditions included an Aminex HPX-87H (300 mm × 7.8 mm) analytical column at a temperature of 40 °C; the photodiode array detector was Waters-2996 (Waters, Dublin, Ireland); a detection wavelength of 210 nm; a mobile phase of 0.005 mol/L H_2SO_4; a flow rate of 0.4 mL/min; respective injection volumes of 20 µL; isocratic elution. Qualitative analysis was based on retention time. The quantification occurred according to the peak area.

2.7. Determining the Aroma Compounds via HS-SPME/GC-MS-FID

The volatile aroma compounds were determined using a method delineated by Chen, Chia, and Liu [25]. HS-SPME/GC-MS-FID was used to determine the volatile aroma compounds in the sample. The pH of wine samples was adjusted to 2.5 ± 0.1 using 1 mol/L hydrochloric acid. A 20 mL headspace bottle was filled with 5.0 mL of the wine sample and sealed with a cap with a rubber septum. The injection port temperature was 250 °C, the carrier gas (He) flow rate was 1.0 mL/min, and the splitless injection mode was selected. The temperature programming conditions included an initial temperature of 50 °C for 5 min, which was increased to 230 °C at a rate of 5 °C/min, and maintained for 30 min. The signals were acquired in full scan mode via electron ionization (EI) at 70 eV, an interface temperature of 280 °C, an ion source temperature of 230 °C, a quadrupole temperature of 150 °C, a scanning mass range (m/z) of 35–500 amu, and a scan speed of 5.2 times/S, and a solvent delay of 3 min. The qualitative analysis of aroma components was performed using the NIST 14 mass spectrum library, while the relative percentages of each volatile compound were determined according to the GC-FID peak area. All samples were tested in triplicate.

2.8. Sensory Evaluation

The wine aroma evaluation panel consisted of 12 members who had extensive wine tasting experience. Members awarded the samples scores ranging from 1–5 and evaluated six aromatic qualities of the Cabernet Sauvignon wine (alcohol, floral, fruity, preserved fruit, spicy, and fermented aroma). Furthermore, the participants performed a preference ranking and difference comparison. The samples were randomly numbered and presented to the panelists. The aroma characteristics were characterized via the five-point intensity method (blank—no odor, 1—just detected the odor, 2—weak, 3—medium, 4—strong, 5—intense).

2.9. Statistical Analysis

The mean and standard variance were calculated using the data of triplicate fermentations, and the results were expressed as mean ± standard deviation (SD). The significant differences were carried out using SPSS 21.0 software (SPSS Inc., Chicago, IL, USA) based on a one-way analysis of variance (ANOVA, significance level was 0.05) with Duncan's multiple range test. The Shapiro–Wilk's and Levene's tests were used in the test for the normality and homogeneity of variances. Principal component analysis (PCA) was performed using Matlab R2013b software (MathWorks, Natick, MA, USA).

3. Results

3.1. Selecting a Non-Saccharomyces Yeast with High-Alcoholicity Endurance

The TTC medium chromogenic method was employed to examine the red color intensity of 85 non-*Saccharomyces* strains. The results demonstrated that the chromogenic levels of the strains were significantly different. The strains B21, D1, D6, F26, E31, and H5 displayed a prominent dark red color and had better consistency between the three parallel. Commercial *Saccharomyces cerevisiae* S1 was used as a positive control, and the strain H67 was used as a negative control. The colorimetric results are shown in Figure 1a. Although the TTC medium method could quickly screen out the strains displaying better color, quantitative evaluation was impossible; therefore, six selected yeasts were utilized in the follow-up study to determine their tolerance at different ethanol concentrations. As shown in Table 1, strain D1 exhibited the worst ethanol tolerance and only proliferated at a 4% ethanol concentration. Strains B21 and E31 showed good tolerance of a 6% ethanol concentration. The strains displaying the highest ethanol tolerance were F26, H5, and D6, which could form colonies on the medium containing 8% ethanol.

Table 1. The ethanol tolerance results of the screened strains.

Strain	Ethanol Concentration (% (*v*/*v*))				
	4	6	8	10	12
D1	+	-	-	-	-
D6	+	+	+	-	-
B21	+	+	-	-	-
E31	+	+	-	-	-
H5	+	+	+	-	-
F26	+	+	+	-	-

Note: "+" indicates that the strain can grow colonies on the medium, and "-" indicates that the strain cannot grow on the medium.

To further evaluate the tolerance of the D6, F26, and H5 strains against ethanol, their growth curves at different ethanol concentrations were measured at OD600, as shown in Figure 1b. Commercial *Saccharomyces cerevisiae* S1 was used as the control group. In ethanol conditions, all strains displayed extended lag phases, reduced growth rates, and significantly lower maximal OD values than in non-ethanol conditions ($p < 0.05$). At an ethanol concentration of 6% (v/v), the proliferation of strains D6, F26, and H5 was noticeably inhibited. The absorbance values of strains H5 and F26 at 30 h were only half that of the control group, while strain D6 was reduced to one-fourth. At an ethanol concentration of 8% (v/v), the absorbance values of the three strains when reaching the stationary phase were less than one-fourth of the control group; however, in general, the growth rate of H5 was significantly faster than F26 and D6 at different ethanol concentrations; therefore, H5 was selected for subsequent experiments.

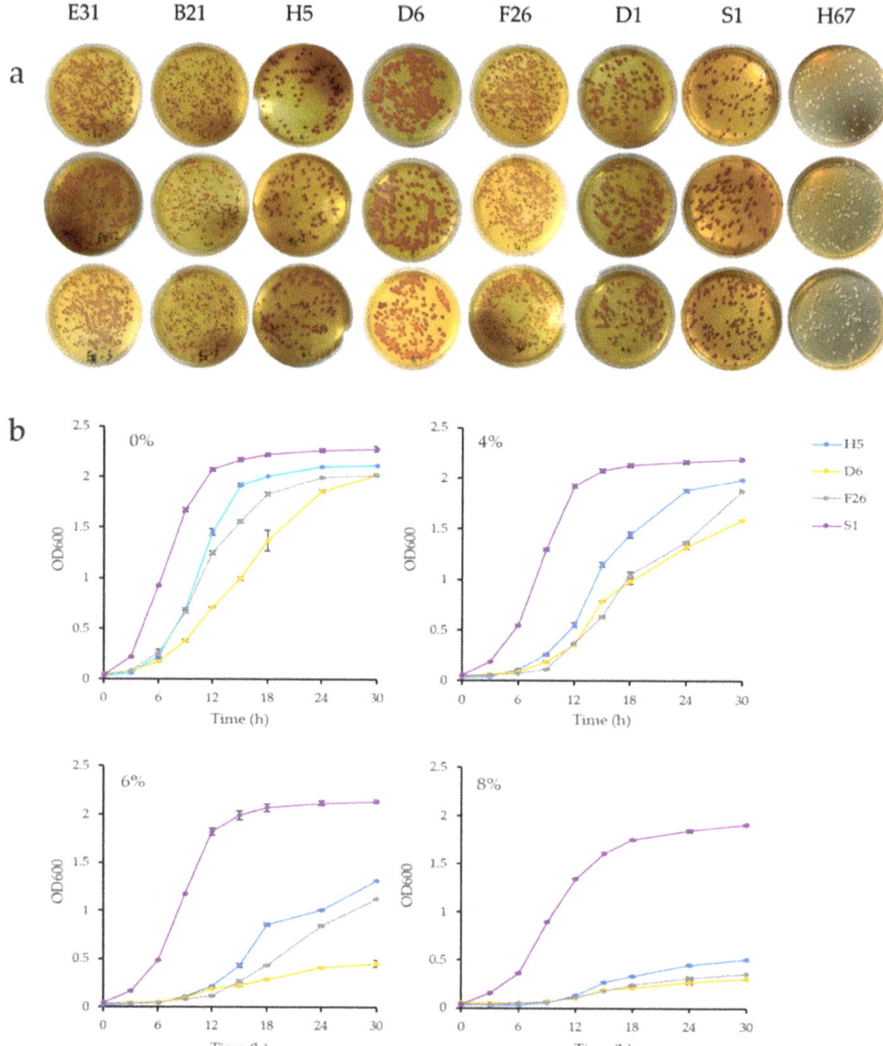

Figure 1. (**a**) The coloration results obtained via the TTC medium chromogenic screening method. The red intensity level of colonies in the culture medium can indirectly reflect the ethanol tolerance of the strains; (**b**) the growth curves of the screened strains at ethanol concentrations of 0%, 4%, 6%, and 8% (v/v).

3.2. Screening for Non-Saccharomyces Yeasts with High Ethanol Tolerance Using DES and UV

The effect of a mutagen dosage on the lethality of strain H5 is shown in Figure 2. The cells cultivated to the logarithmic phase were subjected to UV irradiation and DES mutagenesis. Figure 2a shows that the lethality increased with extended UV irradiation time. The lethality reached 76.28% and 90.29% at UV irradiation times of 40 s and 50 s, respectively. Figure 2b shows that the lethality increased with extended DES mutagenesis time. The lethality reached 69.38% and 83.14% at DES mutagenesis times of 8 min and 10 min, respectively. The treatment condition that produced a lethality of about 80% was set as the mutation concentration [21]; therefore, a UV irradiation time of 40 s (lethality 76.28%)

and a 1% DES mutagenesis time of 10 min (lethality 83.14%) were set as the mutation conditions for strain H5.

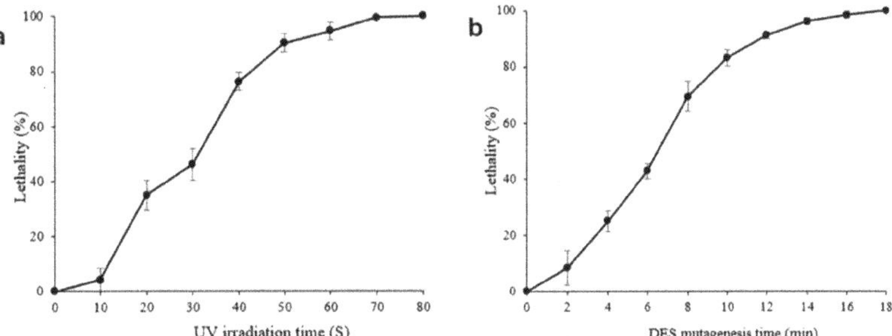

Figure 2. The lethality curve during mutagenesis. (**a**) The lethality curve of strain H5 subjected to UV mutagenesis after exposure to different irradiation times. A UV irradiation time of 40 s was set as the mutation conditions for strain H5; (**b**) the lethality curve of strain H5 subjected to DES mutagenesis after exposure to different times. Treatment with 1% DES for 10 min was set as the mutation conditions for *S. cerevisiae*. The results expressed as mean ± SD.

After treatment with 1% DES for 10 min and UV irradiation for 40 s, 15 mutants (UV1-UV8 and DES1-DES7) displaying a dark red colony color and screened via high-throughput TTC medium were collected. These strains were streaked onto the YPD solid medium containing ethanol to evaluate the ethanol tolerance of the mutant strains. Strain H5 and commercial *Saccharomyces cerevisiae* S1 were used as controls, and the results are shown in Table 2. The ethanol tolerance of the UV5 and UV8 mutants improved significantly after UV irradiation, increasing to 11% and 10%, respectively. After DES treatment, at the same ethanol concentration (8%), only DES7 exhibited better proliferation than strain H5, while the ethanol tolerance of the other strains did not improve.

Table 2. The ethanol tolerance results of mutants H5 and S1 introduced as control strains.

Strain	Ethanol Concentration (% (*v*/*v*))				
	8	**9**	**10**	**11**	**12**
S1	+++	+++	++	++	++
H5	+	-	-	-	-
UV1	-	-	-	-	-
UV2	-	-	-	-	-
UV3	+	-	-	-	-
UV4	-	-	-	-	-
UV5	+++	++	++	+	-
UV6	-	-	-	-	-
UV7	+	-	-	-	-
UV8	+++	++	++	-	-
DES1	-	-	-	-	-
DES2	-	-	-	-	-
DES3	-	-	-	-	-
DES4	I				
DES5	+	-	-	-	-
DES6	-	-	-	-	-
DES7	++	-	-	-	-

Note: "+++" indicates that the strain grows well on the medium, "++" indicates that the strain grows generally, "+" indicates that the strain can grow, and "-" indicates that the strain cannot grow.

To further explore the ethanol tolerance of the mutants, their growth curves were measured at different ethanol concentrations, using strain H5 and commercial *Saccharomyces cerevisiae* S1 as controls, as shown in Figure 3. The growth rates of strain UV5 and UV8 were higher than that of strain H5 at different ethanol concentrations, but they did not exceed that of commercial *Saccharomyces cerevisiae* S1. At an ethanol concentration of 6%, the absorbance values of UV5 and UV8 exceeded 2.00 at 30 h, while that of strain H5 was 0.84. At an ethanol concentration of 8%, the absorbance values of UV5 and UV8 remained higher than 1.00 after 30 h, while that of strain H5 was only 0.46. When the ethanol concentration reached 10%, strain H5 was completely inhibited, and the OD value did not increase, while strains UV5 and UV8 continued to proliferate, and the absorbance values exhibited significant differences ($p < 0.05$); however, the growth of strain DES7 at 6% and 8% ethanol concentrations was slightly superior to that of strain H5, while the absorbance values were significantly different ($p < 0.05$) at 30 h. DES7 showed no growth at 10% ethanol concentration, which was consistent with the results in the solid YPD medium. The findings indicated that the ethanol tolerance of strains UV5 and UV8 surpassed that of strain H5.

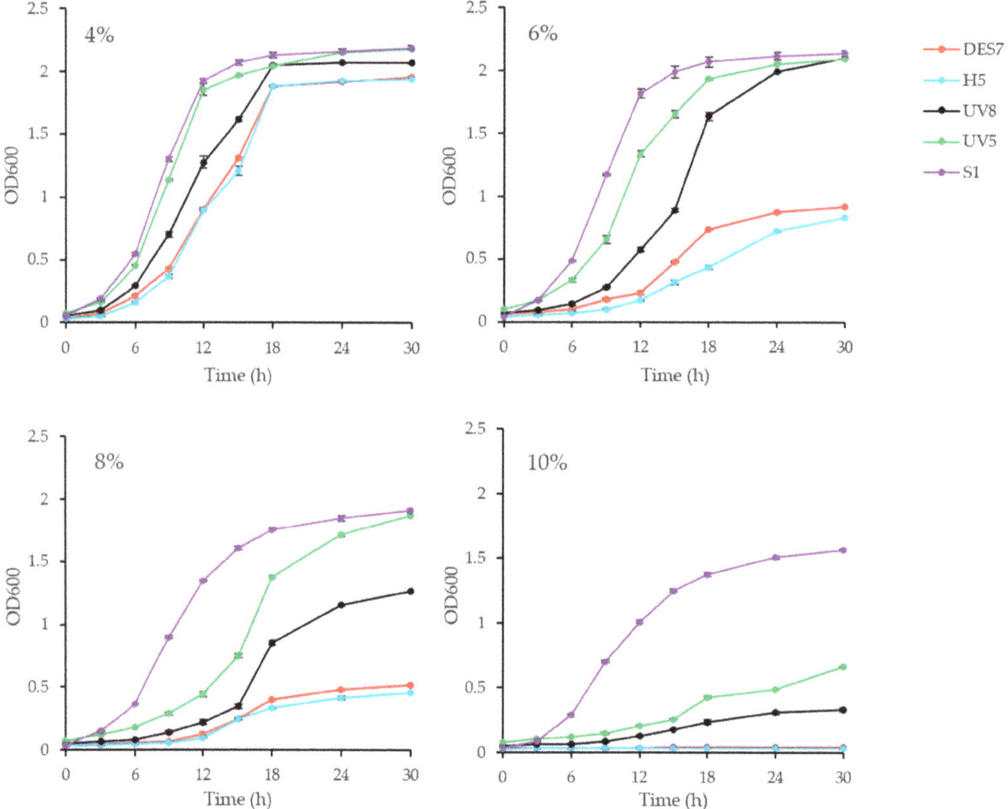

Figure 3. The growth curves of the mutants at ethanol concentrations of 4%, 6%, 8%, and 10% (*v*/*v*). Strain H5 and S1 served as controls at various ethanol concentrations.

3.3. *Evaluation of the Fermentation Ability of the Mutagenized Strains*

3.3.1. The Performance of the Mutants during Fermentation

The yeast cell dynamics during the fermentation process are shown in Figure 4a, while the °Brix change is shown in Figure 4b. As shown in Figure 4a, the yeasts continued to proliferate as the fermentation progressed, reaching their maximum count at 7 d and maintaining high cell viability until the end of the process. The °Brix value results showed that neither strain H5 nor the three mutants could fully utilize the sugar content in grape must.

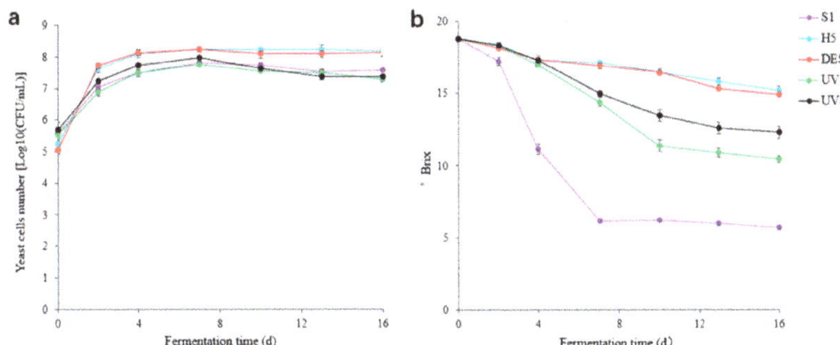

Figure 4. (**a**) The changes in the colony numbers during fermentation; (**b**) changes in the °Brix values of the different fermentation groups during fermentation.

The impact of the yeast metabolism on the wine composition was evaluated, quantifying the main analytical components of the grape must and wine at the end of fermentation. The chemical analysis results are summarized in Table 3. Except for commercial *Saccharomyces cerevisiae* S1, the remaining four yeast strains still contained residual sugar that could not be ignored at 16 d fermentation (>0.4 g/L). This confirmed that they could not complete the wine fermentation process independently, which was consistent with the °Brix value test results. The ability of strain UV5 to utilize glucose and fructose improved significantly compared with strain H5, reducing the content of these substances to 15.17 g/L and 40.10 g/L, respectively. Strain UV8 also used more glucose, reducing the residual glucose content to 31.83 g/L after fermentation. At the end of fermentation, the ethanol and glycerol content produced by the three mutant strains and strain H5 were significantly lower than those of strain S1. The ethanol yield of strain UV5 was 3.20 times that of the H5, while the glycerol yield increased by 16.27%. Although strain UV8 consumed more glucose than strain H5, the ethanol production only increased by 0.39% (v/v), while no significant differences were evident in glycerol production ($p < 0.05$). The glucose, fructose, ethanol, and glycerol levels indicated that the fermentation performance of UV5 and UV8 was surpassed that of strain H5, while the fermentation performance of UV5 was stronger. The data indicated that UV irradiation improved the fermentation performance of strain H5, which was not significantly different from that of strain DES7 ($p < 0.05$).

Table 3. The physicochemical indexes in the grape must and wine.

	Grape Must (Day 0)	Wine (Day 16)				
		S1	H5	DES7	UV5	UV8
°**Brix**	18.77 ± 0.05 [a]	5.67 ± 0.09 [e]	15.20 ± 0.28 [b]	14.90 ± 0.08 [b]	10.43 ± 0.25 [d]	12.27 ± 0.42 [c]
Ethanol (% v/v)	N.D.	11.32 ± 0.12 [a]	2.05 ± 0.10 [d]	2.28 ± 0.07 [cd]	6.55 ± 0.06 [b]	2.44 ± 0.17 [c]
Glycerol (g/L)	N.D.	10.45 ± 0.08 [a]	7.07 ± 0.18 [c]	7.62 ± 0.31 [bc]	8.22 ± 0.51 [b]	6.81 ± 0.75 [c]
Sugar (g/L)						
Glucose	105.08 ± 0.76 [a]	N.D.	61.18 ± 2.53 [b]	62.78 ± 1.26 [b]	15.17 ± 2.34 [d]	31.83 ± 3.58 [c]
Fructose	106.05 ± 1.00 [a]	2.16 ± 0.08 [e]	63.39 ± 2.14 [c]	69.12 ± 2.82 [b]	40.10 ± 1.59 [d]	68.59 ± 2.80 [b]
Acid (g/L)						
Citric acid	0.81 ± 0.02 [b]	0.75 ± 0.02 [c]	0.67 ± 0.02 [d]	0.58 ± 0.02 [e]	0.77 ± 0.01 [bc]	0.88 ± 0.03 [a]
Tartaric acid	3.00 ± 0.05 [a]	2.28 ± 0.08 [c]	3.06 ± 0.00 [a]	2.96 ± 0.02 [a]	1.87 ± 0.03 [d]	2.51 ± 0.11 [b]
Malic acid	4.29 ± 0.08 [a]	2.86 ± 0.07 [d]	2.29 ± 0.03 [f]	2.59 ± 0.03 [e]	3.53 ± 0.07 [b]	3.30 ± 0.08 [c]
Succinic acid	3.59 ± 0.12 [c]	4.42 ± 0.13 [a]	3.82 ± 0.31 [bc]	3.66 ± 0.09 [c]	4.15 ± 0.03 [ab]	2.94 ± 0.21 [d]
Lactic acid	N.D.	N.D.	1.05 ± 0.22 [a]	0.86 ± 0.14 [a]	N.D.	N.D.
Acetic acid	0.06 ± 0.01 [c]	0.11 ± 0.00 [b]	0.15 ± 0.02 [a]	0.11 ± 0.01 [b]	0.09 ± 0.00 [b]	0.15 ± 0.01 [a]

Note: Different letters in the same row indicate that the content levels of the measured substances are significantly different, $p < 0.05$. N.D. denotes "not determined".

Organic acids are important compounds found in grape must and wine. After alcoholic fermentation, the citric acid content of strain UV8 was the highest at 0.88 g/L, while that of DES7 was the lowest at 0.58 g/L. Tartaric acid represents the dominant organic acid in wine, primarily contributing to its acidity and tart taste [26]. The tartaric acid content of strains UV5, UV8, and S1 was significantly reduced after fermentation compared with grape must. This was attributed to a decline in the solubility and precipitation of tartaric acid crystals due to alcohol accumulation during fermentation, decreasing the tartaric acid content [27,28]. Malic acid provides a green, astringent taste that can be synthesized or degraded by yeast during alcoholic fermentation [26,29]. Significant differences were apparent between the malic acid levels of the five experimental groups. Compared with grape must, the malic acid content declined substantially after fermentation. The lactic acid in wine is mainly produced during malolactic fermentation. Lactic acid bacteria convert part of the malic acid into lactic acid, rendering the taste of the wine softer and less harsh. Only some yeast strains can produce trace amounts of lactic acid during the fermentation process due to the lack of an effective lactic acid pathway [30]. During this experiment, lactic acid was only detected in strains H5 and DES7, providing milky and buttery flavors to the wine. The results suggested that UV irradiation affected the metabolic capacity of the organic acids in yeasts.

3.3.2. The Influence of Mutants on the Volatile Aroma Compounds

The volatile metabolites in grape must and wine samples were identified via HS-SPME/GC-MS-FID. A total of 102 volatile aroma compounds were detected, which could be divided into six categories: alcohols, acids, esters, aldehydes and ketones, terpenes, and other compounds. The volatile aroma compounds in the samples were semi-quantitatively analyzed using the area normalization method. The peak area and relative content of each substance are shown in Table S2.

Twenty-six aroma compounds were detected in the grape must, including eight alcohols, one acid, one ester, eight aldehydes and ketones, one terpene, and seven other aroma compounds. After fermentation, the aroma compound diversity detected in each sample was significantly increased compared with the grape must. Fifty-five volatile aroma compounds were identified in the S1 fermentation samples, 59 in strain H5, 56 in DES7, 54 in UV5, and 48 in UV8. The aroma compounds detected during the fermentation processes mainly consisted of esters, alcohols, and acids. Although the total number of aroma compounds in each sample was similar, the types and relative content were significantly different. Compared with the S1 group, the unique volatile compounds detected in other groups included citronellol, 2-methylpropanol, benzyl alcohol, isovaleric acid, hexyl caproate, methyl salicylate, etc. Compared with strain H5, mutant DES7 displayed fewer aroma compound species and relative peak areas; however, although the aroma compound types decreased in UV5 and UV8, the relative peak areas were higher.

As vital aroma compounds produced during alcoholic fermentation, esters present strong floral, fruity flavors and significantly contribute to the fruity aroma of young wines [31,32]. The total peak area of the esters in group S1 was 2357.37×10^6, which was significantly higher than in other groups. The total peak area of the esters in group H5 was 294.87×10^6, accounting for 58.74% of the total aroma content, among which ethyl caproate was the most abundant, accounting for 19.04%, providing tropical fruit aromas, such as bananas and pineapples. The total peak area of the esters in the DES7 group was 202.57×10^6, accounting for 49.56% of the total aroma content, of which ethyl caproate was the highest, accounting for 22.83%. The relative ester content in UV5 was significantly higher than in H5. The total peak area was 1790.31×10^6, accounting for 88.19% of the total aroma content. Furthermore, the ethyl caprate, ethyl caprylate, and ethyl dodecanoate were higher, accounting for 32.22%, 15.47%, and 15.47%, respectively. Ethyl decanoate provided oily, floral, grape, and other aromas, while ethyl caprylate was responsible for pleasant fruity, buttery, mushroom, and other aromas. Ethyl dodecanoate presented a strong aroma of peanuts [33]. The total peak area of the esters in the UV8 group was 938.43 × 10^6, accounting for 91.85% of the total aroma content, of which ethyl acetate was the most abundant, accounting for 74.85%. The peak area of the ethyl acetate in the UV8 group was significantly higher than in other groups, possibly indicating that UV8 can produce high ethyl acetate concentrations. Although ethyl acetate can add grape, cherry, and other fruit characteristics, when its content exceeds 100 mg/L, the wine presents a solvent/nail varnish-like and other unpleasant aromas [34].

Higher alcohols are aroma compounds produced by yeast via amino acid metabolism and can enhance the wine aroma complexity to a certain extent [35]. Compared with the S1 group, more types of higher alcohols were evident in the other groups. Fourteen alcohols were detected in the H5 group, the relative content accounting for 32.44%. Although more than ten alcohols were detected in each mutant group, the relative alcohol content in the UV5 and UV8 groups were significantly reduced, at 3.69% and 5%, respectively. Phenethyl alcohol and isoamyl alcohol, representing higher alcohols synthesized by yeast, were the most abundant in each sample after fermentation, mainly providing aromas of roses and fruit, such as bananas. In addition, compared with strain H5, the unique alcohols detected in the DES7 group were 1,2-butanediol, trans-2-hexenol, octanol, while amyl alcohol was found in the UV5 group, and geraniol and 2, 6-octanediol in the UV8 group.

High acid levels may cause undesirable flavors in wine [36]. This experiment showed low acid content levels in each group after fermentation. Aldehydes and ketones are metabolites found during the early fermentation phase and are generally abundant in grape must. In this study, eight aldehydes and ketones were detected in the grape must, among which hexanal, hexenal, and β-damascenone displayed higher levels. After fermentation, the relative aldehyde and ketone content in each group was below 2%. These results indicated that the metabolic ability of the volatile aroma compounds of the strains was modified by DES mutagenesis and UV irradiation.

PCA was performed on the 43 primary aroma compounds to delineate the differences between these components in the different samples more intuitively. The results are illustrated in Figure 5a, showing that the variance contribution rate was 43.42% in the first principal component (PC1), and 22.48% in the second principal component (PC2), while the cumulative variance contribution rate was 65.90%. The aromas presented on the positive semi-axis of PC1 is mainly included fermented notes provided by alcohols, esters, and acids. UV5 and S1 were designated to the positive semi-axis of the PC1, while H5, DES7, UV8, and grape must were assigned to the negative semi-axis of PC1. The aroma mainly included that of the original grape must variety and some fermented aromas provided by hexanal, trans-2-hexenol, damascene, phenylacetaldehyde, citronellol, and phenethyl acetate. This may be related to the fermentation ability of the five strains, of which only S1 completed the alcoholic fermentation process. Of the remaining four strains, the fermentation ability of UV5 was the highest.

The positive semi-axis area of PC2 included ethyl caprylate, ethyl nonanoate, nerolid alcohol, lauric acid, n-hexanol, and hexyl acetate, while the negative semi-axis area was represented by nonyl acetate and caprylic acid, isoamyl esters, isoamyl alcohol, citronellol, phenethyl alcohol, and ethyl caproate. The five groups of samples were divided into four categories according to the scores of the two principal components. UV5 appeared in the upper right quadrant, S1 in the lower right quadrant, grape must in the upper left quadrant, and H5, DES7, and UV8 in the lower left quadrant. Although UV8 was in the third quadrant with H5 and DES7, as shown in Table S2, this strain contained a higher level of ethyl acetate, providing the wine with the unpleasant smell of nail polish. In addition, the grape must and UV5 were designated to the positive semi-axis of the Y-axis according to PC2, indicating that this strain retained the original grape must aroma within a certain range.

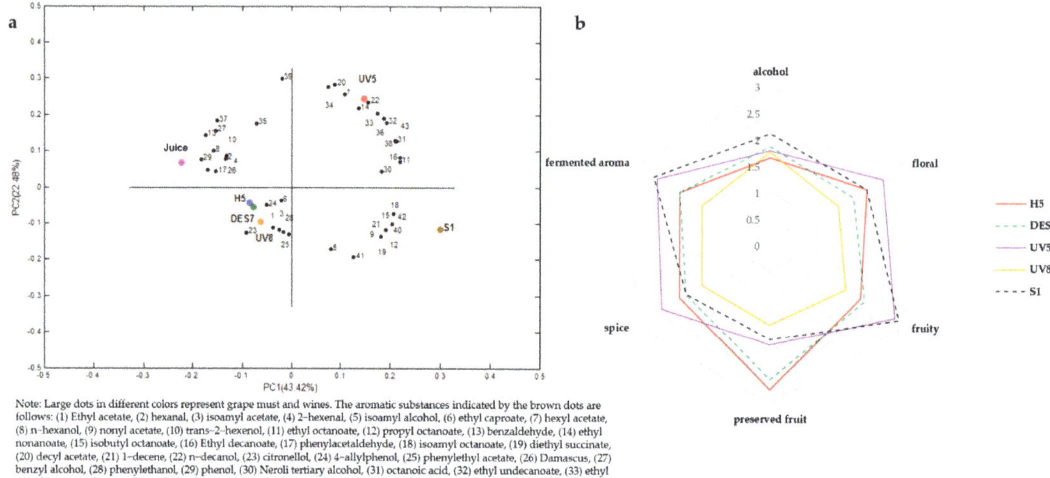

Note: Large dots in different colors represent grape must and wines. The aromatic substances indicated by the brown dots are follows: (1) Ethyl acetate, (2) hexanal, (3) isoamyl acetate, (4) 2-hexenal, (5) isoamyl alcohol, (6) ethyl caproate, (7) hexyl acetate, (8) n-hexanol, (9) nonyl acetate, (10) trans-2-hexenol, (11) ethyl octanoate, (12) propyl octanoate, (13) benzaldehyde, (14) ethyl nonanoate, (15) isobutyl octanoate, (16) Ethyl decanoate, (17) phenylacetaldehyde, (18) isoamyl octanoate, (19) diethyl succinate, (20) decyl acetate, (21) 1-decene, (22) n-decanol, (23) citronellol, (24) 4-allylphenol, (25) phenylethyl acetate, (26) Damascus, (27) benzyl alcohol, (28) phenylethanol, (29) phenol, (30) Neroli tertiary alcohol, (31) octanoic acid, (32) ethyl undecanoate, (33) ethyl tetradecanoate, (34) tetradecanol, (35) nonanoic acid, (36) capric acid, (37) thymol, (38) ethyl hexadecenoate, (39) 2,4-di-tert-butylphenol, (40) ethyl stearate, (41) ethyl linoleate, (42) ethyl oleate, (43) lauric acid.

Figure 5. (**a**) PCA of selected volatile compounds in Cabernet Sauvignon must and wine; (**b**) the aroma radar map of the sensory evaluation.

The aroma radar map of the sensory evaluation is shown in Figure 5b, indicating that the volatile profile of the UV5 group presented prominent floral, fruity, fermented, and spicy aromas. The S1 group displayed higher alcohol, fermented, and fruity aromas scores. The volatile profiles of H5 and DES7 were similar, with the preserved fruit aroma scoring the highest. Previous studies have shown that the β-glucosidase extracted from *Pichia terricola* can significantly increase the phenol and isoprenoid concentrations in wine, providing preserved fruit and raisin aromas [37]. The UV8 group displayed the smallest volatile profile and presented an unpleasant nail polish smell. The sensory evaluation and HS-SPME/GC-MS-FID results could be mutually verified.

3.4. Identification of Strain UV5

Strain UV5 was identified via molecular taxonomy via sequence similarity analysis of the ITS region of the ribosomal DNA gene. The sequence of the UV5 ITS region was consistent with the type of strain of *Pichia terricola* (CBS 2617), and the similarity percentage with the ITS sequence of the type strain was 99.51%; therefore, strain UV5 was identified as *Pichia terricola*.

4. Discussion

The aroma properties in wine are complex and essential for evaluating its flavor resulting from the interaction between various volatile components, such as esters, alcohols, aldehydes and ketones, acids, and terpenes [38]. Alcohols and esters mostly add rich floral and fruity characteristics to the wine. Aldehydes provide a floral, green apple aroma, while some acids and volatile phenols enhance wine aroma complexity; however, excessive amounts of these volatile aroma substances can cause undesirable and unanticipated odors [39]. Furthermore, wine can present different aroma characteristics and styles due to variation in metabolic pathways, enzymatic activity, and yeast strain metabolite type and content [40]. For example, the mono-culture fermentation of *Torulaspora delbrueckii* produces lower levels of volatile acids than *Saccharomyces cerevisiae* [41]. *Metschnikowia pulcherrima* generates more ethyl caprylate, providing the wine with pineapple or pear aromas [42]. Yeasts are in continuous and extensive competition during the alcoholic fermentation process. Temperature, alcohol, oxygen, SO_2, carbon source, nitrogen source, and other factors affect this competitive relationship as fermentation progresses, impacting the taste and aroma characteristics of wine [43]. Toxic metabolites mediate ethanol inhibition. The toxic effect of ethanol on yeast is mainly reflected in the impact of cell morphology and physiological cell activity. Morphological cell modifications are primarily evident in cell enlargement and skeleton evacuation. Changes in the physiological cell activity are mainly represented by the destruction of the cell membrane structure, affecting the synthesis and metabolism of biological macromolecules, and modifying the activity of related enzymes during the glycolysis process [44,45]. Different yeast species display varying ethanol tolerance levels. Compared with *Saccharomyces cerevisiae*, non-*Saccharomyces* yeasts display a weaker tolerance to alcohol. Studies show that *Saccharomyces cerevisiae* 101, ITB, and AAV2 can tolerate 15% ethanol, while *Issatchenkia occidentalis* APC and *Issatchenkia orientalis* cl1132 can only tolerate 10% [46]. The present study showed that *Pichia terricola* H5 could tolerate 8% ethanol. Strains D6 (*Wickerhamomyces anomalus*) and F26 (*Meyerozyma caribbica*) were tolerant to an 8% ethanol concentration. This was consistent with the results obtained by Kim et al. [47], who showed that *W. anomalus* and *M. caribbica* could proliferate at 8% ethanol.

The potential of genetically modified yeasts has been greatly weakened due to the severe restriction of wine markets to the application of genetically modified organisms. Non-genetically modified organism techniques have been used to improve the performance of yeasts to meet the demands of the wine industry [48]. To improve the ethanol tolerance of yeasts, the performance of the strains was modified via mutagenetic breeding and genetic recombination. Some studies constructed hybrids between *Kluyveromyces fragilis* and *Saccharomyces cerevisiae* using protoplast fusion technology to directionally change the ethanol tolerance of *Kluyveromyces fragilis* [49]. Compared with genetic recombination, mutation breeding presents advantages, such as simple operation, low technology and equipment requirements, and high mutation frequency; therefore, mutation breeding is widely used for yeast cultivation. Pattanakittivorakul et al. [50] obtained a thermotolerant ethanol-fermenting yeast strain via UV radiation and ethyl methane sulfonate mutagenesis that effectively improved the efficiency of ethanol production. Watanabe et al. [51] treated *Pichia stipitis* with UV irradiation for 25 min to obtain the *Pichia stipitis* NBRC1687 mutant after screening. It displayed significant ethanol tolerance, with the ethanol content produced via fermentation in the same conditions was 1.4 times that of the parental strain. In this experiment, *Pichia terricola* H5 was used as the mutagenic material and subjected to UV irradiation and DES mutagenesis to improve its ethanol tolerance. The results showed that the ethanol tolerance of UV5 and UV8 increased significantly after UV irradiation, reaching 11% and 10%, respectively, which was consistent with previous studies indicating that UV irradiation could modify the ethanol tolerance of yeast. Of the DES mutants, only the ethanol tolerance of DES7 was slightly higher than the parental H5, differing from the findings in existing literature [52]. This may be attributed to the DES concentration and the processing time used during the experiment. Less strain variability is evident at a lower mutagen concentration and shorter mutagenesis time, reducing the likelihood of changes in the strain performance [53].

This study selected three mutants for alcoholic fermentation with commercial *Saccharomyces cerevisiae* S1 and strain H5 after analyzing their ethanol tolerance while evaluating the fermentation performance and volatile aromatic substances. The fermentation results showed that the three mutants and H5 could not complete alcoholic fermentation independently, while their fermentation ability was weaker than commercial *Saccharomyces cerevisiae*, which was consistent with the research results of Clemente-Jimenez et al. [42]. Although the residual sugar content exceeded 4 g/L when only *Pichia terricola* was used for alcoholic fermentation, the fermentation capacity of the UV5 mutant was improved. Compared with strain H5, the residual sugar content in the UV5 group was reduced by 32.83%, while the ethanol content was 3.75 times that of the H5 group. UV irradiation improved the ethanol tolerance and alcohol production of yeast, which was consistent with previous studies [54]. Furthermore, *Pichia terricola* could reduce acid levels by degrading malic acid and citric acid to improve the taste of the wine [55]. This study showed that the citric and malic acid levels were significantly lower in the H5 strain and DES7 mutant than in S1 while exhibiting higher levels in the UV5 group. It is speculated that while the UV-irradiated yeast develops ethanol resistance, it may also affect the metabolic cell pathways related to citric and malic acid; however, the specific reasons for this require further analysis. Clemente-Jimenez et al. [42] indicated that *Pichia terricola* displayed low fermentation ability while producing high ethyl acetate levels, causing the wine to present an unpleasant odor. The H5 group did not exhibit high ethyl acetate content, while the sensory evaluation results did not indicate the presence of an unpleasant odor; however, high ethyl acetate levels were evident in the UV8 mutant group, displaying a severely unpleasant odor, which was consistent with the literature. This may be because UV irradiation not only induces the ability of cells to resist ethanol stress but also promotes the expression of genes related to ethyl acetate synthesis in cells; however, further research is necessary to provide confirmation. Moreover, the aroma determination results indicated that the fermentation of strain UV5 with improved ethanol tolerance produced higher levels of ethyl caprate, ethyl caprylate, and ethyl dodecanoate, providing a stronger floral, fruity fragrance.

Since this study only measured the ethanol tolerance of the strain at the macro level, the related stress resistance pathways and mechanisms require further exploration. Studies have shown that 359 genes in *Saccharomyces cerevisiae* cells are associated with ethanol tolerance [56]. The mutant genotypes and mutation sites require further exploration to clarify the differences in ethanol tolerance, fermentation characteristics, and aroma characteristics between the mutant and strain H5.

5. Conclusions

This research aims to improve the ethanol tolerance of non-*Saccharomyces* yeast and apply it to wine fermentation. An appropriate UV irradiation dose increased the ethanol tolerance of strain H5 and improved its fermentation performance. In particular, the production of ethyl caprate, ethyl caprylate, and other volatile aroma compounds in strain UV5 were significantly higher, which increased the aroma structure complexity of the wine.

Supplementary Materials: The following supporting information can be downloaded at: https://www.mdpi.com/article/10.3390/fermentation8040149/s1, Table S1: The region and vineyards of strains used; Table S2: The average peak area and relative peak area of volatile aroma compounds in each sample.

Author Contributions: Conceptualization, W.H. and J.Z.; Formal analysis, J.G., X.H. and Y.Y.; Funding acquisition, W.H. and J.Z.; Investigation, Y.Y.; Methodology, J.G. and X.H.; Project administration, J.G. and X.H.; Supervision, W.H. and J.Z.; Writing—review and editing, J.G., Y.Y. and J.Z. All authors have read and agreed to the published version of the manuscript.

Funding: This work was supported by the National "Thirteenth Five-Year" Plan for Science and Technology Support (2016YFD0400500).

Institutional Review Board Statement: Not applicable.

Informed Consent Statement: Not applicable.

Data Availability Statement: All data are available within the article and Supplementary Materials.

Conflicts of Interest: The authors declare no conflict of interest.

References

1. Gobbi, M.; De Vero, L.; Solieri, L.; Comitini, F.; Oro, L.; Giudici, P.; Ciani, M. Fermentative Aptitude of Non-*Saccharomyces* Wine Yeast for Reduction in the Ethanol Content in Wine. *Eur. Food Res. Technol.* **2014**, *239*, 41–48. [CrossRef]
2. Prior, K.J.; Bauer, F.F.; Divol, B. The Utilisation of Nitrogenous Compounds by Commercial Non-*Saccharomyces* Yeasts Associated with Wine. *Food Microbiol.* **2019**, *79*, 75–84. [CrossRef] [PubMed]
3. Ciani, M.; Comitini, F.; Mannazzu, I.; Domizio, P. Controlled Mixed Culture Fermentation: A New Perspective on the Use of Non-*Saccharomyces* Yeasts in Winemaking. *FEMS Yeast Res.* **2010**, *10*, 123–133. [CrossRef] [PubMed]
4. Padilla, B.; Gil, J.V.; Manzanares, P. Past and Future of Non-*Saccharomyces* Yeasts: From Spoilage Microorganisms to Biotechnological Tools for Improving Wine Aroma Complexity. *Front. Microbiol.* **2016**, *7*, 1–20. [CrossRef]

5. Jolly, N.P.; Varela, C.; Pretorius, I.S. Not Your Ordinary Yeast: Non-*Saccharomyces* Yeasts in Wine Production Uncovered. *FEMS Yeast Res.* **2014**, *14*, 215–237. [CrossRef]
6. Sadoudi, M.; Tourdot-Marechal, R.; Rousseaux, S.; Steyer, D.; Gallardo-Chacon, J.J.; Ballester, J.; Vichi, S.; Guerin-Schneider, R.; Caixach, J.; Alexandre, H. Yeast-Yeast Interactions Revealed by Aromatic Profile Analysis of Sauvignon Blanc Wine Fermented by Single or Co-Culture of Non-*Saccharomyces* and *Saccharomyces* Yeasts. *Food Microbiol.* **2012**, *32*, 243–253. [CrossRef]
7. Cordero-Bueso, G.; Esteve-Zarzoso, B.; Cabellos, J.M.; Gil-Diaz, M.; Arroyo, T. Biotechnological Potential of Non-*Saccharomyces* Yeasts Isolated During Spontaneous Fermentations of Malvar (*Vitis Vinifera Cv. L.*). *Eur. Food Res. Technol.* **2013**, *236*, 193–207. [CrossRef]
8. Giorello, F.; Valera, M.J.; Martin, V.; Parada, A.; Salzman, V.; Camesasca, L.; Farina, L.; Boido, E.; Medina, K.; Dellacassa, E.; et al. Genomic and Transcriptomic Basis of *Hanseniaspora vineae*'s Impact on Flavor Diversity and Wine Quality. *Appl. Environ. Microb.* **2019**, *85*, e01959-18. [CrossRef]
9. Whitener, M.E.B.; Carlin, S.; Jacobson, D.; Weighill, D.; Divol, B.; Conterno, L.; Du Toit, M.; Vrhovsek, U. Early Fermentation Volatile Metabolite Profile of Non-*Saccharomyces* Yeasts in Red and White Grape Must: A Targeted Approach. *LWT Food Sci. Technol.* **2015**, *64*, 412–422. [CrossRef]
10. Anfang, N.; Brajkovich, M.; Goddard, M.R. Co-Fermentation with *Pichia Kluyveri* Increases Varietal Thiol Concentrations in Sauvignon Blanc. *Aust. J. Grape Wine Res.* **2009**, *15*, 1–8. [CrossRef]
11. Liu, P.T.; Lu, L.; Duan, C.Q.; Yan, G.L. The Contribution of Indigenous Non-*Saccharomyces* Wine Yeast to Improved Aromatic Quality of Cabernet Sauvignon Wines by Spontaneous Fermentation. *LWT Food Sci. Technol.* **2016**, *71*, 356–363.
12. Heard, G.M.; Fleet, G.H. The Effects of Temperature and Ph on the Growth of Yeast Species During the Fermentation of Grape Juice. *J. Appl. Bacteriol.* **1988**, *65*, 23–28. [CrossRef]
13. Addo-Quaye, C.; Tuinstra, M.; Carraro, N.; Weil, C.; Dilkes, B.P. Whole-Genome Sequence Accuracy Is Improved by Replication in a Population of Mutagenized Sorghum. *G3 Genes Genomes Genet.* **2018**, *8*, 1079–1094.
14. Zhang, G.Q.; Lin, Y.P.; Qi, X.N.; Wang, L.X.; He, P.; Wang, Q.H.; Ma, Y.H. Genome Shuffling of the Nonconventional Yeast *Pichia anomala* for Improved Sugar Alcohol Production. *Microb. Cell Fact.* **2015**, *14*, 112. [CrossRef] [PubMed]
15. Long, L.J.; Lee, P.H.; Small, E.M.; Hillyer, C.; Guo, Y.; Osley, M.A. Regulation of UV Damage Repair in Quiescent Yeast Cells. *DNA Repair* **2020**, *90*, 102861. [CrossRef] [PubMed]
16. Ikehata, H.; Ono, T. The Mechanisms of UV Mutagenesis. *J. Radiat. Res.* **2011**, *52*, 115–125. [PubMed]
17. Hoffmann, G.R. Genetic Effects of Dimethyl Sulfate, Diethyl Sulfate, and Related-Compounds. *Mutat. Res.* **1980**, *75*, 63–129. [CrossRef]
18. Abrol, V.; Kushwaha, M.; Arora, D.; Mallubhotla, S.; Jaglan, S. Mutation, Chemoprofiling, Dereplication, and Isolation of Natural Products from *Penicillium oxalicum*. *ACS Omega* **2021**, *6*, 16266–16272. [CrossRef]
19. Pang, Z.W.; Liang, J.J.; Qin, X.J.; Wang, J.R.; Feng, J.X.; Huang, R.B. Multiple Induced Mutagenesis for Improvement of Ethanol Production by *Kluyveromyces marxianus*. *Biotechnol. Lett.* **2010**, *32*, 1847–1851. [CrossRef]
20. Yi, S.; Zhang, X.; Li, H.X.; Du, X.X.; Liang, S.W.; Zhao, X.H. Screening and Mutation of *Saccharomyces cerevisiae* UV-20 with a High Yield of Second Generation Bioethanol and High Tolerance of Temperature, Glucose and Ethanol. *Indian J. Microbiol.* **2018**, *58*, 440–447. [CrossRef]
21. Kang, L.Z.; Han, F.; Lin, J.F.; Guo, L.Q.; Bai, W.F. Breeding of New High-Temperature-Tolerant Strains of *Flammulina velutipes*. *Sci. Hortic.* **2013**, *151*, 97–102.
22. Hesham, A.E.L.; Wambui, V.; Jo, H.O.; Maina, J.M. Phylogenetic Analysis of Isolated Biofuel Yeasts Based on 5.8S-ITS rDNA and D1/D2 26S rDNA Sequences. *J. Genet. Eng. Biotechnol.* **2014**, *12*, 37–43. [CrossRef]
23. Kurtzman, C.P.; Fell, J.W.; Boekhout, T.; Robert, V. Methods for Isolation, Phenotypic Characterization and Maintenance of Yeasts. In *The Yeasts, a Taxonomic Study*, 5th ed.; Kurtzman, C.P., Fell, J.W., Boekhout, T., Eds.; Elsevier: Amsterdam, The Netherlands, 2011; pp. 87–110.
24. Sun, X.Y.; Zhao, Y.; Liu, L.L.; Jia, B.; Zhao, F.; Huang, W.D.; Zhan, J.C. Copper Tolerance and Biosorption of *Saccharomyces cerevisiae* During Alcoholic Fermentation. *PLoS ONE* **2015**, *10*, e0128611. [CrossRef] [PubMed]
25. Chen, D.; Chia, J.Y.; Liu, S.Q. Impact of Addition of Aromatic Amino Acids on Non-Volatile and Volatile Compounds in Lychee Wine Fermented with *Saccharomyces cerevisiae* Merit.Ferm. *Int. J. Food Microbiol.* **2014**, *170*, 12–20. [CrossRef]
26. Volschenk, H.; Van Vuuren, H.J.J.; Viljoen-Bloom, M. Malic Acid in Wine: Origin, Function and Metabolism During Vinification. *S. Afr. J. Enol. Vitic.* **2006**, *27*, 123–136. [CrossRef]
27. Peinado, R.A.; Mauricio, J.C.; Moreno, J. Aromatic Series in Sherry Wines with Gluconic Acid Subjected to Different Biological Aging Conditions by *Saccharomyces cerevisiae* Var. Capensis. *Food Chem.* **2006**, *94*, 232–239. [CrossRef]
28. Lelova, Z.; Ivanova-Petropulos, V.; Masar, M.; Lisjak, K.; Bodor, R. Optimization and Validation of a New Capillary Electrophoresis Method with Conductivity Detection for Determination of Small Anions in Red Wines. *Food Anal. Methods* **2018**, *11*, 1457–1466. [CrossRef]
29. Redzepovic, S.; Orlic, S.; Majdak, A.; Kozina, B.; Volschenk, H.; Viljoen-Bloom, M. Differential Malic Acid Degradation by Selected Strains of *Saccharomyces* During Alcoholic Fermentation. *Int. J. Food Microbiol.* **2003**, *83*, 49–61. [CrossRef]
30. Ferreira, A.M.; Mendes-Faia, A. The Role of Yeasts and Lactic Acid Bacteria on the Metabolism of Organic Acids During Winemaking. *Foods* **2019**, *9*, 1231. [CrossRef]
31. Younis, O.S.; Stewart, G.G. Sugar Uptake and Subsequent Ester and Higher Alcohol Production by *Saccharomyces cerevisiae*. *J. Inst. Brew.* **1998**, *104*, 255–264. [CrossRef]
32. Lambrechts, M.; Pretorius, I. Yeast and Its Importance to Wine Aroma. *S. Afr. J. Enol. Vitic.* **2000**, *21*, 97–129. [CrossRef]
33. Sumby, K.M.; Grbin, P.R.; Jiranek, V. Microbial Modulation of Aromatic Esters in Wine: Current Knowledge and Future Prospects. *Food Chem.* **2010**, *121*, 1–16. [CrossRef]

34. Pérez-Martín, F.; Seseña, S.; Izquierdo, P.M.; Palop, M.L. Esterase Activity of Lactic Acid Bacteria Isolated from Malolactic Fermentation of Red Wines. *Int. J. Food Microbiol.* **2013**, *163*, 153–158. [CrossRef]
35. Lilly, M.; Styger, G.; Bauer, F.F.; Lambrechts, M.G.; Pretorius, I.S. The Effect of Increased Branched-Chain Amino Acid Transaminase Activity in Yeast on the Production of Higher Alcohols and on the Flavour Profiles of Wine and Distillates. *FEMS Yeast Res.* **2006**, *6*, 726–743. [PubMed]
36. Swiegers, J.; Bartowsky, E.; Henschke, P.; Pretorius, I. Yeast and Bacterial Modulation of Wine Aroma and Flavour. *Aust. J. Grape Wine Res.* **2005**, *11*, 139–173. [CrossRef]
37. De Ovalle, S.; Cavello, I.; Brena, B.M.; Cavalitto, S.; González-Pombo, P. Production and Characterization of a β-Glucosidase from *Issatchenkia terricola* and Its Use for Hydrolysis of Aromatic Precursors in Cabernet Sauvignon Wine. *LWT Food Sci. Technol.* **2018**, *87*, 515–522. [CrossRef]
38. Mauriello, G.; Capece, A.; D'Auria, M.; Garde-Cerdán, T.; Romano, P. SPME-GC Method as a Tool to Differentiate VOC Profiles in *Saccharomyces cerevisiae* Wine Yeasts. *Food Microbiol.* **2009**, *26*, 246–252. [CrossRef]
39. Dubourdieu, D.; Torninaga, T.; Masneuf, I.; Des Gachons, C.P.; Murat, M.L. The Role of Yeasts in Grape Flavor Development During Fermentation: The Example of Sauvignon Blanc. *Am. J. Enol. Vitic.* **2006**, *57*, 81–88.
40. Romano, P.; Fiore, C.; Paraggio, M.; Caruso, M.; Capece, A. Function of Yeast Species and Strains in Wine Flavour. *Int. J. Food Microbiol.* **2003**, *86*, 169–180. [CrossRef]
41. Renault, P.; Miot-Sertier, C.; Marullo, P.; Hernández-Orte, P.; Lagarrigue, L.; Lonvaud-Funel, A.; Bely, M. Genetic Characterization and Phenotypic Variability in *Torulaspora delbrueckii* Species: Potential Applications in the Wine Industry. *Int. J. Food Microbiol.* **2009**, *134*, 201–210. [CrossRef]
42. Clemente-Jiménez, J.M.; Mingorance-Cazorla, L.; Martinez-Rodríguez, S.; Las Heras-Vázquez, F.J.; Rodríguez-Vico, F. Molecular Characterization and Oenological Properties of Wine Yeasts Isolated During Spontaneous Fermentation of Six Varieties of Grape Must. *Food Microbiol.* **2004**, *21*, 149–155. [CrossRef]
43. Liu, Y.Z.; Rousseaux, S.; Tourdot-Marechal, R.; Sadoudi, M.; Gougeon, R.; Schmitt-Kopplin, P.; Alexandre, H. Wine Microbiome: A Dynamic World of Microbial Interactions. *Crit. Rev. Food Sci.* **2017**, *57*, 856–873.
44. Bai, F.W.; Chen, L.J.; Zhang, Z.; Anderson, W.A.; Moo-Young, M. Continuous Ethanol Production and Evaluation of Yeast Cell Lysis and Viability Loss under Very High Gravity Medium Conditions. *J. Biotechnol.* **2004**, *110*, 287–293. [CrossRef]
45. Teixeira, M.C.; Raposo, L.R.; Mira, N.P.; Lourenco, A.B.; Sa-Correia, I. Genome-Wide Identification of *Saccharomyces cerevisiae* Genes Required for Maximal Tolerance to Ethanol. *Appl. Environ. Microbiol.* **2009**, *75*, 5761–5772. [CrossRef] [PubMed]
46. Archana, K.M.; Ravi, R.; Anu-Appaiah, K.A. Correlation between Ethanol Stress and Cellular Fatty Acid Composition of Alcohol Producing Non-*Saccharomyces* in Comparison with *Saccharomyces cerevisiae* by Multivariate Techniques. *J. Food Sci. Technol.* **2015**, *52*, 6770–6776. [CrossRef]
47. Kim, D.H.; Lee, S.B.; Jeon, J.Y.; Park, H.D. Development of Air-Blast Dried Non-*Saccharomyces* Yeast Starter for Improving Quality of Korean Persimmon Wine and Apple Cider. *Int. J. Food Microbiol.* **2019**, *290*, 193–204. [PubMed]
48. Pérez-Torrado, R.; Querol, A.; Guillamón, J.M. Genetic Improvement of non-GMO Wine Yeasts: Strategies, Advantages and Safety. *Trends Food Sci. Technol.* **2015**, *45*, 1–11. [CrossRef]
49. Farahnak, F.; Seki, T.; Ryu, D.D.Y.; Ogrydziak, D. Construction of Lactose-Assimilating and High-Ethanol-Producing Yeasts by Protoplast Fusion. *Appl. Environ. Microbiol.* **1986**, *51*, 362–367. [CrossRef]
50. Pattanakittivorakul, S.; Lertwattanasakul, N.; Yamada, M.; Limtong, S. Selection of Thermotolerant *Saccharomyces cerevisiae* for High Temperature Ethanol Production from Molasses and Increasing Ethanol Production by Strain Improvement. *Antonie Leeuwenhoek* **2019**, *112*, 975–990. [CrossRef]
51. Watanabe, T.; Watanabe, I.; Yamamoto, M.; Ando, A.; Nakamura, T. A UV-Induced Mutant of *Pichia stipitis* with Increased Ethanol Production from Xylose and Selection of a Spontaneous Mutant with Increased Ethanol Tolerance. *Bioresour. Technol.* **2011**, *102*, 1844–1848.
52. Dong, Y.; Zhang, N.; Lu, J.H.; Lin, F.; Teng, L.R. Improvement and Optimization of the Media of *Saccharomyces cerevisiae* Strain for High Tolerance and High Yield of Ethanol. *Afr. J. Microbiol. Res.* **2012**, *6*, 2357–2366.
53. Liang, M.H.; Liang, Y.J.; Wu, X.N.; Zhou, S.S.; Jiang, J.G. Mutation Breeding of *Saccharomyces cerevisiae* with Lower Methanol Content and the Effects of Pectinase, Cellulase and Glycine in Sugar Cane Spirits. *J. Sci. Food Agric.* **2015**, *95*, 1949–1955. [CrossRef] [PubMed]
54. Sridhar, M.; Kiran Sree, N.; Venkateswar, R.L. Effect of UV Radiation on Thermotolerance, Ethanol Tolerance and Osmotolerance of *Saccharomyces cerevisiae* VS1 and VS3 Strains. *Bioresour. Technol.* **2002**, *83*, 199–202. [CrossRef]
55. Wen, L.K.; Wang, L.F.; Wang, G.Z. Degradation of L-Malic and Critic Acids by *Issatchenkia terricola*. *Food Sci.* **2011**, *32*, 220–223.
56. Yoshikawa, K.; Tanaka, T.; Furusawa, C.; Nagahisa, K.; Hirasawa, T.; Shimizu, H. Comprehensive Phenotypic Analysis for Identification of Genes Affecting Growth under Ethanol Stress in *Saccharomyces cerevisiae*. *FEMS Yeast Res.* **2009**, *9*, 32–44. [CrossRef]

Article

The Impact of Must Nutrients and Yeast Strain on the Aromatic Quality of Wines for Cognac Distillation

Charlie Guittin [1], Faïza Maçna [1], Isabelle Sanchez [2], Adeline Barreau [3], Xavier Poitou [3], Jean-Marie Sablayrolles [1], Jean-Roch Mouret [1,]* and Vincent Farines [1]

1 SPO, Université de Montpellier, INRAE, Institut Agro, 34060 Montpellier, France; charlie.guittin@inrae.fr (C.G.); faiza.macna@inrae.fr (F.M.); jean-marie.sablayrolles@inrae.fr (J.-M.S.); vincent.farines@inrae.fr (V.F.)
2 MISTEA, INRAE, Institut Agro, 34060 Montpellier, France; isabelle.sanchez@inrae.fr
3 R&D Department, Jas Hennessy & Co., 16100 Cognac, France; abarreau@moethennessy.com (A.B.); xpoitou@moethennessy.com (X.P.)
* Correspondence: jean-roch.mouret@inrae.fr; Tel.: +33-4-9961-2274

Abstract: In order to understand the influence of nitrogen and lipid nutrition on the aromatic quality of wines for cognac distillation, we developed a transdisciplinary approach that combined statistical modeling (experimental central composite design and response surface modeling) with metabolomic analysis. Three *Saccharomyces cerevisiae* strains that met the requirements of cognac appellation were tested at a laboratory scale (1 L) and a statistical analysis of covariance was performed to highlight the organoleptic profile (fermentative aromas, terpenes, alcohols and aldehydes) of each strain. The results showed that nitrogen and lipid nutrients had an impact on the aromatic quality of cognac wines: high lipid concentrations favored the production of organic acids, 1-octen-3-ol and terpenes and inhibited the synthesis of esters. Beyond this trend, each yeast strain displayed its own organoleptic characteristics but had identical responses to different nutritional conditions.

Keywords: nutrients; central composite design; *Saccharomyces cerevisiae*; wine; strain effect; aromas

Citation: Guittin, C.; Maçna, F.; Sanchez, I.; Barreau, A.; Poitou, X.; Sablayrolles, J.-M.; Mouret, J.-R.; Farines, V. The Impact of Must Nutrients and Yeast Strain on the Aromatic Quality of Wines for Cognac Distillation. *Fermentation* **2022**, *8*, 51. https://doi.org/10.3390/fermentation8020051

Academic Editors: Bin Tian and Jicheng Zhan

Received: 3 January 2022
Accepted: 20 January 2022
Published: 25 January 2022

Publisher's Note: MDPI stays neutral with regard to jurisdictional claims in published maps and institutional affiliations.

Copyright: © 2022 by the authors. Licensee MDPI, Basel, Switzerland. This article is an open access article distributed under the terms and conditions of the Creative Commons Attribution (CC BY) license (https://creativecommons.org/licenses/by/4.0/).

1. Introduction

Cognac is a French protected designation of origin (PDO) whose quality is recognized worldwide. This wine spirit is obtained after a double distillation using almost exclusively the Ugni blanc grape variety. The geographical area of production is divided into six crus around the city of Cognac, approximately 75,000 hectares, encircling the Charente River. Despite this, cognac spirits can be distinguished by very different and specific characteristics. Indeed, the quality of the final product not only depends on the terroir but also on viticultural and oenological practices, including blending and aging [1]. The character, delicacy and roundness that the Charente eaux-de-vie will acquire are mainly due to its aromas resulting from a great variety of volatile substances. Fermentative aromas (higher alcohols, esters, aldehydes, acetyls, fatty acids, etc.) are synthesized by yeast during alcoholic fermentation [2] while varietal aromas, such as terpenes or aliphatic aldehydes, originate from the grape berries. For both these aromatic classes, the proportions found in wines are influenced, modified or even disturbed during alcoholic fermentation by different environmental parameters [3]. Thus, yeast assimilable nitrogen (YAN) and lipids, which are the main nutrients in grape musts [4], can have a significant impact on the production of volatile compounds.

During recent years, many studies have focused on the impact of YAN must supplementations on the production of volatile compounds [3,5–12]. The results are partly contradictory because they vary depending on the type of must (synthetic or natural), the nature of the added assimilable nitrogen (mineral and/or organic) and the yeast strain used for fermentation. Nevertheless, global trends can be considered. Generally, the concentration of higher alcohols goes through a maximum as a function of must YAN concentration [5,8,13,14] while that of esters increases continuously [5,11,13,15]. The effect of lipid nutrition on aroma synthesis has been less well-studied. Lipids can be synthesized by yeast in the presence of oxygen; however, this mechanism is limited during oenological fermentation [16]. Therefore, yeasts primarily use the lipids (phytosterols) present in the solid particles of the must [17]. Most studies have concluded that phytosterols have an overall positive effect on higher alcohol synthesis and a negative impact on ester synthesis [9,18–20].

In this study, an experimental design was implemented to understand the influence of YAN and lipid nutrients, and to model their effect on aroma production. Compared to previous studies [7,9,13], this work has several specificities: (1) the alcoholic fermentations were carried out with natural musts containing very high lipid concentrations due to the high turbidities encountered in cognac (500 to 2000 NTU compared to 100 to 150 NTU in conventional oenology); (2) a metabolomic analysis including several families of volatile compounds was carried out in order to obtain a general overview of aroma production, not limited to a specific category of molecules; (3) three yeast strains of *Saccharomyces cerevisiae* commonly used in the cognac appellation area were included in the statistical modeling design to compare their metabolic response to nutrients and identify their specificities in terms of aromatic profiles.

2. Materials and Methods

2.1. Fermentations

Fermentations were performed in 1.2 L cylindrical glass fermenters containing 1 L of must with continuous magnetic stirring (150 rpm) at 23 °C. CO_2 release was measured by an accurate, automatic online monitoring of weight loss every 20 min [21]. The fermenters were inoculated with 20 g/hL of active dry yeast (Lalvin FC9®, Lallemand SA, Montreal, QC, Canada; Fermivin 7013® and Fermivin SM102®, Erbslöh S.A.S., Servian, France) previously rehydrated for 30 min at 37 °C in a 50 g/L glucose solution. The must (Ugni blanc grape variety) was harvested in the Cognac region of France. The must characteristics were as follows: 183 g/L total sugars, 114 mg/L assimilable nitrogen, pH 3.32 and a turbidity of 100 NTU. Ugni blanc grapes were pressed, and the must was settled for 24 h at 4 °C in the presence of 3 mL/hL enzymes (MYZYM SPIRIT, Institut Oenologique de Champagne, Mardeuil, France). A final turbidity of 100 NTU was achieved for the must and the sludge was collected separately. A correlation between the sludge concentration and turbidity was previously performed (R^2 = 0.999) by adding different amounts of solid particles to final volumes (100 mL) of must:

$$Turbidity = 75 \times [solids\ particles] + 100 \quad (1)$$

with the concentration of added solid particles ([*solid particles*]) in % (*v*/*v*) and the *turbidity* in NTU.

Turbidity was measured with a 2100 N turbidimeter (Hach®, Lognes, France).

Based on the determined relationship (Equation (1)) between turbidity and the percentage of solid particles, 59, 100, 200, 300 and 340 mL of sludge were added to a final volume of 1 L of must, and turbidities of 500, 820, 1600, 2380 and 2700 NTU, respectively, were obtained. Each corresponding volume of must was removed from the fermenter to be substituted by the sludge. Assimilable nitrogen was adjusted to 115, 140, 200, 260 or 285 mg N_{ass}/L with a solution of amino acids and NH_4Cl, respecting the proportions of 30% mineral nitrogen (NH_4^+) and 70% organic nitrogen (a mix of amino acids) found in the initial Ugni blanc must. The free amino acid content of the initial Ugni blanc must was determined by cation-exchange chromatography (see Section 2.3). The composition of the amino acid solution added was as follows (in g/L): tyrosine,1.77; tryptophan, 1.63; isoleucine, 0.74; aspartate, 16.48; glutamate, 17.2; arginine, 46.65; leucine, 1.81; threonine, 4.11; glycine, 0.23; glutamine, 14.18; alanine, 7.17; valine, 4.17; methionine, 0.44; phenylalanine, 3.59; serine, 6.42; histidine, 1.60; lysine, 0.53; asparagine, 0.82; and proline, 42.98. A solution of NH_4Cl (80.24 g/L) was used as an ammonium source. To obtain 115, 140, 200, 260 and 285 mg/L of assimilable nitrogen in the must, 0.03, 0.76, 2.51, 4.27 and 5 mL of the amino acid stock solution and 0.04, 0.91, 3.02, 5.12 and 6 mL of an NH_4Cl solution were added to 1 L of must, respectively.

2.2. *Quantification of Sterols and Fatty Acids in Grape Solids*

2.2.1. Dry Matter

A total of 200 mL of must were centrifuged for 10 min at 10,000 rpm to concentrate the grape solids. The supernatant was removed and the grape solids were washed consecutively three times with an NaCl solution (10 mM) to remove sugars. The final pellet was freeze-dried overnight to recover the dry matter (DM).

2.2.2. Lipid Composition

Total lipids were extracted from lyophilized grape solids (aliquot of 1 g) overnight with methanol/chloroform (2:1, *v*/*v*), and the solid residue was then extracted for 2 h with methanol/chloroform/water (2:1:0.8, *v*/*v*/*v*). The organic extracts were dried over Na_2SO_4 and concentrated to dryness using a rotatory evaporator. Phytosterols (campesterol, stigmasterol and β-sistosterol) and main fatty acids were determined in the remaining solids according to the protocol used by Grison et al. (2015), as adapted by Casalta et al. (2019). Total fatty acid concentration was 37.28 mg/g in the DM (Table 1) C18 unsaturated acids represented approximately 48% of the total fatty acid content, and the most abundant saturated fatty acid was palmitic acid (22%). The concentration of phytosterols was 8.43 mg/g in the DM (Table 2), i.e., within the limits of the concentrations described for other grape varieties [17]. The main phytosterol was β-sitosterol (83%), while campesterol and stigmasterol accounted for approximately 11% of the total phytosterols.

Table 1. Fatty acid composition of Ugni blanc grape solids.

Name	Formula	mg/g of Dry Matter	mg/L of Grape Solids
Lauric	C12:0	0.03	0.82
Myristic	C14:0	0.12	3.24
Pentadecylic	C15:0	0.06	1.46
Palmitic	C16:0	8.29	215.07
Palmitoleic	C16:1	0.14	3.51
Margaric	C17:0	0.08	2.04
Stearic	C18:0	0.93	24.23
Oleic	C18:1	2.95	76.49
Linoleic	C18:2	17.69	458.83
Linolenic	C18:3	4.43	114.84
Arachidic	C20:0	0.36	9.24
Gondoic	C20:1	0.07	1.94
Heneicosylic	C21:0	0.12	3.12
Behenic	C22:0	1.00	25.85
Tricosylic	C23:0	0.13	3.34
Lignoceric	C24:0	0.32	8.30
Total		**37.28**	**897.57**

Table 2. Phytosterols composition of Ugni blanc grape solids.

Name	mg/g of Dry Matter	mg/L of Grape Solids
Campesterol	0.51	13.24
Stigmasterol	0.39	10.12
β-sitosterol	7.01	181.80
Sitostanol	0.24	6.34
Unidentified sterols	0.28	7.1
Total	**8.43**	**218.60**

Using the lipid composition of the solid particles from the dry matter (Tables 1 and 2) and Equation (1), we calculated turbidities of 500, 820, 1600, 2380 and 2700 NTU corresponding to 12.9, 21.9, 43.8, 65.6 and 74.5 mg/L sterols, respectively. These turbidities (500, 820, 1600, 2380 and 2700 NTU) also corresponded to fatty acid concentrations of 57.1, 96.8, 193.5, 290.3 and 329.9 mg/L, respectively.

2.3. *Analytical Methods*

Ammonium concentration was determined enzymatically (R-Biopharm AG™, Darmstadt, Germany). The free amino acid content of the must was measured by cation-exchange chromatography with post-column ninhydrin derivatization (Biochrom 30, Biochrom™, Cambridge, UK), as described by [22].

Ethanol, glucose, fructose, glycerol, succinic acid, alpha-ketoglutaric acid and acetic acid concentrations were determined by High Performance Liquid Chromatography (HPLC) (HPLC 1260 Infinity, Agilent™ Technologies, Santa Clara, CA, USA) on a Phenomenex Rezex ROA column (Phenomenex™, Le Pecq, France) at 60 °C. The column was eluted with 0.005 N H_2SO_4 at a flow rate of 0.6 mL/min. Organic acids were analyzed with a UV detector (Agilent™ Technologies, Santa Clara, CA, USA) at 210 nm; the concentrations of the other compounds were quantified with a refractive index detector (Agilent™ Technologies, Santa Clara, CA, USA). The analysis was carried out with the Agilent™ OpenLab CDS 2.x software package (Santa Clara, CA, USA).

The concentrations of ethyl acetate, ethyl propanoate, ethyl 2-methylpropanoate, ethyl butanoate, ethyl 2-methylbutanoate, ethyl 3-methylbutanoate, ethyl hexanoate, ethyl octanoate, ethyl decanoate, ethyl dodecanoate, ethyl lactate, diethyl succinate, 2-methylpropyl acetate, 2-methylbutyl acetate, 3-methyl butyl acetate, 2-phenylethyl acetate, 2-methylpropanol, 2-methylbutanol, 3-methylbutanol, hexanol, 2-phenylethanol, propanoic acid, butanoic acid, 2-methylpropanoic acid, 2-methylbutanoic acid, 3-methylbutanoic acid, hexanoic acid, octanoic acid, decanoic acid and dodecanoic acid were measured in the liquid phase after sample pretreatment by double liquid-liquid extraction with dichloromethane in the presence of deuterated standards (ethyl butanoate-D5, ethyl decanoate-D5, ethyl hexanoate-D5, ethyl octanoate-D5 and phenylethanol-D4) [9]. The samples were analyzed with a Hewlett Packard (Agilent™ Technologies, Santa Clara, CA, USA) 6890 gas chromatograph equipped with a CTC Combi PAL autosampler AOC-5000 (Shimadzu™, Columbia, SC, USA) and coupled to a Hewlett Packard 5973 mass spectrometry detector (Agilent™ Technologies, Santa Clara, CA, USA).

The total liquid concentration of acetaldehyde was precisely measured using a commercial enzymatic test kit (Ref. 984347, ThermoFischer Scientific™, Waltham, MA, USA) and a Thermo Scientific™ Gallery™ automated photometric analyzer.

The metabolomic analysis was realized by solid-phase microextraction (SPME).

Before SPME analysis, 5 mL of wine were placed into a 20 mL headspace amber glass vial and diluted with 5 mL of ultra-pure water. The samples were spiked with 100 µL of 2-octanol (internal standard solution at 1 mg/L in ethanol) and saturated with sodium chloride before closure with a PTFE-faced silicone septum-aluminum crimp cap. A 50/30 µm DVB/CAR/PDMS fiber (divinylbenzene/carboxene/polydimethylsiloxane;

Supelco, Bellefonte, PA, USA) was used for volatile extraction. Before the analysis, the fiber was conditioned into the GC injector at 270 °C for 30 min to prevent contamination. The sample was pre-incubated for 5 min at 40 °C. Adsorption lasted for 30 min at 40 °C with stirring at 250 rpm (10 s ON and 1 s OFF). Next, desorption took place in the injector in splitless mode for 600 s at 250 °C. All analyses were made in triplicate. The injections were fully automated using an MPS robotic autosampler (Gerstel) operated by Maestro software (Gerstel). An Agilent 8890 gas chromatograph (Agilent Technologies, Palo Alto, CA, USA), coupled to an Agilent 5977B mass spectrometer were used. The sample was analyzed on a DB-HeavyWAX column (60 m × 0.25 mm i.d., 0.25 µm film thickness, Agilent Technologies). The injector and MSD transfer line temperatures were fixed at 250 °C. The oven temperature was held at 40 °C for 10 min, raised to 240 °C at a rate of 3 °C per minute and then held for 5 min. The column carrier gas was helium at a constant flow rate of 1 mL/min. The splitless injection mode was used. The temperature of the ion source was set at 230 °C while the electron impact mass spectra were recorded at 70 eV ionization energy. The GC-MS analysis was carried out in the scanning mode (SCAN); the mass range was from 35 to 350 m/z.

2.4. Statistical Analysis

2.4.1. Experimental Central Composite Design

The statistical analysis was performed with R software, version 3.6.2 (R Development Core Team 2012), and the rsm library [23].

In order to study and understand the impact of yeast nutrition in cognac, we used a central composite design (CCD), the most common design used to fit quadratic models, described by [24]. CCD combines a two-level factorial design with axial points (star points) and at least one point at the center of the experimental region in order to fit quadratic polynomials. The center points are usually repeated to obtain a good estimate of the experimental error (pure error). If the distance from the center of the design space to a factorial point is ± 1 unit for each factor, the distance from the center of the design space to a star point is $\pm\alpha$, where $|\alpha| > 1$. The star points represent new extreme values (low and high) for each factor in this design. The variables included the concentration of assimilable nitrogen and turbidity coded levels, $-\alpha$, -1, 0, +1, +α (the value of α is 1.41 rotatable design), and are shown in Table 3 as well as the actual levels of the variables in the CCD experiments.

The effect of the two independent variables on each volatile compound (Y) was modelled by a polynomial response surface:

$$Y = \beta_0 + \beta_{1x1} + \beta_{2x2} + \beta_{12x1x2} + \beta_{11x1^2} + \beta_{22x2^2} + \varepsilon \tag{2}$$

where $x1$ and $x2$ represent the coded values of the initial nitrogen content and turbidity, respectively, Y is the predicted response, β_0 is the intercept term, β_i is the linear coefficient, β_{ii} is the quadratic coefficient and β_{ij} is the interaction coefficient. When necessary, a simplified model was fitted for some compounds by suppressing the interactive terms of the equation according to validity criteria.

In addition, the normality of residual distributions and the homogeneity of variance were studied with standard diagnostic graphs; no violation of the assumptions was detected.

The accuracy and general ability of each polynomial model described above was evaluated by a lack of fit test, the Fisher test and the adjusted coefficient of determination (adjR2).

The reliability of the fitted models was very good overall: each polynomial model produced a non-significant lack of fit test at a 0.05 threshold, a significant Fisher test at a 0.05 threshold and a range of adjR2 between 35% and 99%.

A CCD was performed for each strain independently.

For graphical representations of the surface responses, a custom version of the persp() function was used.

Table 3. The experimental plan of fermentation conditions for each yeast used in spirit production.

	Independent Variable			
Experiments	**Assimilable Nitrogen (mgN/L)**		**Turbidity (NTU)**	
	Coded Level	**Uncoded Level**	**Coded Level**	**Uncoded Level**
1	−1	140	−1	820
2	+1	260	−1	820
3	−1	140	+1	2380
4	+1	260	+1	2380
5 [a]	0	200	0	1600
6 [a]	0	200	0	1600
7 [a]	0	200	0	1600
8 [a]	0	200	0	1600
9	−1.41	115	0	1600
10	+1.41	285	0	1600
11	0	200	−1.41	500
12	0	200	+1.41	2700

[a] Four replicates included at the center of the experimental domain.

2.4.2. Analysis of Covariance (ANCOVA)

ANCOVA is a general linear model with a continuous outcome variable and two or more predictor variables, where at least one is continuous (covariate) and at least one is categorical (factor). In addition to studying the impacts of nitrogen and lipid nutrition on the synthesis of fermentative aromas, an analysis of covariance was performed in order to demonstrate a difference in production between strains, which we will call the "strain effect". This analysis did not allow us to determine which strain was different but evidenced any overall difference.

2.4.3. Analysis of Correlation Tree and PCA

The statistical evaluations of the metabolomic analysis were performed using the JMP (SAS Institute JMP, Brie Comte Robert, France) software, version 15.0.0.

3. Results and Discussion

The impacts of YAN and turbidity on the (1) kinetic parameters, (2) metabolites of central carbon metabolism and (3) production of volatile compounds (fermentative aromas, aldehydes and terpenes) were studied. A central composite design was used to minimize the number of factor combinations required to evaluate the effects of these two factors. Twelve fermentations were performed for each of the three strains with five different assimilable nitrogen concentrations (from 115 to 285 mgN/L) and five different turbidities (from 500 to 2700 NTU). The conditions are summarized in Table 3.

The central composite design enabled us to plot the concentrations of volatile compounds as a function of the two environmental parameters and to build an associated mathematical model. With this model, a "weight" was assigned to each parameter (assimilable nitrogen and turbidity); thus, a classification of these two parameters according to their influence on the synthesis of each volatile compound or on the fermentative parameters could be performed. The significance of this effect was determined from its expressed p-value (p). Based on the sign of the coefficients (Equation (2)) of the polynomial surface response, the positive or negative effect of the fermentative parameter on the production of the studied volatile compound was determined. The interactions between the parameters (i.e., the effect of one factor as a function of the value of another) were also studied, as well as the non-linear effects of these parameters, also called quadratic effects.

3.1. Fermentation Kinetics

In all the fermentations, the sugars were almost exhausted. The residual sugar content for all fermentations performed was lower than or equal to 5 g/L. The YAN was completely consumed at the end of the growth phase; therefore, assimilable nitrogen was the limiting nutrient in these experiments. The three yeast strains displayed similar fermentation kinetics (Figure 1), although the effects of the nutrients were slightly different. Under identical nutritional conditions, the maximum rates of CO_2 production were identical for the three strains, while the fermentation durations were noticeably different: strain SM102® was 50 h faster at low nitrogen concentration than strain 7013® and 30 h faster than strain FC9® (Figure 1).

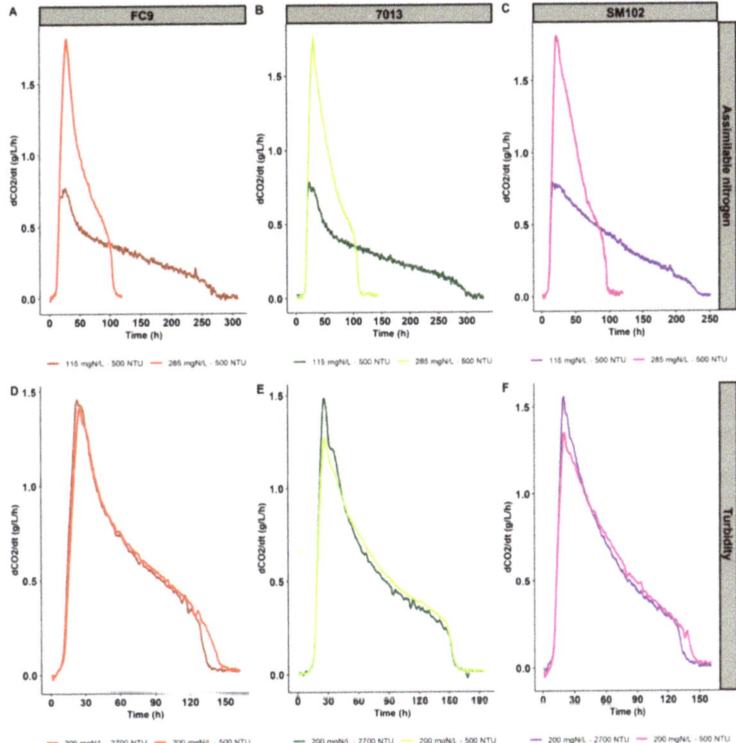

Figure 1. The impact of assimilable nitrogen concentration in the must (**A**–**C**) and turbidity (**D**–**F**) on fermentation kinetics at 23 °C for three yeast strains: FC9® (**A**,**D**), 7013® (**B**,**E**) and SM102® (**C**,**F**).

For all strains, a very strong positive impact of assimilable nitrogen was noted on the maximum CO_2 production rate, resulting in a shorter fermentation time, in agreement with previous works [25–27]. The maximum CO_2 production rate was also positively impacted by turbidity, but to a much lesser extent (Table 4).

Table 4. The effects of nutrient parameters on maximum CO_2 production rate and central carbon metabolism (MCC) compounds for FC9®, 7013® and SM102® strains described by the CCP model with ANCOVA analysis to identify a "strain effect".

Strain	Effect	Variable	Kinetic Parameters		MCC Compounds		
			Vmax	**Time of Fermentation**	**Glycerol**	**Succinate**	**α-Ketoglutarate**
FC9®	*Simple effects*	N_{ass}	↗	↘	↗	↘	↘
		NTU	↗		↗	↗	↗
	Interaction effects	N_{ass}:NTU			↗	↗	
	Quadratic effects	N_{ass}	↘	↗			
		NTU					
7013®	*Simple effects*	N_{ass}	↗	↘	↗		↘
		NTU	↗				↗
	Interaction effects	N_{ass}:NTU					
	Quadratic effects	N_{ass}	↘	↗			
		NTU					
SM102®	*Simple effects*	N_{ass}	↗		↗	↘	↘
		NTU	↗		↗	↗	↗
	Interaction effects	N_{ass}:NTU	↗		↗	↗	
	Quadratic effects	N_{ass}	↘		↘	↘	
		NTU	↘		↘	↘	
ANCOVA (p-value): « strain effect »			+++	+++	+++	+++	+++

■: $p < 0.001$, ■: $p < 0.01$, ■: $p < 0.05$, ■: $p < 0.001$, ■: $p < 0.01$, ■: $p < 0.05$, ■: $p < 0.001$, ■: $p < 0.01$, ■: $p < 0.05$, ■: not significant, (↗): positive effect, (↘): negative effect, (+++): ANCOVA $p < 0.05$, N_{ass}: assimilable nitrogen, NTU: turbidity, V_{max}: maximum rate of CO_2 production.

3.2. Compounds of Central Carbon Metabolism

The amount of ethanol produced in all conditions was similar, whereas significant differences were observed in the concentrations of glycerol, succinic acid, acetic acid and α-ketoglutaric acid, which are metabolic markers that result from central carbon metabolism. The effects of the nutrients were different depending on the compound, although succinate and α-ketoglutarate generally reacted in the same way (Table 4). Firstly, only very small quantities of acetic acid (<0.015 g/L) were found, compared to the concentrations of 0.34 to 1.21 g/L generally present in wine [3]. This observation might be explained by the fact that our working range of phytosterol concentration is much higher (from 13 to 75 mg/L) than the range commonly used in previous studies [9,28,29]; indeed, the production of acetic acid during alcoholic fermentation is low for high concentrations of phytosterols [28,30,31]. Therefore, there was almost no demand for acetyl-CoA, the precursor of lipid biosynthesis. In these conditions, the pool of cytoplasmic acetyl-CoA not used to produce lipids can be converted to citrate by the citrate synthase Cit2p. Then, citrate can be transported to the mitochondria and participate in succinic acid synthesis via the TCA cycle [9]. Therefore, the synthesis of α-ketoglutaric and succinic acids is promoted by the intake of lipids (Table 4). On the other hand, and consistent with bibliographic data, the concentrations of α-ketoglutaric and succinic acids decreased when the YAN supply increased [5,7,9,11]. Meanwhile, the turbidity and assimilable nitrogen positively impacted glycerol production (Table 4). When nitrogen is the limiting nutrient face to lipids, as was the case in this work, glycerol production increases with lipid content [30]. This can be explained by the important role of glycerol in maintaining the redox balance of the cell. When the production of organic acids is increased after lipid addition, an excess of NADH will be synthesized that will then be oxidized by the glycerol synthesis pathway [32]. In this study, the glycerol concentration was also favored by YAN supply (Table 4). This quite surprising result differs from the literature [9,14,32,33]; however, it is still consistent with our previous study [7]. Finally, a highly significant strain effect was observed for all central metabolism compounds. Indeed, strain FC9® produced less organic acids and glycerol in all conditions, with average concentrations of 0.09 ± 0.03; 0.81 ± 0.08 and 6.81 ± 0.33 g/L of α-ketoglutaric acid, succinic acid and glycerol, respectively, whereas the other two strains (7013® and SM102®) produced 0.12 ± 0.05 and 0.14 ± 0.04 g/L of α-ketoglutaric acid, respectively; 1.08 ± 0.07 and 1.09 ± 0.1 g/L of succinic acid, respectively; and 8.18 ± 0.42 and 8.14 ± 0.37 g/L of glycerol, respectively.

3.3. *Aromas*

Volatile compounds present in distilled beverages have been reported in the literature [34]. Aroma compounds involved in olfactory perception belong to different chemical classes, such as alcohols, esters, aldehydes, norisoprenoids and terpenes. In this work, a metabolomic analysis was applied to the study of the impact of YAN and lipids on the production of fermentative aromas (higher alcohols and esters), terpenes, aldehydes (acetaldehyde) and more atypical alcohols present in wines intended for charentaise batch distillation.

3.3.1. Fermentative Aromas

Since all the higher alcohols (isobutanol, isoamyl alcohol, hexanol, methionol and 2-phenylethanol) and acetate esters (isobutyl acetate, 2-methylbutyl acetate, isoamyl acetate, hexyl acetate and 2-phenylethyl acetate) reacted in the same way to the nutrients, we considered them as sums and by compound families (Table 5).

Table 5. The effects of nutrient parameters on fermentative aromas: sum of higher alcohols and acetate esters for FC9®, 7013® and SM102® strains described by the CCP model with ANCOVA analysis to identify a "strain effect".

Strain	Effect	Variable	Sum of Higher Alcohols	Sum of Acetate Esters
FC9®	*Simple effects*	N_{ass}	↘	↗
		NTU	↗	↘
	Interaction effects	N_{ass}:NTU	↗	
	Quadratic effects	N_{ass}		
		NTU		
7013®	*Simple effects*	N_{ass}	↘	↗
		NTU	↗	↘
	Interaction effects	N_{ass}:NTU		
	Quadratic effects	N_{ass}	↘	
		NTU	↘	↗
SM102®	*Simple effects*	N_{ass}	↘	↗
		NTU	↗	↘
	Interaction effects	N_{ass}:NTU		
	Quadratic effects	N_{ass}		
		NTU		↗
ANCOVA (*p*-value): « strain effect »			+++	+++

■: $p < 0.001$, ■: $p < 0.01$, ■: $p < 0.05$, ■: $p < 0.001$, ■: $p < 0.01$, ■: $p < 0.05$, ■: $p < 0.001$, ■: $p < 0.01$, ■: $p < 0.05$, ■: not significant, (↗): positive effect, (↘): negative effect, (+++): ANCOVA $p < 0.05$, N_{ass}: assimilable nitrogen, NTU: turbidity.

For the three yeast strains, the production trend of higher alcohols decreased with an increasing supply of assimilable nitrogen (p-value < 0.001) but was promoted by enhanced turbidity (p-value < 0.001), whereas opposite effects of the nutrients were observed for acetate esters (Table 5 and Figure 2A,B). These data are consistent with previous findings [7]. The observed effects can be explained by the regulation of acetyltransferases that catalyze the conversion of higher alcohols into their corresponding esters. The positive impact of nitrogen on acetate ester [7,9] production can be explained by the higher expression of genes coding for alcohol acetyltransferases during nitrogen supplementation of the must [10,35]. In this case, the alcohol/ester balance changes in favor of ester accumulation with increasing doses of YAN in the medium. The effect of turbidity is, however, quite surprising because there is no metabolic connection between the anabolic pathways of alcohol production and the lipid pathway. This effect can also be explained by the activity of alcohol acetyl transferases, which decreases in the presence of lipids. Indeed, the expression of ATF1 (the major gene encoding this enzymatic activity) is repressed by lipids [36,37]. In the presence of increased lipids, the repression of acetate ester synthesis is to the benefit of the accumulation of higher alcohols. Although the strains responded similarly to the nutrients in the synthesis of higher alcohols and acetate esters, their production levels were significantly different (Table 5), indicating a probable strain effect on the contribution to the wine's fruity and floral character. YAN supply resulted in complex effects on the synthesis of ethyl esters depending on the involved *Saccharomyces cerevisiae* strain for fermentations (Table 6).

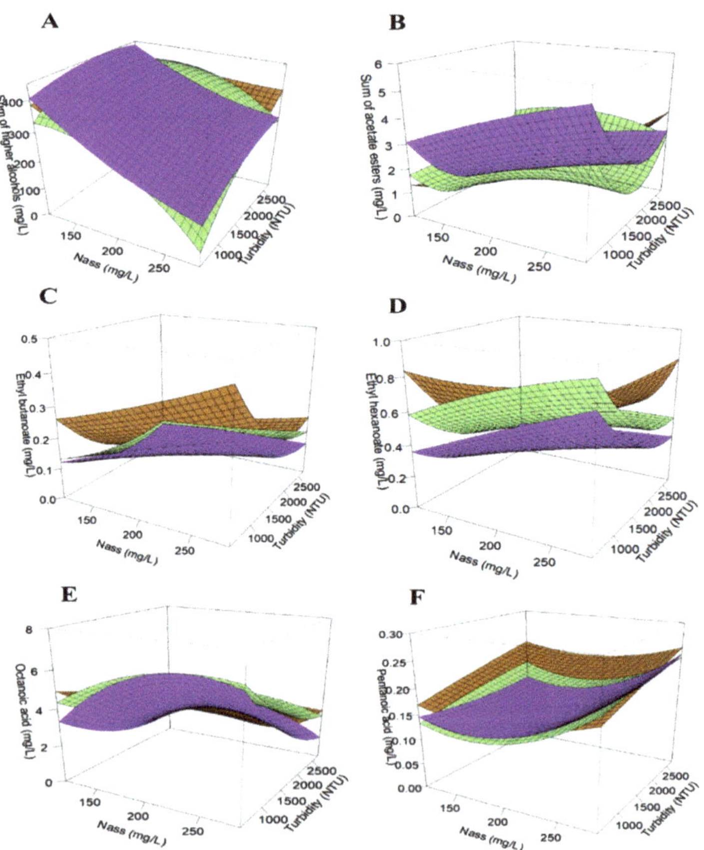

Figure 2. Response surfaces of fermentative aromas: sum of higher alcohols (**A**), sum of acetate esters (**B**), ethyl esters (ethyl butanoate (**C**) and ethyl hexanoate (**D**)) and acids (octanoic acid (**E**) and pentanoic acid (**F**)) for all three strains: FC9® (—), 7013® (—) and SM102® (—).

Table 6. The effects of nutrient parameters on fermentative aromas: ethyl esters for FC9®, 7013® and SM102® strains described by the CCP model with ANCOVA analysis to identify a "strain effect".

Strain	Effect	Variable	Ethyl Isobutanoate	Ethyl Butanoate	Ethyl Hexanoate	Ethyl Octanoate	Ethyl Decanoate
FC9®	*S.E*	N_{ass}	↘				
		NTU		↘	↘		
	I.E	N_{ass}:NTU					
	Q.E	N_{ass}	↗				
		NTU		↗	↗		
7013®	*S.E*	N_{ass}	↘	↗	↗		↗
		NTU	↗	↘	↘	↘	↘
	I.E	N_{ass}:NTU		↘			
	Q.E	N_{ass}	↗				
		NTU		↗	↗	↗	
SM102®	*S.E*	N_{ass}		↗			
		NTU	↗	↘	↘		↘
	I.E	N_{ass}:NTU			↘		
	Q.E	N_{ass}					
		NTU		↗	↗		
ANCOVA (*p*-value): « strain effect »			+++	+++	+++	+++	+++

■: $p < 0.001$, ■: $p < 0.01$, ■: $p < 0.05$, ■: $p < 0.001$, ■: $p < 0.01$, ■: $p < 0.05$, ■: $p < 0.001$, ■: $p < 0.01$, ■: $p < 0.05$, ■: not significant, (↗): positive effect, (↘): negative effect, (+++): ANCOVA $p < 0.05$, N_{ass}: assimilable nitrogen, NTU: turbidity, S.E: Simple effects, I.E: Interaction effects, Q.E: quadratic effects.

Ethyl butanoate increased with assimilable nitrogen in 7013® and SM102® strains, while only 7013® reacted positively to this nutrient for the synthesis of ethyl hexanoate and ethyl decanoate. The production of these previous esters was not impacted by nitrogen addition for strain FC9®. Lastly, an opposing behavior was noticed with a decrease in ethyl isobutanoate production induced by nitrogen supply (FC9® and 7013®).

This complex YAN effect has already been observed in several studies, certainly due to the fact that the synthesis pathways of these esters are not linked to nitrogen metabolism [7,9,12,14] and depend on the yeast strain used [38]. More generic trends were noticed for the impact of turbidity, with a positive effect on the synthesis of ethyl isobutanoate (7013® and SM102®) and a negative effect on almost all other ethyl esters (Table 6). In addition to this negative impact, a positive quadratic effect (i.e., the bending of the curve) was also visible for ethyl butanoate and hexanoate (anise seed, apple-like aroma), where a minimum production of these flavors was obtained close to 1500 NTU (Figure 2C,D). Observing a similar effect of phytosterols on ethyl esters for the three strains is rather logical because these compounds are produced from the lipid metabolism whose regulation appears to be relatively conserved within the *Saccharomyces cerevisiae* species. Moreover, the negative effect of this nutrient is in accordance with the data obtained by [19]), showing that the addition of unsaturated fatty acids to the fermentation medium results in a global decrease in ethyl ester production. As for ethyl esters, the acids did not all respond in the same way to assimilable nitrogen (Table 7).

The production of C:6, C:8 and C:10 acids increased with YAN addition (7013® and SM102®). The production of theses acids is favored by the contribution of assimilable nitrogen in the must and passes through a maximum of concentration (quadratic effect) at 200 mg/L of nitrogen (Figure 2E), as previously observed by [7,39]. An opposing effect was observed for propionic acid, C:3 (only for SM102®) and pentanoic acid, C:5 (all three strains), with decreased synthesis when the nitrogen concentration was increased and a minimum concentration (quadratic effect) at 200 mg/L of nitrogen (Figure 2F). As observed for ethyl esters, lipid nutrition provided a more consensual response in acid production with a globally decreased synthesis (Table 7). First, the impact of nitrogen on acid synthesis can vary depending on the *Saccharomyces cerevisiae* strain, as do the intrinsic production levels [39,40]. Second, the pairing of the production of esters from their acid precursors was not systematic and not all of the pairs of compounds reacted in the same way to nutrient changes. This latter observation indicates that, for ethyl esters, the key factor is not only the availability of the precursors (fatty acids) but also the enzymatic activity (Eht1p and Eeb1p) responsible for this bioconversion [41].

Table 7. The effects of nutrient parameters on fermentative aromas: acids for FC9®, 7013® and SM102® strains described by the CCP model with ANCOVA analysis to identify a "strain effect".

Strain	Effect	Variable	Propionic Acid	Pentanoic Acid	Octanoic Acid	Decanoic Acid	Hexanoic Acid
FC9®	S.E	N_{ass}		↘			
		NTU	↘		↘	↘	
	I.E	N_{ass}:NTU					
	Q.E	N_{ass}		↗			
		NTU			↗		
7013®	S.E	N_{ass}		↘	↗	↗	↗
		NTU	↘		↘		↘
	I.E	N_{ass}:NTU					↘
	Q.E	N_{ass}		↗	↘	↘	
		NTU	↗		↗		↗
SM102®	S.E	N_{ass}	↘	↘	↗	↗	↗
		NTU					↘
	I.E	N_{ass}:NTU					↘
	Q.E	N_{ass}	↗	↗	↘	↗	
		NTU			↗		↗
ANCOVA (p-value): « strain effect »			+++	+++	+++	+++	+++

■: $p < 0.001$, ■: $p < 0.01$, ■: $p < 0.05$, ■: $p < 0.001$, ■: $p < 0.01$, ■: $p < 0.05$, ■: $p < 0.001$, ■: $p < 0.01$, ■: $p < 0.05$, ■: not significant, (↗): positive effect, (↘): negative effect, (+++): ANCOVA $p < 0.05$, N_{ass}: assimilable nitrogen, NTU: turbidity, S.E: Simple effects, I.E: Interaction effects, Q.E: quadratic effects.

3.3.2. Terpenes

While the contribution of terpenoids has been extensively studied in grapes and wines [42,43], some significant works have also been specially dedicated to mono- and sesquiterpenes in brandies and cognac [1,44–46]. Indeed, terpenes are some of the compounds that impart a fruity and floral character to cognac [47]. Terpenes are generally known and described as varietal compounds occurring in free as well as bound forms, depending on the odor active molecule (aglicone) that may or may not be bound to a sugar moiety. The majority of free monoterpenes are found in the grape berry skin as well as in the pulp, while the forms combined with sugar are mainly found in the juice [48]. Bound forms are commonly known as aroma precursors since they undergo hydrolysis easily, generating the active odor molecule and free sugar. Studies have shown that β-glucosidase activity is low or non-existent in *Saccharomyces cerevisiae* strains [49,50] because the fermentation conditions (acidic pH and high sugar concentration) are unfavorable for the activity of this enzyme.

Therefore, the terpenes studied are considered to be the free forms when initially found in the musts [48] and the glycolized forms as those released during wine distillation [51]. In wines made from Ugni blanc grapes, the monoterpenes found include trans-nerolidol (rose aroma), citronellol (shade of lemongrass), α-terpineol (lily of the valley aroma) and linalool (floral aroma). Firstly, it should be noted that increasing the concentrations of assimilable nitrogen reduced wine terpene levels (trans-nerolidol, citronellol and linalool) for all strains (Table 8).

Table 8. Effects of assimilable nitrogen and turbidity on varietal flavors, including terpenes and ketone compounds for FC9®, 7013® and SM102® strains described by the CCP model with ANCOVA analysis.

Strain	Effect	Variable	Terpenes				Ketone
			Trans-Nerolidol	Citronellol	α-Terpineol	Linalool	Acetophenone
FC9®	*Simple effects*	N_{ass}				↘	
	Interaction effects	N_{ass}:NTU				↗	
	Quadratic effects	N_{ass}				↗	
7013®	*Simple effects*	N_{ass}	↘	↘		↘	↗
		NTU		↗			↘
	Quadratic effects	N_{ass}		↘			↘
		NTU		↘			
SM102®	*Simple effects*	N_{ass}	↘	↘		↘	↗
		NTU		↗		↗	↘
	Interaction effects	N_{ass}:NTU					↘
	Quadratic effects	N_{ass}					↘
ANCOVA (p-value): « strain effect »			+++	+++	+++	+++	+++

■: $p < 0.001$, ■: $p < 0.01$, ■: $p < 0.05$, ■: $p < 0.001$, ■: $p < 0.01$, ■: $p < 0.05$, ■: $p < 0.001$, ■: $p < 0.01$, ■: $p < 0.05$, ■: not significant, (↗): positive effect, (↘): negative effect, (+++): ANCOVA $p < 0.05$, N_{ass}: assimilable nitrogen, NTU: turbidity.

These results are partly comparable to the ones presented by [3], who reported that nitrogen supplementation (DAP) significantly decreased citronellol and nerol concentrations, while linalool was present in higher concentrations at moderate DAP concentrations (350 mgN/L) [52]. Secondly, the changes in nitrogen content had no effect on the synthesis of α-terpineol (Table 5A), likely suggesting that the low pH of the medium led to its formation through the chemical transformation of linalool, rather than through the involvement of a cyclase enzyme, as also demonstrated by [52]. Thus, the effect of nitrogen on terpenes is difficult to interpret and further studies will be necessary.

On the other hand, we observed that turbidity favors the presence of citronellol and linalool in wines (Table 8). This result can be explained by the fact that increasing turbidity results in the addition of important quantities of solid particles (from 6 to 34%), composed of the skin and pulp of grapes [17], that are important sources of free monoterpenes, such as citronellol or linalool [17]. Moreover, it has been proposed that terpene and sterol biosynthesis are related [53] and that terpenes can also be synthesized by *Saccharomyces cerevisiae* [52,54,55]. Anaerobic conditions were suggested to inhibit several essential steps in ergosterol biosynthesis, including squalene epoxidation and the oxidative demethylation/dehydrogenation of lanosterol, which are essential steps in the formation of ergosterol [56]. In this study, a large excess of phytosterols (13 to 75 mg/L) likely favored terpene biosynthesis from geranyl pyrophosphate (GPP) by the yeast. Finally, a strain effect was identified on terpene synthesis (Table 8). Strain SM102® produced more linalool while the 7013® strain synthesized more citronellol (Figure 3A). Although the link between the varietal aromas and the impact of fermentation on their production remains unclear, it has previously been shown that linalool production can vary depending on the *Saccharomyces cerevisiae* strain used, because the presence of polyphenolic and aromatic fractions in grapes exerts a strong influence on yeast metabolism [57]. In our case, the yeast strains' metabolism did not always favor the production of linalool, nerolidol or citronellol from the precursor geranyl PP (GPP) [52]. In the FC9® strain, linalool synthesis (followed by α-terpineol) was very low. These two aromas only accounted for 5% of the terpenes produced, instead of about 25% for the other two strains (Figure 3). These data confirm that, in the case of wines dedicated to cognac distillation, the proportion of terpenes of varietal origin is essential; however, the proportion linked to yeast metabolism is also important. In fact, high turbidities provide the geraniol precursor in large quantities, and yeast becomes a tool to modulate the terpene-linked aromatic fraction of wines.

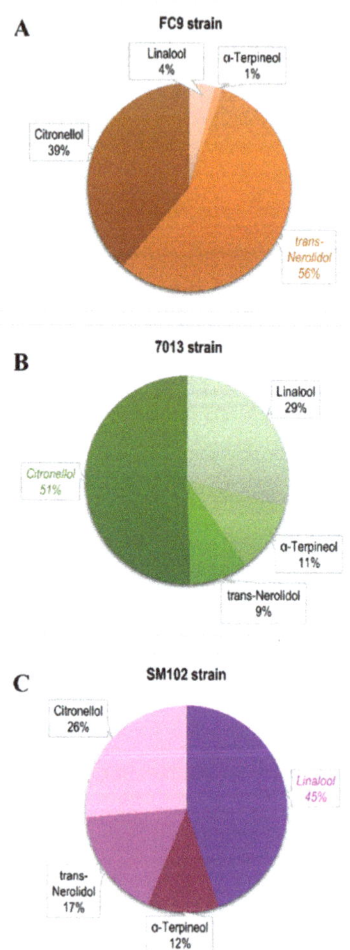

Figure 3. The distribution of terpene synthesis for strain (**A**) FC9® (—), (**B**) 7013® (—) and (**C**) SM102® (—), including all fermentative conditions.

3.3.3. 1-octen-3-ol

1-octen-3-ol is a natural chemical derived from linoleic acid, which was first isolated from the matsutake pine mushroom and thereafter from plants and other fungi. This alcohol is a well-known compound associated with fresh mushroom odors in grapes and wines [58] and is associated with rotten grapes, particularly due to *Botrytis cinerea* [59]. 1-octen-3-ol has a relatively low perception threshold and, because its odor persists after alcoholic fermentation [60], it may be responsible for defects in wine. As shown in Table 9, the presence of 1-octen-3-ol seemed to be favored by the presence of solid particles in high quantities.

Table 9. Effects of assimilable nitrogen and turbidity on other alcohols for FC9®, 7013® and SM102® strains described by the CCP model with ANCOVA analysis.

Strain	Effect	Variable	1-Octen-3-ol	2-Nonanol
FC9®	*Simple effects*	N_{ass}		
		NTU		
	Interaction effects	N_{ass}:NTU	↗	
	Quadratic effects	N_{ass}		↗
		NTU		
7013®	*Simple effects*	N_{ass}		
		NTU	↗	↗
	Interaction effects	N_{ass}:NTU		↘
	Quadratic effects	N_{ass}		↗
		NTU		↗
SM102®	*Simple effects*	N_{ass}		
		NTU	↗	
	Interaction effects	N_{ass}:NTU		↘
	Quadratic effects	N_{ass}		
		NTU	↘	
ANCOVA (p-value): « strain effect »			/	+++

■: $p < 0.001$, ■: $p < 0.01$, ■: $p < 0.05$, ■: $p < 0.001$, ■: $p < 0.01$, ■: $p < 0.05$, ■: $p < 0.001$, ■: $p < 0.01$, ■: $p < 0.05$, ■: not significant, (↗): positive effect, (↘): negative effect, (+++): ANCOVA $p < 0.05$, N_{ass}: assimilable nitrogen, NTU: turbidity.

However, the vineyard from which the must was obtained was considered as healthy and was not affected by *Botrytis cinerea*. In our case, 1-octen-3-ol biosynthesis resulted from the aerobic oxidation of linoleic acid through a specific enzymatic reaction [61]. In the elaboration of cognac, the collection of musts as well as the pre-fermentation stages are carried out in the total absence of SO_2. Oxidative phenomena, such as the oxidation of linoleic acid, are therefore favored. 1-octen-3-ol is thus a marker of this oxidative phenomenon, which was all the more marked as the musts were rich in solid particles and therefore in lipids.

3.3.4. Acetaldehyde

Acetaldehyde is a key compound in yeast metabolism and an intermediate of glycolysis that is produced during alcoholic fermentation, where pyruvate can be converted to acetaldehyde and carbon dioxide by pyruvate decarboxylase (PDC) and the acetaldehyde can then be reduced to ethanol through the action of alcohol dehydrogenase (ADH). In addition, acetaldehyde is a powerful aromatic compound that can be found in many food matrices [62]; however, excess acetaldehyde in cognac is not desired as it is responsible for a poor aromatic profile [47]. During the aging of brandies, acetaldehyde reacts with ethanol to form diethyl acetal, which can impart a strong apple smell when present in too high a concentration. In wine, free acetaldehyde can form more or less stable combinations with other molecules; we can then speak of combined or bound acetaldehyde, the sum of the two forms being the total acetaldehyde amount. In this study, only total acetaldehyde was discussed. The final acetaldehyde concentration appeared to be favored by a high concentration of assimilable nitrogen (Table 10).

Table 10. Effects of assimilable nitrogen and turbidity on aliphatic aldehydes for FC9®, 7013® and SM102® strains described by the CCP model with ANCOVA analysis.

Strain	Effect	Variable	Acetaldehyde	Benzaldehyde
FC9®	*Simple effects*	N_{ass}	↗	
		NTU		
	Interaction effects	N_{ass}:NTU		
	Quadratic effects	N_{ass}		
		NTU		
7013®	*Simple effects*	N_{ass}	↗	↘
		NTU	↘	↗
	Interaction effects	N_{ass}:NTU	↘	
	Quadratic effects	N_{ass}	↘	
		NTU		
SM102®	*Simple effects*	N_{ass}	↗	↘
		NTU	↘	
	Interaction effects	N_{ass}:NTU	↘	
	Quadratic effects	N_{ass}		↗
		NTU		
ANCOVA (*p*-value): « strain effect »			+++	/

■: $p < 0.001$, ■: $p < 0.01$, ■: $p < 0.05$, ■: $p < 0.001$, ■: $p < 0.01$, ■: $p < 0.05$, ■: $p < 0.001$, ■: $p < 0.01$, ■: $p < 0.05$, ■: not significant, (↗): positive effect, (↘): negative effect, (+++): ANCOVA $p < 0.05$, N_{ass}: assimilable nitrogen, NTU: turbidity.

A high ammonium consumption can be associated with an increased concentration of final acetaldehyde [5], whereas a high concentration of amino acids in the must significantly decreases the acetaldehyde content [63]. A strain effect is superimposed on this nitrogen effect (mineral versus organic): in some strains, the residual acetaldehyde at the end of fermentation will be lower for a high nitrogen concentration, while for others the residual acetaldehyde will be higher [39]. The strain effect on the production of this molecule was also highlighted in this study (Table 9), which was in agreement with the fact that *Saccharomyces cerevisiae* strains have a different potential to produce this aroma [64]. Unlike nitrogen, an increase in lipid content in the musts reduced acetaldehyde synthesis. Presumably, when there is an excess of lipids, the yeast does not need to produce acetyl-CoA for lipid biosynthesis; therefore, the yeast does not need to overproduce and therefore accumulate acetaldehyde to switch to the acetate (lipid precursor) metabolism pathway. Nevertheless, the impact of sterols on acetaldehyde has been poorly studied and makes comparisons difficult. One study showed that the most turbid musts had significantly higher concentrations of acetaldehyde during the first 4 days of fermentation; however, at the end, the difference was statistically non-significant [65]. Therefore, the impact of turbidity remains difficult to analyze and is not well understood at this time [65].

The other compounds mentioned, benzaldehyde, 2-nonanol and acetophenone, are recognized for their importance to the aromatic quality of cognac [1].

3.4. *Aromatic Typicity of the Strains*

Regardless of the yeast strain, the synthesis of aroma compounds is similarly affected by the main fermentation parameters, such as YAN and lipids. However, notable differences were obtained in the production levels for the three strains tested. A fruity wine or, on the contrary, a softer or rounder wine, can be obtained by prior management of nutritional parameters that will impact aroma synthesis, and also by the choice of yeast strain. A strong interaction between these two control levers of fermentation was demonstrated here: the FC9® strain displayed a greater synthesis of fruity aromas, such as esters (Figure 4A).

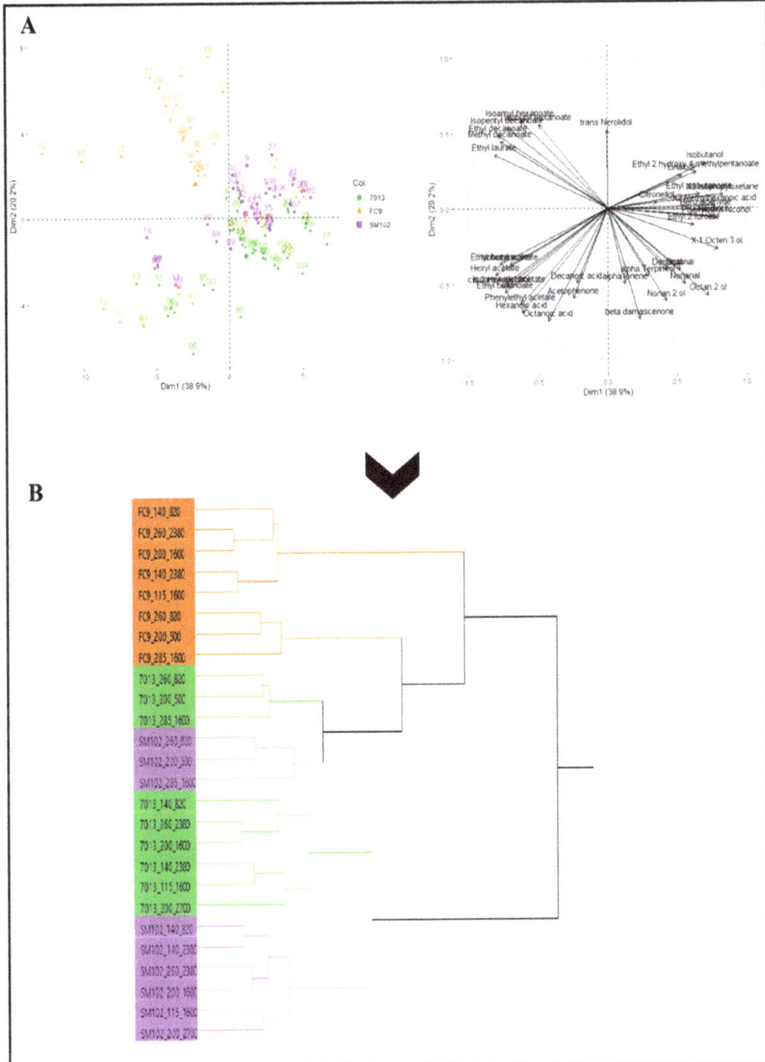

Figure 4. Principal component analysis of aroma production (fermentative aromas, terpenes, aldehydes, noriseprenoids and alcohols) for the three strains (FC9® (—), 7013® (—) and SM102® (—) under all conditions of fermentation (**A**) and its correlation with the hierarchical classification of the aromatic composition of wines (**B**).

The other two strains, 7013® and SM102®, seemed to be quite similar, although they each had their own aromatic character. This result was also supported by the decision tree, which demonstrated the aromatic typicity of strain FC9® (Figure 4B). The other strains had a similar strong interaction effect with the fermentation parameters: the same aroma fingerprint for SM102® and 7013® was correlated to the same levels of YAN and lipid nutrition parameters. Beyond the classic aromas resulting from the well-known fermentation metabolism, the *Saccharomyces cerevisiae* strains differed in their terpene fraction through the distribution of monoterpene production (Figure 3). Strain FC9® contributed a rose aroma note (trans-nerolidol), strain 7013® had lemon grass nuances (citronellol) and strain SM102® produced a flower aroma (linalool). Under high lipid level conditions, the metabolic pathway of terpene biosynthesis is activated and should not be neglected compared to the varietal source.

4. Conclusions

In conclusion, this study allowed us to analyze the impact of the two main fermentation parameters, nitrogen and turbidity, on the production of fermentation aromas and also on a wider range of aromas through a metabolomic analysis. First, it appeared that the results for the CCM metabolites were consistent with the literature: at a high initial lipid content, the acetate synthesis pathway was repressed while that of glycerol was favored. Moreover, in the working range of assimilable nitrogen in this study, the production of higher alcohols decreased, while their synthesis was favored by the presence of phytosterols and fatty acids. The effects of these nutrients on acetate esters were the complete opposite, indicating that the key element in the regulation of their synthesis is not the availability of their precursor. Finally, the metabolism of ethyl esters and their corresponding acids remained more complex to understand, although the lipid concentration was decreased. This study also supported the fact that terpenes are not only aromas of varietal origin, and an important part of their production originates from yeast metabolism under conditions of excess lipids in the medium, and moreover, in a different way, depending on the strains used. In addition, acetaldehyde, which plays a central role in *Saccharomyces cerevisiae* metabolism, seemed to be strongly impacted by the nitrogen and lipid supply. A forthcoming study will detail the chronology of its synthesis based on data obtained from an online monitoring system.

Author Contributions: Conceptualization, X.P., J.-R.M. and V.F.; methodology, C.G., F.M. and A.B.; formal analysis, C.G., I.S. and A.B.; investigation, C.G., A.B. and V.F.; writing—original draft preparation, C.G., I.S., A.B. and V.F.; writing—review and editing, C.G., X.P., J.-R.M., J.-M.S. and V.F. All authors have read and agreed to the published version of the manuscript.

Funding: This research received no external funding.

Institutional Review Board Statement: Not applicable.

Informed Consent Statement: Not applicable.

Data Availability Statement: The datasets generated during and/or analyzed in the current study are available from the corresponding author upon reasonable request.

Acknowledgments: We would like to thank the aroma analysis platform (PTV, UMR SPO, INRAE Montpellier) for their training, patience and availability for the analysis of aromas and the ChemoSens platform (CSGA, INRAE Dijon) for the analysis of lipids. Finally, we thank Thérèse Marlin for the analysis of amino acids (ADEL Team, UMR SPO, INRAE Montpellier) and Mélanie Veyret (UE Pech Rouge, INRAE) for her help with the acetaldehyde assays.

Conflicts of Interest: All authors declare no conflict of interest.

References

1. Ferrari, G.; Lablanquie, O.; Cantagrel, R.; Ledauphin, J.; Payot, T.; Fournier, N.; Guichard, E. Determination of Key Odorant Compounds in Freshly Distilled Cognac Using GC-O, GC-MS, and Sensory Evaluation. *J. Agric. Food Chem.* **2004**, *52*, 7. [CrossRef] [PubMed]
2. Lurton, L.; Snakkers, G.; Roulland, C.; Galy, B.; Versavaud, A. Influence of the fermentation yeast strain on the composition of wine spirits. *J. Sci. Food Agric.* **1995**, *67*, 485–491. [CrossRef]
3. Vilanova, M.; Siebert, T.E.; Varela, C.; Pretorius, I.S.; Henschke, P.A. Effect of ammonium nitrogen supplementation of grape juice on wine volatiles and non-volatiles composition of the aromatic grape variety Albariño. *Food Chem.* **2012**, *133*, 124–131. [CrossRef]
4. Casalta, E.; Cervi, M.F.; Salmon, J.M.; Sablayrolles, J.M. White wine fermentation: Interaction of assimilable nitrogen and grape solids: Interaction of assimilable nitrogen and grape solids on alcoholic fermentation under oenological conditions. *Aust. J. Grape Wine Res.* **2013**, *19*, 47–52. [CrossRef]
5. Beltran, G.; Esteve-Zarzoso, B.; Rozès, N.; Mas, A.; Guillamón, J.M. Influence of the timing of nitrogen additions during synthetic grape must fermentations on fermentation kinetics and nitrogen consumption. *J. Agric. Food Chem.* **2005**, *53*, 996–1002. [CrossRef] [PubMed]

6. Gobert, A.; Tourdot-Maréchal, R.; Sparrow, C.; Morge, C.; Alexandre, H. Influence of nitrogen status in wine alcoholic fermentation. *Food Microbiol.* **2019**, *83*, 71–85. [CrossRef] [PubMed]
7. Guittin, C.; Maçna, F.; Sanchez, I.; Poitou, X.; Sablayrolles, J.-M.; Mouret, J.-R.; Farines, V. Impact of high lipid contents on the production of fermentative aromas during white wine fermentation. *Appl. Microbiol. Biotechnol.* **2021**, *105*, 6435–6449. [CrossRef]
8. Jiménez-Martí, E.; Aranda, A.; Mendes-Ferreira, A.; Mendes-Faia, A.; lí del Olmo, M. The nature of the nitrogen source added to nitrogen depleted vinifications conducted by a Saccharomyces cerevisiae strain in synthetic must affects gene expression and the levels of several volatile compounds. *Antonie Leeuwenhoek* **2007**, *92*, 61–75. [CrossRef]
9. Rollero, S.; Bloem, A.; Camarasa, C.; Sanchez, I.; Ortiz-Julien, A.; Sablayrolles, J.-M.; Dequin, S.; Mouret, J.-R. Combined effects of nutrients and temperature on the production of fermentative aromas by Saccharomyces cerevisiae during wine fermentation. *Appl. Microbiol. Biotechnol.* **2015**, *99*, 2291–2304. [CrossRef]
10. Seguinot, P.; Rollero, S.; Sanchez, I.; Sablayrolles, J.-M.; Ortiz-Julien, A.; Camarasa, C.; Mouret, J.-R. Impact of the timing and the nature of nitrogen additions on the production kinetics of fermentative aromas by Saccharomyces cerevisiae during winemaking fermentation in synthetic media. *Food Microbiol.* **2018**, *76*, 29–39. [CrossRef]
11. Torrea, D.; Varela, C.; Ugliano, M.; Ancin-Azpilicueta, C.; Leigh Francis, I.; Henschke, P.A. Comparison of inorganic and organic nitrogen supplementation of grape juice—Effect on volatile composition and aroma profile of a Chardonnay wine fermented with Saccharomyces cerevisiae yeast. *Food Chem.* **2011**, *127*, 1072–1083. [CrossRef] [PubMed]
12. Torrea, D.; Henschke, P.A. Ammonium supplementation of grape juice-effect on the aroma profile of a Chardonnay wine. *TechRev* **2004**, *150*, 59–63.
13. Mouret, J.R.; Camarasa, C.; Angenieux, M.; Aguera, E.; Perez, M.; Farines, V.; Sablayrolles, J.M. Kinetic analysis and gas–liquid balances of the production of fermentative aromas during winemaking fermentations: Effect of assimilable nitrogen and temperature. *Food Res. Int.* **2014**, *62*, 1–10. [CrossRef]
14. Vilanova, M.; Ugliano, M.; Varela, C.; Siebert, T.; Pretorius, I.S.; Henschke, P.A. Assimilable nitrogen utilisation and production of volatile and non-volatile compounds in chemically defined medium by Saccharomyces cerevisiae wine yeasts. *Appl. Microbiol. Biotechnol.* **2007**, *77*, 145–157. [CrossRef]
15. Miller, A.C.; Wolff, S.R.; Bisson, L.F.; Ebeler, S.E. Yeast Strain and Nitrogen Supplementation: Dynamics of Volatile Ester Production in Chardonnay Juice Fermentations. *Am. J. Enol. Vitic.* **2007**, *58*, 470–483.
16. Andreasen, A.A.; Stier, T.J.B. Anaerobic nutrition of Saccharomyces cerevisiae. I. Ergosterol requirement for growth in a defined medium. *J. Cell. Physiol.* **1953**, *41*, 23–36. [CrossRef]
17. Casalta, E.; Salmon, J.-M.; Picou, C.; Sablayrolles, J.-M. Grape Solids: Lipid Composition and Role during Alcoholic Fermentation under Enological Conditions. *Am. J. Enol. Vitic.* **2019**, *70*, 147–154. [CrossRef]
18. Rollero, S.; Mouret, J.-R.; Sanchez, I.; Camarasa, C.; Ortiz-Julien, A.; Sablayrolles, J.-M.; Dequin, S. Key role of lipid management in nitrogen and aroma metabolism in an evolved wine yeast strain. *Microb. Cell Factories* **2016**, *15*, 32. [CrossRef]
19. Saerens, S.M.G.; Delvaux, F.; Verstrepen, K.J.; Van Dijck, P.; Thevelein, J.M.; Delvaux, F.R. Parameters Affecting Ethyl Ester Production by Saccharomyces cerevisiae during Fermentation. *Appl. Environ. Microbiol.* **2008**, *74*, 454–461. [CrossRef]
20. Yunoki, K.; Yasui, Y.; Hirose, S.; Ohnishi, M. Fatty acids in must prepared from 11 grapes grown in Japan: Comparison with wine and effect on fatty acid ethyl ester formation. *Lipids* **2005**, *40*, 361–367. [CrossRef]
21. Sablayrolles, J.M.; Barre, P.; Grenier, P. Design of a laboratory automatic system for studying alcoholic fermentations in anisothermal enological conditions. *Biotechnol. Tech.* **1987**, *1*, 181–184. [CrossRef]
22. Crépin, L.; Nidelet, T.; Sanchez, I.; Dequin, S.; Camarasa, C. Sequential Use of Nitrogen Compounds by Saccharomyces cerevisiae during Wine Fermentation: A Model Based on Kinetic and Regulation Characteristics of Nitrogen Permeases. *Appl. Environ. Microbiol.* **2012**, *78*, 8102–8111. [CrossRef] [PubMed]
23. Lenth, R.V. Response-Surface Methods in R, Using rsm. *J. Stat. Softw.* **2009**, *32*, 1–17. [CrossRef]
24. Box, G.E.P.; Wilson, K.B. On the Experimental Attainment of Optimum Conditions. *J. R. Stat. Soc. Ser. B* **1951**, *13*, 1–38. [CrossRef]
25. Manginot, C.; Sablayrolles, J.M.; Roustan, J.L.; Barre, P. Use of constant rate alcoholic fermentations to compare the effectiveness of different nitrogen sources added during the stationnary phase. *Enz. Microbiol. Tech.* **1997**, *20*, 373–380. [CrossRef]
26. Blateyron, L.; Sablayrolles, J. Stuck and Slow Fermentations in Enology: Statistical Study of Causes and Effectiveness of Combined Additions of Oxygen and Diammonium Phosphate. *J. Biosci. Bioeng.* **2001**, *91*, 184–189. [CrossRef]
27. Bely, M.; Sablayrolles, J.M.; Barre, P. Description of Alcoholic Fermentation Kinetics: Its Variability and Significance. *Am. J. Enol. Vitic.* **1990**, *41*, 319–324.
28. Ochando, T.; Mouret, J.-R.; Humbert-Goffard, A.; Sablayrolles, J.-M.; Farines, V. Impact of initial lipid content and oxygen supply on alcoholic fermentation in champagne-like musts. *Food Res. Int.* **2017**, *98*, 87–94. [CrossRef]
29. Deroite, A.; Legras, J.-L.; Rigou, P.; Ortiz-Julien, A.; Dequin, S. Lipids modulate acetic acid and thiol final concentrations in wine during fermentation by Saccharomyces cerevisiae × Saccharomyces kudriavzevii hybrids. *AMB Expr.* **2018**, *8*, 130. [CrossRef]
30. Luparia, V.; Soubeyrand, V.; Berges, T.; Julien, A.; Salmon, J.-M. Assimilation of grape phytosterols by Saccharomyces cerevisiae and their impact on enological fermentations. *Appl. Microbiol. Biotechnol.* **2004**, *65*, 25–32. [CrossRef]
31. Rollero, S. Impact des paramètres environnementaux sur la synthèse des arômes fermentaires par Saccharomyces cerevisiae en fermentation oenologique. Ph.D. Thesis, Montpellier SupAgro, Montpellier, France, 2015.
32. Albers, E.; Larsson, C.; Lidén, G.; Niklasson, C.; Gustafsson, L. Influence of the nitrogen source on Saccharomyces cerevisiae anaerobic growth and product formation. *Appl. Environ. Microbiol.* **1996**, *62*, 3187–3195. [CrossRef] [PubMed]

33. Remize, F.; Sablayrolles, J.M.; Dequin, S. Re-assessment of the influence of yeast strain and environmental factors on glycerol production in wine. *J. Appl. Microbiol.* **2000**, *88*, 371–378. [CrossRef] [PubMed]
34. Nykanen, I. *Aroma of Beer, Wine and Distilled Alcoholic Beverages*; Kluwer Academic Publishers: Berlin, Germany, 1983.
35. Verstrepen, K.J.; Laere, S.D.M.V.; Vanderhaegen, B.M.P.; Derdelinckx, G.; Dufour, J.-P.; Pretorius, I.S.; Winderickx, J.; Thevelein, J.M.; Delvaux, F.R. Expression Levels of the Yeast Alcohol Acetyltransferase Genes ATF1, Lg-ATF1, and ATF2 Control the Formation of a Broad Range of Volatile Esters. *Appl. Environ. Microbiol.* **2003**, *69*, 5228–5237. [CrossRef] [PubMed]
36. Fujii, T.; Kobayashi, O.; Yoshimoto, H.; Furukawa, S.; Tamai, Y. Effect of aeration and unsaturated fatty acids on expression of the Saccharomyces cerevisiae alcohol acetyltransferase gene. *Appl. Environ. Microbiol.* **1997**, *63*, 910–915. [CrossRef] [PubMed]
37. Fujiwara, D.; Yoshimoto, H.; Sone, H.; Harashima, S.; Tamai, Y. Transcriptional co-regulation of Saccharomyces cerevisiae alcohol acetyltransferase gene, ATF1 and Δ-9 fatty acid desaturase gene, OLE1 by unsaturated fatty acids. *Yeast* **1998**, *14*, 711–721. [CrossRef]
38. Torrea, D. Production of volatile compounds in the fermentation of chardonnay musts inoculated with two strains of Saccharomyces cerevisiae with different nitrogen demands. *Food Control* **2003**, *14*, 565–571. [CrossRef]
39. Ugliano, M.; Travis, B.; Francis, I.L.; Henschke, P.A. Volatile Composition and Sensory Properties of Shiraz Wines as Affected by Nitrogen Supplementation and Yeast Species: Rationalizing Nitrogen Modulation of Wine Aroma. *J. Agric. Food Chem.* **2010**, *58*, 12417–12425. [CrossRef] [PubMed]
40. Varela, C.; Torrea, D.; Schmidt, S.A.; Ancin-Azpilicueta, C.; Henschke, P.A. Effect of oxygen and lipid supplementation on the volatile composition of chemically defined medium and Chardonnay wine fermented with Saccharomyces cerevisiae. *Food Chem.* **2012**, *135*, 2863–2871. [CrossRef]
41. Saerens, S.M.G.; Delvaux, F.R.; Verstrepen, K.J.; Thevelein, J.M. Production and biological function of volatile esters in *Saccharomyces cerevisiae*. *Microb. Biotechnol.* **2010**, *3*, 165–177. [CrossRef]
42. Black, C.A.; Parker, M.; Siebert, T.E.; Capone, D.L.; Francis, I.L. Terpenoids and their role in wine flavour: Recent advances: Terpenoids: Role in wine flavour. *Aust. J. Grape Wine Res.* **2015**, *21*, 582–600. [CrossRef]
43. Ribereau-Gayon, P.; Dubourdieu, D.; Glories, Y.; Maujean, A. *Traitéd'oenologie: Chimie Du Vin, Stabilisation et Traitements*; Dunod: Malakoff, France, 2017.
44. Malfondet, N.; Gourrat, K.; Brunerie, P.; Le Quéré, J.-L. Aroma characterization of freshly-distilled French brandies; their specificity and variability within a limited geographic area: Aroma characterization of freshly-distilled French brandies. *Flavour Fragr. J.* **2016**, *31*, 361–376. [CrossRef]
45. Schreier, P.; Drawert, F.; Winkler, F. Composition of neutral volatile constituents in grape brandies. *J. Agric. Food Chem.* **1979**, *27*, 365–372. [CrossRef]
46. Thibaud, F.; Courregelongue, M.; Darriet, P. Contribution of Volatile Odorous Terpenoid Compounds to Aged Cognac Spirits Aroma in a Context of Multicomponent Odor Mixtures. *J. Agric. Food Chem.* **2020**, *68*, 13310–13318. [CrossRef] [PubMed]
47. Eliseev, M.; Gribkova, I.; Kosareva, O.; Alexeyeva, O. Effect of organic compounds on cognac sensory profile. *Foods Raw Mater.* **2021**, 244–253. [CrossRef]
48. Wilson, B.; Strauss, C.R.; Williams, J. The Distribution of Free and Glycosidically-Bound Monoterpenes among Skin, Juice, and Pulp Fractions of Some White Grape Varieties. *Am. J. Enol. Vitic.* **1986**, *37*, 107–111.
49. Charoenchai, C.; Fleet, G.H.; Henschke, P.A.; Todd, B.E.N.T. Screening of non-*Saccharomyces* wine yeasts for the presence of extracellular hydrolytic enzymes. *Aust. J. Grape Wine Res.* **1997**, *3*, 2–8. [CrossRef]
50. Strauss, M.L.A.; Jolly, N.P.; Lambrechts, M.G.; van Rensburg, P. Screening for the production of extracellular hydrolytic enzymes by non-Saccharomyces wine yeasts. *J. Appl. Microbiol.* **2001**, *91*, 182–190. [CrossRef]
51. Awad, P.; Athès, V.; Decloux, M.E.; Ferrari, G.; Snakkers, G.; Raguenaud, P.; Giampaoli, P. Evolution of Volatile Compounds during the Distillation of Cognac Spirit. *J. Agric. Food Chem.* **2017**, *65*, 7736–7748. [CrossRef]
52. Carrau, F.M.; Medina, K.; Boido, E.; Farina, L.; Gaggero, C.; Dellacassa, E.; Versini, G.; Henschke, P.A. De novo synthesis of monoterpenes by *Saccharomyces cerevisiae* wine yeasts. *FEMS Microbiol. Lett.* **2005**, *243*, 107–115. [CrossRef] [PubMed]
53. Lynen, F. The pathway from activated acetic acid to the terpenes and fatty acids. *Proc. R. Caroline Inst.* **1964**, 103–138.
54. Gamero, A.; Manzanares, P.; Querol, A.; Belloch, C. Monoterpene alcohols release and bioconversion by Saccharomyces species and hybrids. *Int. J. Food Microbiol.* **2011**, *145*, 92–97. [CrossRef] [PubMed]
55. Steyer, D. Genetic analysis of geraniol metabolism during fermentation. *Food Microbiol.* **2013**, *7*, 228–234. [CrossRef] [PubMed]
56. Vaudano, E.; Moruno, E.G.; Stefano, R. Modulation of Geraniol Metabolism during Alcohol Fermentation. *J. Inst. Brew.* **2004**, *110*, 213–219. [CrossRef]
57. Denat, M.; Pérez, D.; Heras, J.M.; Querol, A.; Ferreira, V. The effects of Saccharomyces cerevisiae strains carrying alcoholic fermentation on the fermentative and varietal aroma profiles of young and aged Tempranillo wines. *Food Chem. X* **2021**, *9*, 100116. [CrossRef] [PubMed]
58. Takeoka, G.R.; Güntert, M.; Engel, K.-H. (Eds.) *Aroma Active Compounds in Foods: Chemistry and Sensory Properties*; ACS Symposium Series; American Chemical Society: Washington, DC, USA, 2001; Volume 794, ISBN 978-0-8412-3694-3.
59. Pallotta, U.; Castellari, M.; Piva, A.; Baumes, R.; Bayonove, C. Effects of Botrytis cinerea on must composition of three italian grape varieties. *Wein-Wissenschaft* **1998**, *53*, 32–36.
60. La Guerche, S.; Dauphin, B.; Pons, M.; Blancard, D.; Darriet, P. Characterization of Some Mushroom and Earthy Off-Odors Microbially Induced by the Development of Rot on Grapes. *J. Agric. Food Chem.* **2006**, *54*, 9193–9200. [CrossRef]

61. Wurzenberger, M.; Grosch, W. The formation of 1-octen-3-ol from the 10-hydroperoxide isomer of linoleic acid by a hydroperoxide lyase in mushrooms (*Psalliota bispora*). *Biochim. Biophys. Acta Lipids Lipid Metab.* **1984**, *794*, 25–30. [CrossRef]
62. Liu, S.-Q.; Pilone, G.J. An overview of formation and roles of acetaldehyde in winemaking with emphasis on microbiological implications. *Int. J. Food Sci. Technol.* **2000**, *35*, 49–61. [CrossRef]
63. Garde-Cerdán, T.; Ancín-Azpilicueta, C. Effect of the addition of different quantities of amino acids to nitrogen-deficient must on the formation of esters, alcohols, and acids during wine alcoholic fermentation. *LWT Food Sci. Technol.* **2008**, *41*, 501–510. [CrossRef]
64. Wucherpfenning, K.; Semmler, G. Formation of acetaldehyde during fermentation in relation to pH-value and to individual vitamins. *Z. Lebensm. Unters. Forsch.* **1972**, *148*, 138–145.
65. Bosso, A.; Guaita, M. Study of some factors involved in ethanal production during alcoholic fermentation. *Eur. Food Res. Technol.* **2008**, *227*, 911–917. [CrossRef]

Article

Aroma Perception of Rose Oxide, Linalool and α-Terpineol Combinations in Gewürztraminer Wine

Mildred Melina Chigo-Hernandez, Aubrey DuBois and Elizabeth Tomasino *

Department of Food Science and Technology, Oregon State University, Corvallis, OR 97331, USA; mildredmelina.chigohernndez@oregonstate.edu (M.M.C.-H.); aubrey.dubois@oregonstate.edu (A.D.)
* Correspondence: elizabeth.tomasino@oregonstate.edu

Abstract: *Cis*-Rose oxide was found to be an important chiral compound in Gewürztraminer wine, with an enantiomeric ratio range from 76 to 58%. The enantiomeric ratio showed an important influence on white wine aroma when other monoterpenes were present. The aim of this study was to evaluate rose oxide at different ratios and changes to aroma perception, and the interaction of rose oxide with linalool and α-terpineol. A wine model was made based on Gewürztraminer wine. Twelve models were created with different ratios of rose oxide and concentrations of linalool and α-terpineol. Triangle tests, check-all-that-apply (CATA) and descriptive analysis were used to evaluate the aroma of the wines. Results show that the rose oxide ratios of 70:30 and 65:35 were statistically different. Additional descriptive analysis showed that the ratios altered aroma when linalool and α-terpineol were at low and medium concentrations. At high concentrations, linalool and α-terpineol masked any influence from rose oxide. Understanding how monoterpenes alter aroma perception of white wine when at different combinations and concentrations is important to achieving desired wine qualities and helps provide information on how flavor chemistry results can be interpreted without having to run sensory analysis.

Keywords: monoterpenes; triangle test; check-all-that-apply; correspondence analysis; Cochran's Q-test

Citation: Chigo-Hernandez, M.M.; DuBois, A.; Tomasino, E. Aroma Perception of Rose Oxide, Linalool and α-Terpineol Combinations in Gewürztraminer Wine. *Fermentation* **2022**, *8*, 30. https://doi.org/10.3390/fermentation8010030

Academic Editors: Bin Tian and Jicheng Zhan

Received: 23 December 2021
Accepted: 11 January 2022
Published: 13 January 2022

Publisher's Note: MDPI stays neutral with regard to jurisdictional claims in published maps and institutional affiliations.

Copyright: © 2022 by the authors. Licensee MDPI, Basel, Switzerland. This article is an open access article distributed under the terms and conditions of the Creative Commons Attribution (CC BY) license (https://creativecommons.org/licenses/by/4.0/).

1. Introduction

The aroma of wine is an important aspect of wine quality [1] and is normally described using pleasant aromas, such as floral, green fruit, citrus fruit, stone fruit, tropical fruit, red fruit, black fruit, dry fruit, herbaceous, herbal, spices, and more [2,3]. Aroma characteristics in wine are of interest to consumers, especially for white wine, as it reflects the typicity for a specific region or terroir [4,5]. One of the most important factors in determining wine typicity and quality is which compounds are most important to aroma perception [4,6].

Aroma is one of the first quality aspects assessed when evaluating wine [7]. The aroma compounds travel from the wine, into the air and through the nose, until they bind to the receptors of the olfactory bulb [8]. The olfactory bulb is the first relay station of the central olfactory system in the mammalian brain, and contains a few thousand glomeruli on its surface. Individual glomeruli represent a single type of odorant receptor [9], and each receptor can detect a limited number of odorants substances [10].

Many olfactory receptors are specific to different enantiomers [11]. Enantiomers are a type of isomer that display a quality called chirality, meaning that they have same molecular formula and the same connectivity, but they differ in the way that they are oriented in three-dimensional space [12]. The olfactory receptors react differently with the two enantiomeric forms of a chiral odorant, leading to differences in odor strength and quality [11]. A slight modification of the chemical structure of a stimulus molecule can lead to large changes in the odor impression. Linalool is an example of this characteristic, as its (+)-enantiomer displays a sweet, petitgrain aroma, and the (−) enantiomer has lavender notes and an oily, woody aroma [11].

The sensory perception of aroma in wine is complex. A single glass of wine can contain hundreds of aroma compounds [8]. Traditionally, it was thought that only compounds at high concentrations, those above their known perception thresholds, influenced aroma perception [13]. However, it has been shown that low impact odorants may act to change the perception of other odorants in a mixture, may interact synergistically or antagonistically, and can significantly impact aroma perception [5].

Many monoterpenes are chiral [14] and their enantiomeric forms can be found in grapes and wine [15]. There has been a growing awareness and interest in their enantiomer specific properties [16], as aroma characteristics of wine cannot be understood only from the knowledge of aroma composition alone [17]. In some cases, both enantiomers have similar odors and thresholds [12], such as α-pinene [18]. In others, one enantiomer is odorous and some do not have odors [12] like androstenone [11], while others have different detection thresholds, such as *cis*-rose oxide [12,19]. Few studies have evaluated the different chiral monoterpenes and the effect of interactions among them. Monoterpenes are of interest, as monoterpene combinations have been found to influence the aroma of wine, resulting in different aroma qualities [17].

Monoterpenes are present in grapes of all wine varieties, but the highest concentrations occur in aromatic varieties such as Muscat, Gewürztraminer, Irsai Oliver, and Riesling [20]. Considerable research has been done with respect to the identification and contribution of terpene compounds to the muscat aroma of muscat grapes and wines [21]. In general, there are indications that many terpenes, not just linalool and geraniol, contribute to muscat and related aromas [21]. Furthermore, it was found that terpenes interact to such an extent that one component can increase the aroma intensity of another compound, and that a mixture could become more aromatic than the most aromatic single component which belongs to that mixture [21].

The monoterpene content of Pinot gris, Riesling and Muscat wines has been investigated [21–24], including how monoterpenes influence aroma perception [17,25]. However, there is much less information available focusing on Gewürztraminer wines. Gewürztraminer is one of the world's oldest vine varieties found in Europe, America, and Australia [26]. Their aroma is characterized as having a "Traminer" smell or aroma quality reminiscent of the tropical fruit lychee [27]. Other terms used to describe this on Gewürztraminer wine are: spicy, floral (rose petals), fruity (citrus, grapefruit, peach), lychee, cold cream, honey, and jasmine tea [28]. The compounds that are responsible for the overall flavor of Gewürztraminer wines [26,29], and their monoterpene concentration has been investigated [22]. Yeast strain used during fermentation and time of grape harvest has been found to influence the aromatic profile [20,30]. *Cis*-rose oxide has been found to have a significant influence in odor profiles [29].

Cis-rose oxide has been identified as the most characteristic odor compound in Gewurztraminer [27]. Additionally, linalool, geraniol, nerol, citronellol, and α-terpineol are known to be significant compounds, where linalool is considered important in both free and bound form [31]. In spite of this, *cis*-rose oxide is the main impact odorant related to sweet and fruity [11], and it is thought to produce the aroma of lychee in Gewürztraminer wine [32].

Additionally, (−)-rose oxide has shown an important influence in wine aroma when monoterpenes were added in different mixtures in white wine, where the wines that were perceived as different contained (−)-rose oxide [19]. Furthermore, Gewurztraminer wines have reported a range of enantiomeric ratio of *cis*-rose oxide, always in favor of the (−)-enantiomer, going from 58 to 76% [22,29,33], and when combined with *trans*-rose oxide, the percentage has been found from 70 to 85% in favor of the (−)-enantiomer [22]. Additionally, it has been observed that the concentration of monoterpenes changed when using different yeast, and when grapes were harvested later [20,30]. Its enantiomers are found in many mixtures of monoterpenes in wine, but two aspects are unclear. (1) Do the two different ratios of enantiomers produce different aroma qualities in wine? and (2) how does rose oxide alter aroma quality when in mixtures of other monoterpenes? Does it dominate the perceived aroma, or are the differences more subtle? The aim of this work was to evaluate the sensory perception of the different ratios of rose oxide in mixtures with other monoterpenes and assess aroma interactions with other monoterpenes in wine.

2. Materials and Methods

2.1. Wine Base

Wine aroma compounds were removed in the same way as described by Song et al. [17]. LiChrolut EN resin was added to wine at a rate to 1.5 g L^{-1}. Wines with resin were stirred for 24 h before filtering out the resin and storing in Stainless Ball Lock Kegs (AMCYL, Wyoming, MN, USA) with nitrogen at 4 °C for later use. A week prior to sensory analysis, the aroma base (Supplementary Table S1) was added to the dearomatized wine. This model wine was then bottled in 750 mL glass wine bottles with screw caps ((Stelvin Amcor, Zurich, Switzerland) and stored at 4 °C until the sensory panel.

2.2. Chemicals

The following chemical standards were used in the Aroma Base and were obtained from Sigma-Aldrich Co. (St. Louis, MO, USA): Acetaldehyde (≥99%), hexanol (98%), ethyl octanoate (≥98%), isoamyl acetate (≥99%), 2-phenyl acetate (≥99%), ethyl butanoate (≥98%), dyacetyl (97%), methionol (≥98%), isobutyl acetate (≥97%), ethyl 3-methylbutanoate (≥98%), ethyl 2-methylbutanoate (≥98%), isobutanol (≥99%), ethyl acetate (≥99%), phenethyl alcohol (≥99%), butanoic acid (≥99%), ethyl decanoate (≥99%), ethyl hexanoate (≥99%), hexanoic acid (99%), octanoic acid (≥99%) and decanoic acid (≥98%). The terpenes compounds added to the model wine for the treatments were: (−)-rose oxide (Analytical Standard), (−)-rose oxide (Analytical Standard), Linalool (≥97%) and α-terpineol, which were purchased from Sigma-Aldrich Co. (St. Louis, MO, USA). Solvents. Mili-Q water.

2.3. Standards and Wine Treatments

Stock standard solutions of all chemicals used in the aroma base and wine models were prepared in 14% aqueous ethanol. The stock solutions were stored at −20 °C. Composite standards were created from the stock standards so that fewer additions were made when aroma base or treatments were added. Prior to additions, the working standards were defrosted one day before in a fridge (4.9 °C). Aroma base compounds were added to the model wine 4 days prior to the first sensory analysis (Supplementary Table S1). Treatment standards were added to the model wine 1 h prior to each day sensory analysis. Concentrations of the terpenes in each treatment were chosen according to literature concentrations [20,22,30]. The final concentrations chosen can be found in Table 1.

Table 1. Concentrations (μg/L) of Terpenes in each treatment.

Wines	(−) Rose Oxide	(+) Rose Oxide	Linalool	α-Terpineol
Model 1 [a]	15	5		
Model 2 [b]	14	6		
Model 3 [c]	13	7		
Model 4 [d]	12	8		
Model 5 [e]			50	
Model 6 [f]				50
Model 7 [b,g]	14	6	50	50
Model 8 [c,g]	13	7	50	50
Model 9 [b,h]	14	6	20	15
Model 10 [c,h]	13	7	20	15
Model 11 [b,i]	14	6	100	100
Model 12 [c,i]	13	7	100	100

The wine models that had rose oxide had 4 ratios: [a] 75:25, [b] 70:30, [c] 65:35 and [d] 60:40. [e] Just linalool, [f] Just α-terpineol, [g] Medium concentration of linalool and α-terpineol, [h] low concentration of linalool and α-terpineol, [i] high concentration of linalool and α-terpineol.

2.4. Sensory Analysis

This study was approved by the Oregon State University Internal Review Board (#8606). All panelists were non-smokers, not pregnant at the time of the study, free of taste deficit disorders, free of oral lesions or sores, and had no piercings on the lips, cheeks, or tongue. Panelists were regular white wine consumers (at least one serving of white wine a week). All panels occurred at Oregon State University in the Arbuthnot Dairy Lab (Corvallis, OR, USA). The room was kept at a constant of 21 °C, with a mix of natural and artificial light. Each panelist had their own individual booth (61 cm × 71 cm center, 61 cm × 65 cm sides) and two air purifiers (Winix, Vernon Hills, IL, USA) were used for air quality maintenance. Results were collected using Compusense Cloud Software® (Version 21.0.773.192939).

2.5. Triangle Test Procedure

A total of 65 participants (49 Women and 16 men, all above 21 years old) were recruited from Oregon State University and the surrounding area, for the triangle tests. The triangle test evaluated models 1 to 4 (Table 1), as the main goal was to determine if the ratio of rose oxide enantiomers elucidated different sensory responses. Testing occurred between 19 April and 23 April 2021; each panelist attended a single session. Each panelist was presented with six triangle tests to cover the comparison between the 4 ratios in all possible combinations.

For each wine, 20 mL of sample was served in black INAO wine glasses (Lehmann glass, Kiyasa Group, New York, NY, USA) with three-digit random codes and covered with plastic lids (Clark Associates, Inc. Lancaster, PA, USA). The treatments were kept at 4 °C until serving and were poured 30 min before each sensory test. Panelists were instructed to smell the samples in the order indicated, from left to right, and choose the sample that was most different. A 1 min break between each test was required to avoid any carryover effects and fatigue.

2.6. CATA

The treatments that were found to be significantly different in triangle tests and the additional terpene combinations were used in CATA (treatments 2, 3, 5, 6, 7, 8, 9, 10, 11 and 12). The CATA included 23 sensory descriptors with two "other" options so that panelists could write in any descriptors that they thought were important, but were not in the terms provided. CATA analysis occurred in the same room as triangle test at OSU on 29 April 2021. Twenty-five wine consumers (20 women and four men, one—chose not to specify, age 21 to 60) participate in one session (1-h sessions). Panelists were instructed to smell the sample and select all the descriptors that were perceived in the sample. A 30 sec break between each test was required to avoid any carryover effects and fatigue.

2.7. Descriptive Analysis for Aroma Intensity

From the CATA results, 13 aroma descriptors were selected for further descriptive analysis. Attributes with their training standards and images used in training can be found in Table 2. Nineteen wine consumers (11 women and 8 men, all over 21 years old) were trained on recognizing 13 different aroma standards (Table 2). Participants were trained through multiple choice odor and image recognition training [17]. During the second training, panelists also evaluated two random aromatic wines using a 100 mm line scale with anchors at 30 mm and 70 mm labeled none and extreme respectively, to help them become familiar with the test.

Table 2. Standards and Images used for Descriptive Panel Training.

Attribute	Amount per Glass	Components	Image
Honey	1 tsp *	Clover Honey [a]	
Honeysuckle	1 drop	Honey suckle essential oil [b]	
Rose	4 drops	Rosewater concentrate [c]	
Dried Fruit [d]	1 tsp	Golden raisins mix with DI water	
Stone Fruit	1 tsp pf each	White peach [e] and apricot puree [e]	
Pome	1 tsp of each	Green apple [e] and Pear puree [e]	
Melon	1 tsp and 1 drop of each	Honeydew puree made [f] and melon essential oil [g]	
Orange	1 tsp	Orange pure made [h]	
Grapefruit	1 tsp	Grapefruit pure made [h]	
Lemon	1 tsp	Lemon pure made [h]	
Lychee	1 tsp	Lychee pure [i]	
Tropical fruit	1 tsp of each	Mango [e] and passion fruit puree [i]	
Ginger	$\frac{1}{8}$ tsp	Ground ginger [j]	

tsp * = teaspoon, The components were sourced from: Hanna's Honey (Salem, OR, USA) [a], Rainbow Abby 2013 (Guangzhou, China) [b], Nielsen-Massey [c] (Waukegan, IL, USA), the perfect purée of Napa Valley (Napa, CA, USA) [e], Silver Cloud Flavors [g] (Belcamp, MD), Funkin pro (Arlington Rd, LDN, UK) [i] and Private selection [j] (Cincinnati, OH, USA). Purees were kept frozen at −23 °C [d] Dried fruit was pureed with the addition of distilled water one day before each sensory session and kept at refrigeration temperatures until analysis. [f] Puree was made two weeks prior sensory and kept frozen at −23 °C; the seeds and rid were removed to make the puree. [h] Purees were made using the whole fruit two weeks prior sensory and kept frozen at −23 °C. All purees were defrosted one day before the first training and kept in refrigeration at 4 °C between sessions.

2.8. Statistical Analysis

Binomial Statistical model was used to determine differences in the triangle test between the models 1 and 4. As part of the binomial model, a Z-test was used to determine if there is a significant difference between the models ($\alpha = 0.05$). For CATA analysis, a Cochran's Q test and correspondence analysis were performed to determine the aroma descriptors to be used in further descriptive analysis, utilizing line scales. For the line intensity scales, the data were analyzed using principal component analysis (PCA) and agglomerative hierarchical clustering (AHC). All statistical analyses were performed using XLstat (2020.1.3).

3. Results

3.1. Triangle Test

The models evaluated were dearomatized white wines with the same aroma base but with different concentrations of the rose oxide enantiomers comparing four different ratios; 75:25, 70:30, 65:35 and 60:40 (Table 1). A significant difference was found when comparing model 2 with model 3 (Table 3).

Table 3. Triangle Test results for the models with different Rose oxide ratios, performed by z-test (n = 65, for all tests) [a].

Comparison	Number of Participants	Number of Correct Responses	p-Value
Model 1 vs. Model 2	65	24	0.3156
Model 1 vs. Model 3	65	25	0.296
Model 1 vs. Model 4	65	17	>0.5
Model 2 vs. Model 3	65	37	<0.001 ***
Model 2 vs. Model 4	65	21	>0.5
Model 3 vs. Model 4	65	17	>0.5

[a] Significant difference Level: *** $p < 0.001$.

3.2. CATA

There were 22 attributes used more than 15% for CATA and included in the statistical analysis (Table 2). The first three variates of correspondence analysis explained 65.74% of the total variance (Figure 1). The terms could be separated into four groups using multiple pairwise comparisons (Sheskin critical difference) (Supplementary Table S2). Pome fruit, honey, stone fruit, and dried fruit had the greatest citation frequencies.

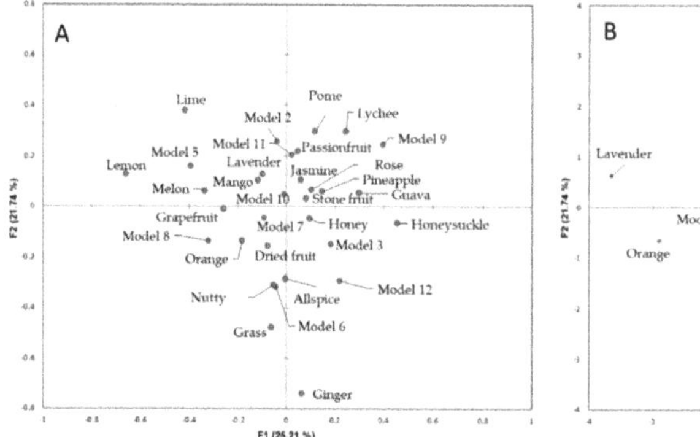

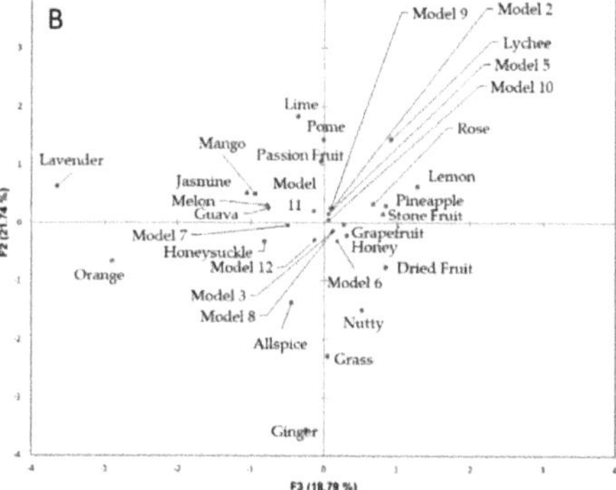

Figure 1. First three dimensions of results from corresponding analysis for CATA data, (**A**) F1 and F2, (**B**) F2 and F3.

3.3. Descriptive Analysis Results

Differences were found in the model wines with the first two variates explained 53.25% of the total variance (Figure 2). A third dimension was not included, since it did not greatly increase the amount of total variance (only explained 15.42% more) and did not elucidate any additional groupings/separation of models. The wine models were classified into 3 clusters using agglomerative hierarchical clustering (AHC) (Supplementary Figure S1. Cluster 1 included models 3 and 10, and was characterized by pome, stone fruit, melon, and grapefruit descriptors. The second cluster included models 6, 8 and 9, and was characterized by lemon aroma. The third cluster included models 2, 5, 7, 11 and 12, and was characterized by lychee, honey, honeysuckle, rose, dried fruit, ginger, orange, and tropical fruit aromas.

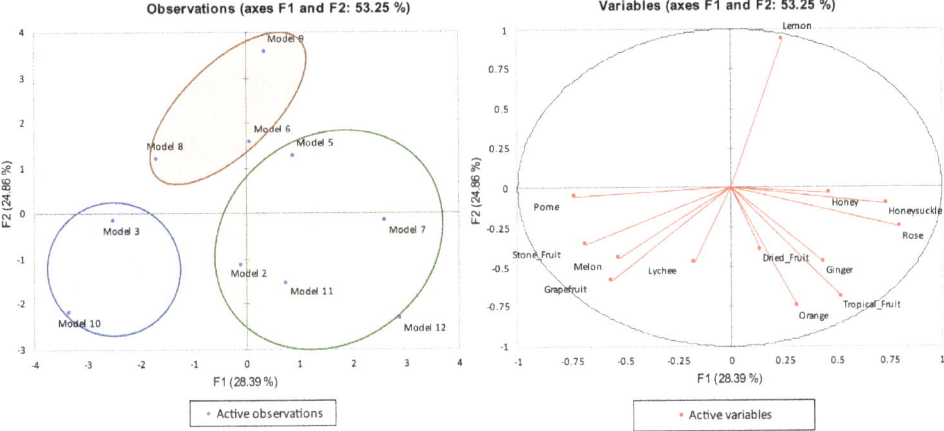

Figure 2. Separation of wines models by monoterpene profile using PCA; circles represented the clusters obtained from dendrogram.

4. Discussion

4.1. Triangle Test

Gewurztraminer wines are known to have different enantiomers of *cis*-rose oxide, with the ratio of enantiomers primarily altered during yeast fermentation [33]. In wine, the (−) enantiomer is dominant [17,33]. The dominant form is important as the enantiomers have different perception thresholds and aroma qualities [34]. Triangle tests (Table 3) show evidence that the ratio of rose oxide (RO) enantiomers altered aroma perception, specifically when comparing the ratio of 65:35 (Model 3) with the ratio of 70:30 (Model 2). This difference is interesting since those are the treatments closer in concentration and unexpectedly, model 2 was not considered to be different from model 4, which had a greater concentration difference in the enantiomer ratio. Unfortunately, model 1 and model 4 were not included in the descriptive analysis, so it is unclear exactly how aroma was altered. However, by looking at where model 2 and model 3 were located in the PCA results (Figure 2), there is the possibility that model 1 and model 4 may be located between these two points in the PCA, and therefore not different enough to show significant differences in the triangle test, something to be investigated in a future study.

4.2. Descriptive Analysis

Rose oxide is not present by itself in wine, and therefore, to understand the impact it has on quality, it is important to appreciate how this compound interacts with other compounds to alter aroma perception. Rose oxide is a monoterpene, and previous work has shown that when one monoterpene is altered, many others are also, as they all share the same biosynthesis pathway, starting with geranyl diphosphate (GPP) [35–37]. This work focused on the interaction between rose oxide, linalool and α-terpineol, based on concentrations determined in previous work [17]. Of the 10 model wines tested, three different clusters were found for descriptive analysis, as shown in Figure 2. How the wines grouped, and which aroma descriptors are related to those wines, clearly show interactions that result in aromatic perception changes.

Clusters are as follows; (1) model 3 and model 10, (2) models 6, 8 and 9, and (3) models 2, 5, 7, 11 and 12. Within these clusters, the rose oxide ratio is clearly a driver of aroma, especially when the other monoterpenes are at low concentrations. Something quite interesting is noted based on the ratios tested. When a ratio of 65:35 of rose oxide is present, this appears to be the main driver of aroma when linalool and α-terpineol are at low concentrations, as models 3 and 10 are in the same cluster (Figure 2). As the concentrations of linalool and α-terpineol increase, the interactions start to alter and change, being dominated by the other two compounds. However, the same does not occur when the rose oxide ratio is 70:30 and wines contain low concentrations of linalool and α-terpineol, as model wines 2 and 9 are not in the same group. This is most likely due to the different thresholds of the two enantiomers (Table 4).

Table 4. Thresholds of Rose oxide enantiomers, Linalool and α-Terpineol in different conditions.

Components	Detection	Double DI Water [38]	Water	Beer [39]	Carbonated Water/Ethanol (5%)	Water/Ethanol (10%)
(−)-*cis*-Rose oxide	50 [34]	-	-	-	-	-
(+)-*cis*-Rose oxide	50 [34]	-	-	-	-	-
cis-Rose oxide	-	-	0.1 [27]	-	-	0.2 [29]
(−)-*trans*-Rose oxide	160 [34]	-	-	-	-	-
(+)-*trans*-Rose oxide	80 [34]	-	-	-	-	-
Linalool	-	(−) 0.8 (+) 0.7	1 [40]	5	3 [41]	15 [29]
α-Terpineol	280–350 [42]	(−) 9180 (+) 6800	460 [37]	2000	450 [41]	-

Of the other monoterpenes being investigated, linalool appears to be greatly altering the aroma of wine, although this is dependent on concentration. As stated previously, linalool by itself is associated with floral aromas [43]. When no linalool is present, or when linalool is at low concentrations, there are no floral aromas associated with the wines (models 2, 3, 9 and 10). As the concentration of linalool increases the aroma of model 7, shifts to honeysuckle and honey, adding a floral characteristic to the wine. The other wines with medium amounts of linalool, model 5 and 8, are primarily associated with lemon and pome. Although model 5 is grouped with model 7, but it is clearly associated with the lemon eigenvector, as model 8 is with pome eigenvector. When linalool is at high concentrations, models 11 and 12, where a more consistent aroma profile is seen, with both wines described as tropical fruit, ginger, etc. The increased concentrations of α-terpineol in models 11 and 12 are altering the aroma of model 7 from floral to more tropical fruit, ginger aromas. A similar trend to linalool can be seen with increasing α-terpineol concentrations. A more systematic study would need to be determined if the main driver of aroma at high concentrations is primarily linalool or α-terpineol and not rose oxide.

It is known that (+)-*cis* rose oxide has floral-green and rose aromas, while (−)-trans has floral-green, herbal (minty) and fruity aromas [34]. Model 3 has a lower (−) enantiomer ratio (65:35) with pome as its main aroma descriptor, and as the (−) enantiomer increases (ratio 70:30, model 2), the aromas that characterized the wine are lychee and dried fruit. This suggests that a slight change in enantiomers is enough to bring out the traditional lychee aroma associated with Gewürztraminer wines [32]. Our results suggest that you need the 70:30 ratio to achieve lychee aroma. The known thresholds of (−)-*cis*-rose oxide, (+)-*trans* rose oxide and (−)-*trans* rose oxide are above the concentration used in our wines (Table 4), although these thresholds were not done in wine, and it is known that monoterpene thresholds decrease when in mixtures [44]. This could mean that the threshold for (−)-rose oxide in wine is lower than these values (Table 4).

A surprising result from the descriptive analysis was the tropical fruit, orange and ginger aromas associated with model 12. Linalool has been found to be an important compound for fresh ginger aroma [45]. Tropical fruit aromas in wine are traditionally associated with esters and volatile thiols, and not with monoterpenes [46]. Although linalool and α-terpineol are important components of passionfruit [47,48]. This combination of high concentrations of the two compounds may be another possible compound combination for tropical fruit aromas, which has been largely overlooked in wine.

The strong association of lemon aroma with model 9 was also unexpected, as lemon aroma is associated with limonene [49]. Limonene was not used in this study. Linalool is known to be a major component of lemon and orange juices [50], but is not considered to be a main driver of these aromas. Additionally, α-terpineol is considered to draw from flavors to aged lemon and citrus products; it is a degradation product of linalool [51]. Based on this information from lemon and orange products, it was a surprise to see such a strong lemon aroma association with combinations of the used terpenes. Although lemon was also noted in models 6 and 8, all three models have different rose oxide enantiomer ratios and concentrations of linalool and α-terpineol. No clear trend or explanation can be seen, and this association should be studied further.

When the ratio of rose oxide changed and the other terpene concentrations remained the same (model 9 versus model 10), the main descriptors changed from lemon to stone fruit, grapefruit and melon. Grapefruit aroma is traditionally associated with volatile thiols, 3-mercaptoheptan-1-ol and 3-mercaptohexanol [15] and not with rose oxide. However, of the other terpenes investigated, linalool is important to grapefruit aroma, when in combination with other monoterpenes [52]. Linalool and α-terpineol have previously been associated with stone fruit aroma in Viognier wines, at similar concentrations to those used in our study (58 and 18.5 µg/L, and 91 and 28 µg/L means, respectively) [53]. Our work supports this aroma association, although only when linalool and α-terpineol are at low concentrations.

5. Conclusions

These results not only suggest that monoterpene interactions impact aroma perception, but that also, the ratios of the enantiomers are crucial on aroma perception when low and medium concentrations of other monoterpenes are present. The low and medium concentrations are the most common concentrations found in wine. These aroma qualities help to anticipate variations in wines when modifying or altering grape and wine practices, with a focus on altering monoterpene content. Further research should be done to tease out the nuances found, and to determine which compound is more influential on aroma perception when higher concentrations are found.

Supplementary Materials: The following are available online at https://www.mdpi.com/article/10.3390/fermentation8010030/s1, Table S1: Final concentration of aroma compounds found in the wine base, Table S2: Multiple pairwise comparisons using the Critical difference (Sheskin) procedure. Figure S1: Differentiation of the wine models in dendrogram of AHC.

Author Contributions: Conceptualization, M.M.C.-H. and E.T.; methodology, A.D., E.T.; software, A.D., E.T.; validation, M.M.C.-H. and E.T.; formal analysis, M.M.C.-H., A.D., and E.T.; investigation, M.M.C.-H. and A.D.; resources, E.T., data curation, A.D. and E.T.; writing—original draft preparation, M.M.C.-H.; writing—review and editing, E.T., A.D.; visualization, M.M.C.-H. and E.T.; supervision, E.T.; project administration, E.T.; funding acquisition, E.T. All authors have read and agreed to the published version of the manuscript.

Funding: This research received no external funding.

Institutional Review Board Statement: The study was conducted according to the guidelines of the Declaration of Helsinki, and approved by the Institutional Review Board (or Ethics Committee) of

Oregon State University (Corvallis, OR, USA), protocol code IRB-2018-8606 and Date of approval: 27 June 2018.

Informed Consent Statement: Informed consent was obtained from all subjects involved in the study.

Data Availability Statement: Not applicable.

Conflicts of Interest: The authors declare no conflict of interest.

References

1. Rocha, S.M.; Coutinho, P.; Coelho, E.; Barros, A.S.; Delgadillo, I.; Coimbra, M.A. Relationships between the Varietal Volatile Composition of the Musts and White Wine Aroma Quality. A Four Year Feasibility Study. *LWT-Food Sci. Technol.* **2010**, *43*, 1508–1516. [CrossRef]
2. Peynaud, E.; Blouin, J. The Taste of smell. In *The Taste of Wine*, 2nd ed.; John Wiley & Sons: New York, NY, USA, 1996; Volume 1, pp. 50–78. [CrossRef]
3. Ruiz, J.; Kiene, F.; Belda, I.; Fracassetti, D.; Marquina, D.; Navascués, E.; Calderón, F.; Benito, A.; Rauhut, D.; Santos, A.; et al. Effects on Varietal Aromas during Wine Making: A Review of the Impact of Varietal Aromas on the Flavor of Wine. *Appl. Microbiol. Biotechnol.* **2019**, *103*, 7425–7450. [CrossRef] [PubMed]
4. Francis, I.L.; Newton, J.L. Determining Wine Aroma from Compositional Data. *Aust. J. Grape Wine Res.* **2005**, *11*, 114–126. [CrossRef]
5. Styger, G.; Prior, B.; Bauer, F.F. Wine Flavor and Aroma. *J. Ind. Microbiol. Biotechnol.* **2011**, *38*, 1145–1159. [CrossRef]
6. Zalazain, A.; Marín, J.; Alonso, G.L.; Salinas, M.R. Analysis of wine primary aroma compounds by stir bar sorptive extraction. *Talanta* **2007**, *71*, 1610–1615. [CrossRef] [PubMed]
7. Jackson, R. *Wine Tasting A Professional Handbook*, 2nd ed.; Academic Press: Berkeley, CA, USA, 2009; Volume 1, pp. 1–22.
8. Grainger, K.; Tattersall, H. Chapter 3 Nose. In *Wine Quality: Tasting and Selection*; John Wiley & Sons, Ltd.: Ames, IA, USA, 2009; pp. 35–39. Available online: https://ebookcentral.proquest.com/lib/osu/detail.action?docID=416557 (accessed on 25 October 2021).
9. Mori, K. Unique Characteristics of the olfactory system. In *The Olfactory System*; Springer: Tokyo, Japan, 2014; pp. 1–18. [CrossRef]
10. Lozano, J.; Santos, J.; Horrillo, M. Classification of White Wine Aromas with an Electronic Nose. *Talanta* **2005**, *67*, 610–616. [CrossRef] [PubMed]
11. Terashini, R.; Buttery, G.; Shahidi, F. Enantioselectivity in Odor Perception. In *Flavor Chemistry Trends and Development*, 3rd ed.; ACS Symposium Series; American Chemical Society: Washington DC, USA, 1989; pp. 155–157.
12. Bentley, R. The Nose as a Stereochemistry. *Enantiomers Odor. Chem. Rev.* **2006**, *106*, 4099–4112. [CrossRef]
13. Buettner, A. Influence of mastication on the concentrations of aroma volatiles—Some aspects of flavour release and flavour perception. *Food Chem.* **2000**, *71*, 347–354. [CrossRef]
14. Ager, D. Terpenes: The expansion of the Chiral Pool. In *Handbook of Chiral Chemicals*; CRC Press: New York, NY, USA, 2005; pp. 59–74.
15. Buettner, A. Wine. In *Springer Handbook of Odor*; Springer: New York, NY, USA, 2017; pp. 143–162.
16. Ganjitabar, H.; Hadidi, R.; Garcia, G.A.; Nahon, L.; Powis, I. Vibrationally-Resolved Photoelectron Spectroscopy and Photoelectron Circular Dichroism of Bicyclic Monoterpene Enantiomers. *J. Mol. Spectrosc.* **2018**, *353*, 11–19. [CrossRef]
17. Tomasino, E.; Song, M.; Fuentes, C. Odor Perception Interactions between Free Monoterpene Isomers and Wine Composition of Pinot Gris Wines. *J. Agric. Food Chem.* **2020**, *68*, 3220–3227. [CrossRef]
18. Laska, M. Olfactory Discrimination Ability of Human Subjects for Ten Pairs of Enantiomers. *Chem. Senses* **1999**, *24*, 161–170. [CrossRef]
19. Song, M.; Xia, Y.; Tomasino, E. Investigation of a Quantitative Method for the Analysis of Chiral Monoterpenes in White Wine by HS-SPME-MDGC-MS of Different Wine Matrices. *Molecules* **2015**, *20*, 7359–7378. [CrossRef] [PubMed]
20. Katarína, F.; Katarína, M.; Katarína, Ď.; Ivan, Š.; Fedor, M. Influence of Yeast Strain on Aromatic Profile of Gewürztraminer Wine. *LWT-Food Sci. Technol.* **2014**, *59*, 256–262. [CrossRef]
21. Marais, J. Terpenes in the Aroma of Grapes and Wines: A Review. *S. Afr. J. Enol. Vitic.* **2017**, *4*, 49–58. [CrossRef]
22. Song, M.; Fuentes, C.; Loos, A.; Tomasino, E. Free Monoterpene Isomer Profiles of Vitis Vinifera, L. Cv. White Wines. *Foods* **2018**, *7*, 27. [CrossRef]
23. Mateo, J.J.; Jiménez, M. Monoterpenes in Grape Juice and Wines. *J. Chromatogr. A* **2000**, *881*, 557–567. [CrossRef]
24. Dziadas, M.; Jeleń, H.H. Analysis of Terpenes in White Wines Using SPE–SPME–GC/MS Approach. *Anal. Chim. Acta* **2010**, *677*, 43–49. [CrossRef]
25. Rapp, A. Volatile Flavour of Wine: Correlation between Instrumental Analysis and Sensory Perception. *Food/Nahrung* **1998**, *42*, 351–363. [CrossRef]
26. Katarína, F.; Bajnociová, L.; Feder, M.; Spanik, I. Investigation of volatile profile of varietal of gewürztraminer wines using two-dimensional gas chromatography. *J. Agric. Nutr. Res.* **2017**, *56*, 73–85.
27. Ong, P.K.C.; Acree, T.E. Similarities in the Aroma Chemistry of Gewürztraminer Variety Wines and Lychee (*Litchi chinesis* Sonn.) Fruit. *J. Agric. Food Chem.* **1999**, *47*, 665–670. [CrossRef]

28. Wiest, R. Apendix A A Quick Reference Guide to Varietal Wines. In *A Guide to the Elite Estates of the Mosel-Saar-Ruwer Wine Region*; Board and Bench Publishing: San Francisco, CA, USA, 1983.
29. Guth, H. Identification of Character Impact Odorants of Different White Wine Varieties. *J. Agric. Food Chem.* **1997**, *45*, 3022–3026. [CrossRef]
30. Lukić, I.; Radeka, S.; Grozaj, N.; Staver, M.; Peršurić, Đ. Changes in physico-chemical and volatile aroma compound composition of Gewürztraminer wine as a result of late and ice harvest. *Food Chem.* **2016**, *196*, 1048–1057. [CrossRef]
31. Vilanova, M.; Genisheva, Z.; Graña, M.; Oliveira, J.M. Determination of Odorants in Varietal Wines from International Grape Cultivars (*Vitis vinífera*) Grown in NW Spain. *S. Afr. J. Enol. Vitic.* **2016**, *34*, 212–222. [CrossRef]
32. Ruiz-García, L.; Hellín, P.; Flores, P.; Fenoll, J. Prediction of Muscat Aroma in Table Grape by Analysis of Rose Oxide. *Food Chem.* **2014**, *154*, 151–157. [CrossRef]
33. Koslitz, S.; Renaud, L.; Kohler, M.; Wüst, M. Stereoselective Formation of the Varietal Aroma Compound Rose Oxide during Alcoholic Fermentation. *J. Agric. Food Chem.* **2008**, *56*, 1371–1375. [CrossRef]
34. Yamamoto, T.; Matsuda, H.; Utsumi, Y.; Hagiwara, T.; Kanisawa, T. Synthesis and odor of optically active rose oxide. *Tetrahedron Lett.* **2002**, *43*, 9077–9080. [CrossRef]
35. Buettner, A. Biosynthesis of Plant-Derived Odorants. In *Springer Handbook of Odor*; Springer: New York, NY, USA, 2017; pp. 13–33.
36. Dunlevy, J.D.; Kalua, C.M.; Keyzers, R.A.; Boss, P.K. The production of flavour and aroma compounds in grape berries. In *Grapevine Molecular Physiology & Biotechnology*, 2nd ed.; Springer: New York, NY, USA, 2009; pp. 293–340.
37. Chen, Q.; Fan, D.; Wang, G. Heteromeric geranyl(geranyl) diphosphate synthase is involved in monoterpene biosynthesis in *Arabidopsis* flowers. *Mol. Plant* **2015**, *8*, 1434–1437. [CrossRef]
38. Padrayuttawat, A.; Yoshizawa, T.; Tamura, H.; Tokunaga, T. Optical isomers and odor thresholds of volatile constituents in Citrus sudachi. *Food Sci. Technol. Int. Tokyo* **1997**, *3*, 402–408. [CrossRef]
39. Hanke, S.; Herrmann, M.; Rückerl, J.; Schönberger, C.; Back, W. Hop volatile compounds (Part II): Transfer rates of hop compounds from hop pellets to wort and beer. *Brew. Sci.* **2008**, *61*, 140–147.
40. Niu, Y.; Sun, X.; Xiao, Z.; Wang, P.; Wang, R. Olfactory impact of terpene alcohol on terpenes aroma expression in Chrysanthemum essential oils. *Molecules* **2018**, *23*, 2803. [CrossRef]
41. Takoi, K.; Koichiro, K.; Itoga, Y.; Katayama, Y.; Shimase, M.; Nakayama, Y.; Watari, J. Biotransformation of Hop-Derived Monoterpene Alcohols by Lager Yeast and Their Contribution to the Flavor of Hopped Beer. *J. Agric. Food Chem.* **2010**, *58*, 5050–5058. [CrossRef] [PubMed]
42. Burdock, G.A.; Fenaroli, G. *Fenaroli's Handbook of Flavor Ingredients*; Taylor & Francis Group: Milton Park, UK, 2009.
43. Buettner, A. Coffee. In *Springer Handbook of Odor*; Springer: New York, NY, USA, 2017; pp. 107–128.
44. Moreno, J.; Peinado, R. Must Aromas. In *Enological Chemistry*; Elsevier Science: Amsterdam, The Netherlands; New York, NY, USA, 2012; pp. 23–40.
45. Sekiwa, Y.; Mizuno, Y.; Yamamoto, Y.; Kubota, K.; Kobayashi, A.; Koshino, H. Isolation of some glucosides as aroma precursors from ginger. *Biosci. Biotechnol. Biochem.* **1999**, *63*, 384–389. [CrossRef] [PubMed]
46. Capone, D.; Barker, A.; Williamson, P.; Francis, I. The role of potent thiols in Chardonnay wine aroma. *Aust. J. Grape Wine Res.* **2018**, *24*, 38–50. [CrossRef]
47. Jordán, M.J.; Goodner, K.L.; Shaw, P.E. Volatile components in tropical fruit essences: Yellow passion fruit (*Passiflora edulis* Sims, *F. flavicarpa* Degner) and banana (*Musa sapientum* L.). *Fla. State Hortic. Soc.* **2000**, *113*, 284–286.
48. Maróstica, M.R.; Pastore, G.M. Tropical fruit flavour. In *Flavours and Fragrances*; Springer: Berlin/Heidelberg, Germany, 2007; pp. 189–201.
49. Vieira, A.J.; Beserra, F.P.; Souza, M.C.; Totti, B.M.; Rozza, A.L. Limonene: Aroma of innovation in health and disease. *Chem.-Biol. Interact.* **2018**, *283*, 97–106. [CrossRef]
50. Buettner, A. Fruits. In *Springer Handbook of Odor*; Springer: New York, NY, USA, 2017; pp. 171–190.
51. Nisperos-Carriedo, M.O.; Shaw, P.E. Comparison of volatile flavor components in fresh and processed orange juices. *J. Agric. Food Chem.* **1990**, *38*, 1048–1052. [CrossRef]
52. Zheng, H.; Zhang, Q.; Quan, J.; Zheng, Q.; Xi, W. Determination of sugars, organic acids, aroma components, and carotenoids in grapefruit pulps. *Food Chem.* **2016**, *205*, 112–121. [CrossRef]
53. Siebert, T.E.; Barter, S.R.; de Barros Lopes, M.A.; Herderich, M.J.; Francis, I.L. Investigation of 'stone fruit'aroma in Chardonnay, Viognier and botrytis Semillon wines. *Food Chem.* **2018**, *256*, 286–296. [CrossRef]

Article

Recycling and Conversion of Yeasts into Organic Nitrogen Sources for Wine Fermentation: Effects on Molecular and Sensory Attributes

Paula Rojas [1], Daniel Lopez [1], Francisco Ibañez [1], Camila Urbina [1], Wendy Franco [2,3], Alejandra Urtubia [1,4] and Pedro Valencia [1,*]

1 Department of Chemical and Environmental Engineering, Universidad Técnica Federico Santa María, Av. España 1680, Valparaíso 2390123, Chile; pau.rojas21@gmail.com (P.R.); daniel.lopezo@alumnos.usm.cl (D.L.); francisco.ibaneze@sansano.usm.cl (F.I.); camila.urbina@usm.cl (C.U.); alejandra.urtubia@usm.cl (A.U.)
2 Department of Chemical and Bioprocess Engineering, Pontificia Universidad Católica de Chile, Av. Vicuña Mackena 4860, Santiago 7820244, Chile; wfranco@ing.puc.cl
3 Department of Health Science, Nutrition and Dietetics, Pontificia Universidad Católica de Chile, Av. Vicuña Mackena 4860, Santiago 7820244, Chile
4 Centro de Biotecnología, Universidad Técnica Federico Santa María, General Bari 699, Valparaíso 2390136, Chile
* Correspondence: pedro.valencia@usm.cl

Citation: Rojas, P.; Lopez, D.; Ibañez, F.; Urbina, C.; Franco, W.; Urtubia, A.; Valencia, P. Recycling and Conversion of Yeasts into Organic Nitrogen Sources for Wine Fermentation: Effects on Molecular and Sensory Attributes. *Fermentation* **2021**, *7*, 313. https://doi.org/10.3390/fermentation7040313

Academic Editors: Bin Tian and Jicheng Zhan

Received: 7 November 2021
Accepted: 13 December 2021
Published: 14 December 2021

Publisher's Note: MDPI stays neutral with regard to jurisdictional claims in published maps and institutional affiliations.

Copyright: © 2021 by the authors. Licensee MDPI, Basel, Switzerland. This article is an open access article distributed under the terms and conditions of the Creative Commons Attribution (CC BY) license (https://creativecommons.org/licenses/by/4.0/).

Abstract: Organic nitrogen plays a significant role in the fermentation performance and production of esters and higher alcohols. This study assessed the use of yeast protein hydrolysate (YPH) as a nitrogen source for grape must fermentation. In this study, we prepared an enzymatic protein hydrolysate using yeasts recovered from a previous fermentation of wine. Three treatments were performed. DAP supplementation was used as a control, while two YPH treatments were used. Low (LDH) and high degrees of hydrolysis (HDH), 3.5% and 10%, respectively, were chosen. Gas chromatography and principal component analysis indicated a significant positive influence of YPH-supplementations on the production of esters and higher alcohols. Significantly high concentrations of 3-methyl-1-penthanol, isoamyl alcohol, isobutanol, and 2-phenylethanol were observed. Significant odorant activity was obtained for 3-methyl-1-pentanol and ethyl-2-hexenoate. The use of YPH as nitrogen supplementation is justified as a recycling yeasts technique by the increase in volatile compounds.

Keywords: yeast protein hydrolysate; nitrogen supplementation; volatile compounds; wine aroma; wine higher alcohols; wine esters

1. Introduction

Despite the variations in wine volume, approximately 270 million hectoliters are produced each year [1], along with the unavoidable generation of wastes. The main byproducts from the wine industry are grape stems, pomace, marc, and lees, which are mainly utilized for landfills, incineration, or animal feed [2]. Yeast is one of the lees' components and contains proteins and carbohydrates, resulting in a potential source of nutrients for the general purposes of the food industry [3]. A wide spectrum of applications of inactive dry yeast preparations in winemaking has been reviewed by Pozo-Bayón et al. [4]. An attractive alternative to increase the bioavailability of the components of yeasts is cell lysis and the enzymatic hydrolysis of proteins [3,5,6]. This process increases the bioavailability of proteins by generating assimilable nitrogen in the form of peptides. Nitrogen supplementation is a key aspect to attempt an effective and profitable fermentation process by affecting the fermentative activity and the formation of metabolites [7] and represents the challenge of achieving an organoleptic profile to satisfy consumers. Four components contribute to the assimilable nitrogen for yeasts during fermentation: ammonium, amino acids, oligopeptides, and proteins [8]. During the fermentation process, nitrogen is required for the synthesis of cell metabolites and structures such as proteins, cell walls, nucleic acids, and, in general, biomass. The assimilation of organic nitrogen in the form of amino acids and peptides allows the direct availability of these molecules, lowering energy consumption [8–10].

The concentration and nature of the nitrogen source directly affect biomass generation and the fermentation rate [11]. Additionally, nitrogen plays a fundamental role in the generation of secondary metabolites, which confer the characteristic flavors and taste to wine. These metabolites are higher alcohols and esters. Higher alcohols are formed in amino acid catabolism by the Ehrlich pathway. Some of these amino acids are direct precursors of higher alcohols, such as isoamyl alcohol from leucine, 2-phenylethanol from phenylalanine, and methionol from methionine [12]. The odors of these compounds are not desirable when isolated. However, they have a minor contribution to the vinous nature of wine and an important role in the formation of esters and aldehydes [8]. Therefore, the production of esters is not directly affected by the nitrogen source. However, nitrogen availability affects the redox balance of the cell and could modify the concentration of acetyl-CoA [13]. Surely, we can conclude that nitrogen has a primary and a secondary role in yeast metabolism and, consequently, in the organoleptic profile of wine.

The most used nitrogen source in the wine industry is diammonium phosphate (DAP), which is an inorganic compound with high bioavailability. It is well documented that nitrogen supplementation with organic sources, such as amino acids, allows an increase in the generation of volatile compounds, enhancing the sensorial perception of wine [13–17]. However, yeasts recovered from the same fermentation are a more economically suitable organic nitrogen source.

A few studies have evaluated the effect of yeast lysates on wine must fermentation. González-Marco et al. [18] studied the generation of biogenic amines after the addition of yeast autolysate to wine. Kevvai et al. [19] studied the growth of *Saccharomyces cerevisiae* in synthetic medium supplemented with yeast hydrolysate. They determined that 40% of the total nitrogen in the fermenting yeasts originated from the yeast hydrolysate. Supplementation with organic nitrogen sources has been studied in more detail in the fermentation of malt worts, especially high-gravity worts. Protein hydrolysates from wheat gluten [20–24], soy [25], and walnut meal [26] were used as supplements. All these studies have confirmed the improvement of fermentation performance in a variety of aspects: increased biomass growth, increased cell viability, increased ethanol content, increased wort fermentability, enhancement of cell membrane integrity, a decrease in radical oxygen species, and increased osmotic and ethanol stress tolerance. In addition, Li et al. [26] observed an improved formation of alcohols and esters and an increased ratio of higher alcohols to esters, which means a better-balanced taste of final beers.

To the best of our knowledge, the protein hydrolysate from recovered yeasts is still unevaluated as a nitrogen source for winemaking. Therefore, we recovered yeasts from previous batches of wine fermentation, hydrolyzed them by enzymatic proteolysis, and used them as a nitrogen supplement for grape must fermentation in a winemaking process involving the whole cycle of yeast usage. The wine production process was conducted in the laboratory; however, the procedure emulated each stage of winemaking in the commercial winery, including bottling before chemical and sensorial analyses. This article evaluates the use of yeast protein hydrolysate (YPH) as a nitrogen source for fermentation and its effect on the molecular and sensorial profiles of wine.

2. Materials and Methods

2.1. Materials

Wine lees were obtained from Villaseñor winery (Maule, Chile; 35°08′37″ S 71°21′35.2″ W), recovered from the fermentation of a Syrah wine. The yeast for fermentation was Lalvin EC1118 (Danstar Ferment AG, Fredericia, Denmark). Alcalase protease manufactured by

Novozymes (Bagsvaerd, Denmark) was used in the protein hydrolysis, corresponding to the endoprotease subtilisin with 24 AU/g. Cabernet Sauvignon grapes obtained from Indomita winery (Casablanca Valley, Chile; 33°21′34.8″ S 71°20′37.8″ W) were used for wine production. Reagents HCl, NaOH, diammonium phosphate (DAP), potassium metabisulfite, and formaldehyde were analytical grade and obtained from Winkler (Santiago, Chile).

2.2. *Preparation of the Nitrogen Source*

2.2.1. Yeast Recovery from Lees

The solid phase from the lees was separated by gravitation and filtration. The wet solid phase was mixed with an equal mass of 1.5% *v*/*v* HCl for 45 min at 50 °C with constant stirring by IKA agitator (Cole Parmer, IL, USA) to dissolve the tartaric salts. Yeasts were recovered by gravitation and filtration, and the liquid phase was discarded. The remaining yeasts were suspended in water at a 1:1 ratio and autoclaved at 121 °C for 15 min in an AMILAB autoclave (Laboratory Instruments, Santiago, Chile). Thermal treatment was used as an inactivating treatment to avoid living yeasts interfering with the fermentation process and as a primary lysate treatment.

2.2.2. Yeast Protein Hydrolysis

Autoclaved yeast suspensions contained 95 g/L dried yeast, corresponding to 31 g/L protein, previously quantified by the Kjeldahl method (AOAC). Next, 300 g of yeast suspension was stirred in a 0.5-L vessel in a water bath at 50 °C (Julabo ED, Seelbach, Germany). The suspension pH was adjusted to 8 with 1.5 N NaOH. The hydrolysis reaction was started by the addition of 15 mAU of Alcalase per g of suspension. The pH was controlled by the addition of 1.5 N NaOH with the G20 Mettler-Toledo auto titrator (Schwerzenbach, Switzerland). Two hydrolysates were prepared at different degrees of hydrolysis (DH), 3.5% and 10%. The DH was controlled based on the volume of NaOH solution added, according to a previous publication [27]. The reaction was stopped by inactivating the protease in a water bath at 85 °C for 30 min. The resulting hydrolysates were aliquoted in 50-mL tubes and frozen at −15 °C (Freezer Daewoo FF09) for later use as a nitrogen source.

2.3. *Wine Production*

The experiments were designed to test the effects of 3 nitrogen sources: diammonium phosphate (DAP), yeast hydrolysate DH 3.5% (LDH), and yeast hydrolysate DH 10% (HDH). Experiments supplemented with DAP were designed as controls, while those supplemented with LDH and HDH corresponded to treatments with low and high degrees of hydrolysis hydrolysates, respectively. Fermentation experiments were performed with 5 replicates for each nitrogen source, resulting in a total of 15 experiments.

2.3.1. Fermentation

Each fermentation replicate consisted of 5 kg of destemmed and crushed Cabernet Sauvignon grapes conditioned with 70 ppm potassium metabisulfite and kept in 6-l open vessels until the first racking. The must contained 219 mg/L nitrogen quantified as free amino groups by the formaldehyde method [28]. Nitrogen sources were added as the necessary amount to achieve 250 mg/L of total nitrogen. The must pH was adjusted to 3.45 with tartaric acid. A total of 4 g of EC1118 yeast was hydrated with 40 mL of equal volumes of must and water at 35 °C and left unstirred for 10 min. Then, the mixture was gently hand-stirred and left to rest for another 15 min. The hydrated yeasts were left at room temperature to let them cool. An adequate aliquot of this suspension was added to each fermentation batch to obtain an initial yeast concentration of 0.05 g/L. The fermentation was followed up for 5 days with withdrawal samples to quantify the specific gravity and sugar concentration using a glass densimeter (TrueBrew, TX, USA) and a digital refractometer Hanna HI96811 (Woonsocket, RI, USA), respectively.

2.3.2. Post-Fermentation

After 8 days of fermentation, each batch was racked individually. The fermentation content was filtered, and the pomace was pressed. The fermented must was kept in closed glass bottles with an airlock for a month. The lees were separated in the second racking, and the fermented must was again kept in closed glass bottles with an airlock for 10 weeks at outdoor temperatures between 8 °C and 12 °C. Tartaric salts precipitated under this condition. After the third racking, tartaric salts were separated, and the wine was corked in 750-mL bottles and maintained at room temperature until molecular characterization was performed.

2.4. Analytical Techniques

The fermentation progress was characterized by the quantification of sugars, acids, and alcohols measured by Alpha Bruker FTIR equipment (Billerica, MA, USA). The working parameters were spectral range 4000–400 cm^{-1} and spectral resolution of 2 cm^{-1}. FTIR equipment was previously calibrated for wine samples, and the analyses were performed by the software Opus 7.0. The Folin–Ciocalteu method was used to quantify the concentration of polyphenols in bottled wines using gallic acid standards between 25 and 125 mg/L [29]. The calibration curve was made measuring absorbance at 760 nm in a Thermo Scientific Orion AquaMate 8000 UV-Vis Spectrophotometer (Fisher Scientific, Göteborg, Sweden). Results of gallic acid calibration are shown in Figure S2 (Supplementary Material). The concentrations of volatile compounds, such as higher alcohols and esters, were quantified to characterize the bottled wines prepared with different nitrogen sources. The procedure consisted of solid-phase microextraction and injection of samples in Shimadzu GCMS QP2010 Ultra gas chromatography-mass spectrometry equipment (Suzhou, China). The mass spectrum of each sample was compared with a NIST-EPA-NIH library consisting of 130,000 spectrums. The quantification was carried out by the internal standard method considering a response factor equal to 1. A total of 23 higher alcohols and 57 esters were analyzed in 15 samples of bottled wines (5 for each supplementation). An analysis scheme is represented in Figure 1.

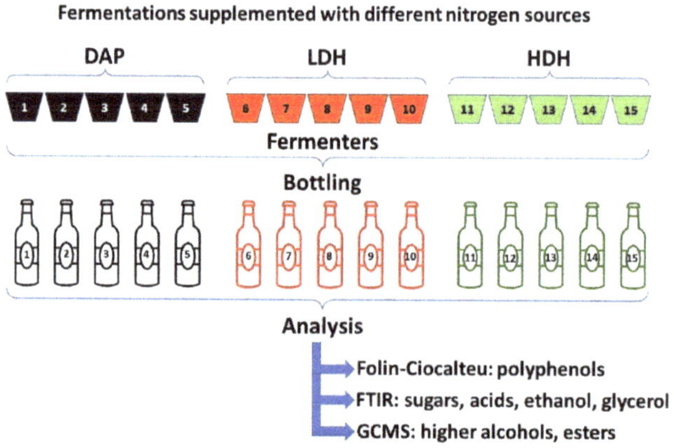

Figure 1. Scheme representing the fermentations, samples, and analysis.

2.5. Sensory Evaluation

A wine tasting session was conducted with 43 participants who were volunteer wine consumers and not wine experts. This test was performed as an initial approach to wine tasting with the objective of obtaining a mere perception of consumers. Three attributes were considered for the sensorial characterization of the wines: sweetness, acidity, and astringency. The first stage consisted of training, where the participants identified the 3 attributes by tasting solutions of glucose, tartaric acid, and tannic acid at concentrations of 20, 0.75, and 1.0 g/L, respectively. The second stage consisted of ordering the attribute intensity of glucose solutions with concentrations of 2, 10, 20, and 30 g/L; tartaric acid solutions with concentrations of 0.25, 0.50, 0.75, and 1.0 g/L; and tannic acid solutions with concentrations of 0.50, 0.75, 1.00, and 1.25 g/L. The aroma was also included as a wine attribute in the test. However, it was not trained and remained as a pure consumer's perception attribute. The third stage consisted of tasting the three wines with different nitrogen supplementation. Two randomly distributed blind samples of each wine were given to the panelists. Finally, the participants were asked to qualify each wine in order of intensity for the three attributes previously trained, the aroma (not trained), and order the wine samples according to their preference. All the processes were performed individually and without contact between participants.

2.6. Statistical Analysis of The Results

The results for chemical analyses performed by FTIR (acids, sugars, ethanol, and glycerol), GCMS (higher alcohols and esters), and polyphenols by the Folin–Ciocalteu method were evaluated for significant differences between 2 samples at 95% confidence. The results from these analytical techniques consisted of 5 replicates for each supplementation. A detailed list of results can be checked in the supplementary material Figures S1 and S2.

Principal component analysis (PCA) was performed to detect the effects of the treatments with different nitrogen sources on the molecular profile based on volatile compounds, including higher alcohols and esters. The raw data for higher alcohols and esters from the 15 wine samples was used as an entry. Higher alcohols and esters were analyzed by PCA in separated procedures for better visualization of the results.

The results obtained with the sensorial panel for each nitrogen source were analyzed through Friedman analysis for a nonparametric statistic [30]. This simple ranking test consisted of the evaluation of the effects of the three nitrogen sources according to H_0: all the effects of the nitrogen source are zero; and H_1: not all the effects of the nitrogen source are zero. Each wine sample was ranked by the panelists, and the ranked sum was compared between samples to evaluate significance at 95%.

3. Results

The progress of fermentation presented in Figure 2 indicated that all batches involving the different nitrogen sources resulted in sugar consumption. The progress profiles and the final density of the musts were similar, as observed in Figure 2a. However, a pattern in the fermentation rate was observed in Figure 2b. The batch supplemented with YPH and HDH had the highest rate, followed by the LDH- and DAP-supplemented batches (Figure 2b). In all cases, fermentation was performed on the sixth day. The same observations were made by measuring °Brix, as shown in Figure 2c. At this point, we can infer that all nitrogen supplementations allowed proper fermentation of the must. The higher the DH, the shorter the lag phase and the faster the fermentation rate.

The bottled wines were characterized by the quantification of acids, sugars, alcohols, and polyphenols. Results were plotted in Figure 3. The highest concentration of acids corresponded to lactic and tartaric acids. High variations in replicates (error bars) were observed for malic and lactic acids. The variations can be attributed to variations in malolactic fermentation, suggesting a lack of homogeneity among replicates. The variations in tartaric acid can be explained by a nonhomogeneous precipitation of this acid among the replicates. The highest acidity was obtained in the wine supplemented with HDH, followed by the LDH- and DAP-supplemented wines. The residual sugars were between 1.2 and 1.4 g/L in the different wines. Glucose presented the same residual concentration, while fructose was different depending on the experiment. The highest fructose concentration was obtained in batches supplemented with DAP. A lower concentration was obtained in the wines supplemented with YPH, consisting of the same value for LDH- and HDH-supplemented wines. Ethanol and glycerol production levels were similar among the different nitrogen supplementations. Polyphenols resulted in different concentrations, depending on the nitrogen source supplemented. Both YPH supplementation, LDH, and HDH resulted in higher concentrations of polyphenols compared with the DAP-supplemented wine. The only difference observed during fermentation was the sugar uptake rate. Faster production of ethanol in batches supplemented with YPH could cause a higher extraction of polyphenols. This inference is not supported by our evidence or by antecedents. Fermentations supplemented with YPH were faster, consumed more sugars, and obtained higher acidity and polyphenols compared to DAP supplementation. After the characterization of wines, analysis of volatile compounds was performed.

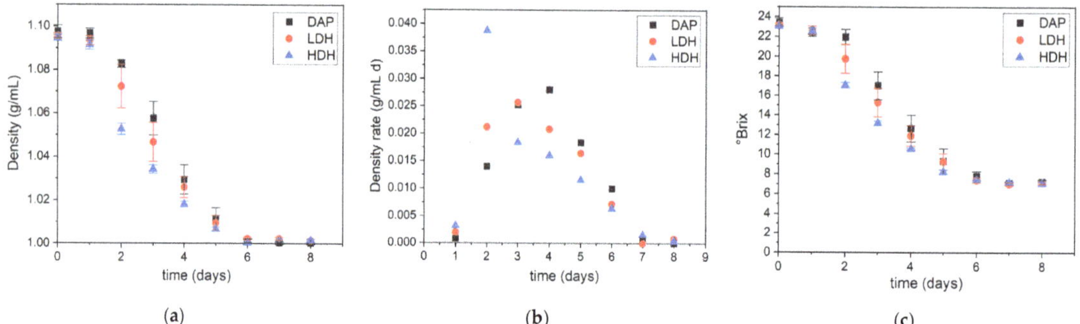

Figure 2. Fermentation progress of Cabernet Sauvignon musts supplemented with different nitrogen sources: (**a**) must density, (**b**) density rate, and (**c**) °Brix versus time.

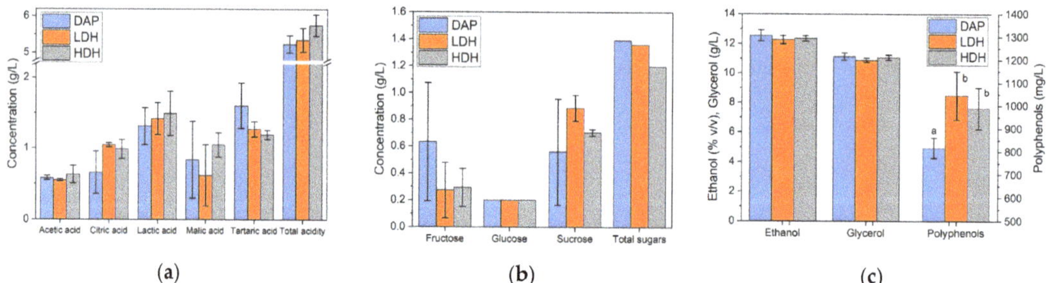

Figure 3. Characterization of Cabernet Sauvignon wines produced by supplemented fermentations with different nitrogen sources: (**a**) acids, (**b**) sugars, and (**c**) alcohols and polyphenols. Different letters indicate significant differences ($p < 0.05$).

The concentrations of higher alcohols and esters are presented in Figure 4. The highest concentration of higher alcohols and esters was obtained in the HDH-supplemented wine. The results agree with previously published data. The production of higher alcohols depends on the availability of amino acids. This evidence shows the feasibility of using YPH as a nitrogen source to promote the formation of higher alcohols. The esters were classified according to their origin, as shown in Figure 4b. In all cases, supplementation with HDH achieved higher concentrations of esters. Comparisons between the pairs DAP/HDH and LDH/HDH resulted in significant differences. Supplementation with LDH resulted in an equal or lower generation of esters compared to DAP-supplemented wine. The necessity of a DH threshold for the hydrolysate to cause an increase in ester production is inferred from this evidence. The requirement of yeast proteases to hydrolyze low-DH peptides means a higher energy cost for yeast, which could impede an improvement of ester production compared to DAP supplementation.

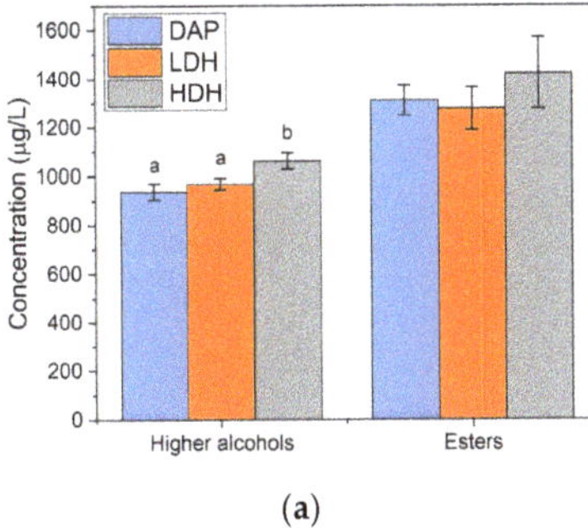

(a)

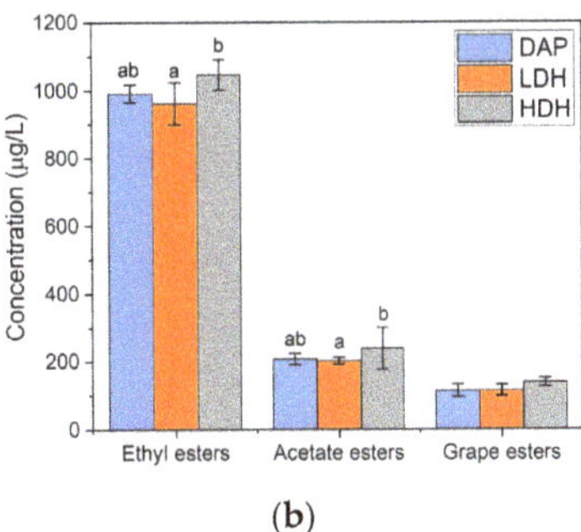

(b)

Figure 4. Characterization of Cabernet Sauvignon wines produced by supplemented fermentations with different nitrogen sources: (**a**) total higher alcohols and total esters, (**b**) ethyl, acetate, and grape esters. Different letters indicate significant differences ($p < 0.05$).

A total of 23 higher alcohols and 57 esters were quantified by gas chromatography. The complete list of higher alcohols and esters can be checked in the worksheet file "Higher alcohols and esters" in the Supplemental Material. A total of 11 higher alcohols and 18 esters presented significant differences among the different supplementations. The list of these volatile compounds is presented in Table 1. The wines supplemented with YPH presented higher volatile compound concentrations than control (DAP) in just 6 higher alcohols and 13 esters. In the rest of the cases, the control (DAP) presented higher values of the compounds. The isoamyl alcohol was the only volatile compound which concentration increased in the order of DAP-, LDH- and HDH-supplementations, presenting 443, 484, and 547 µg/L, respectively. The 2-phenyl ethanol presented a significantly higher concentration in the HDH-supplementation compared to both LDH- and DAP-supplementations. In the case of esters, the compounds ethyl 2-hydroxyhexanoate and ethyl dodecanoate were higher in both YPH-supplementations compared to the control, while isoamyl decanoate was significantly higher in HDH-supplementation compared to both LDH- and DAP-supplementations. The concentrations of all volatile compounds were compared to their odor threshold. Those compounds with concentrations higher than the odor threshold are listed in Table 2. One higher alcohol and 8 esters presented an odor activity value (OAV) higher than 1, which means that these compounds present significant odor among the total aroma components of the sample.

Principal component analysis (PCA) was performed to evaluate the effect of each treatment on each volatile compound. Figures 5 and 6 show the results of PCA for all the treatments and volatile compounds explored. A total of 15 samples were analyzed, corresponding to 5 samples for each nitrogen supplementation. The PCA plot for higher alcohols in Figure 5 presents the component in the x-axis with 35.6% of the variance, and in the y-axis, the other component with 23.4% of the variance. Both represent 59% of the total variance. A difference in the sample's distribution was observed for higher alcohols: samples 1 to 5 (DAP) at the left, samples 6 to 10 (LDH) at the center, and samples 11 to 15 (HDH) at the right. The samples with DAP obtained the highest variability and dispersed along the y-axis but not along the x-axis. Various higher alcohols are located to the right due to their influence contained in PC 1. One of these higher alcohols is the 3-methyl-1-pentanol, which presented a significantly higher concentration in HDH-supplemented wines and significant OAV (Tables 1 and 2).

Table 1. List of significantly different volatile compounds for Cabernet Sauvignon wines with different nitrogen supplementations.

Compound	Concentration (µg/L)					
	DAP		LDH		HDH	
	Mean	Sd	Mean	Sd	Mean	Sd
1-Butanol	3.58 a	0.67	2.80 ab	0.47	2.42 b	0.23
1-Decanol	1.54 a	0.15	1.88 ab	0.28	2.24 b	0.21
1-Heptanol	19.1 ab	2.03	17.38 a	0.79	19.88 b	1.56
1-Hexanol	132.16 a	7.83	115.08 b	6.01	114.86 b	5.85
1-Pentanol	0.62 a	0.04	0.54 ab	0.05	0.52 b	0.04
3-Ethoxy-1-propanol	0.9 a	0.14	0.56 b	0.13	0.36 b	0.09
3-Methyl-1-pentanol	8.52 a	0.86	9.36 ab	0.53	10.56 b	1.18
4-Methyl-1-pentanol	3.78 a	0.26	4.40 b	0.35	4.90 b	0.43
Isoamyl alcohol	443.52 a	10.13	484.02 b	28.19	547.08 c	16.54
Isobutyl alcohol	37.58 a	2.56	40.54 ab	2.73	43.92 b	2.13
2-Phenyl ethanol	255.84 a	15.77	264.72 a	8.70	285.38 b	12.96
Citronellyl acetate	2.16 a	0.32	1.60 b	0.25	1.42 b	0.13
Ethyl 2-hexenoate	3.32 a	0.39	3.70 ab	0.50	4.52 b	0.70
Ethyl 2-hydroxyhexanoate	6.14 a	0.66	8.10 b	1.22	8.32 b	0.73
Ethyl dodecanoate	10.76 a	2.82	15.96 b	2.51	18.46 b	1.74
Ethyl palmitate	0.46 a	0.18	0.78 ab	0.20	0.90 b	0.10
Ethyl propionate	15.62 a	1.67	13.38 ab	1.19	12.58 b	1.15
Ethyl trans-4-decenoate	0.20 a	0.00	0.10 ab	0.00	0.04 b	0.05
Ethyl undecanoate	0.42 a	0.08	0.60 ab	0.07	0.60 b	0.07
Ethyl phenyl lactate	1.36 a	0.13	1.60 ab	0.21	1.78 b	0.24
Hexyl acetate	3.38 a	0.81	2.24 b	0.30	3.52 ab	2.96
Isoamyl butyrate	0.42 a	0.04	0.58 ab	0.20	0.70 b	0.07
Isoamyl decanoate	2.02 a	0.83	3.66 a	1.06	5.42 b	0.91
Isoamyl hexanoate	5.86 a	0.58	6.12 ab	1.13	7.42 b	0.68
Isoamyl isovalerate	0.22 a	0.04	0.30 ab	0.07	0.32 b	0.04
Isoamyl octanoate	8.20 a	1.44	10.48 ab	2.78	12.56 b	1.34
Methyl octanoate	2.30 a	0.19	2.58 ab	0.62	2.76 b	0.13
Octyl formate	6.24 a	0.42	6.86 ab	1.09	7.46 b	0.30
Propyl acetate	1.96 a	0.42	1.16 b	0.13	1.78 ab	1.58

Different letters indicate significant differences ($p < 0.05$).

Table 2. List of volatile compounds with significant OAV for the different nitrogen supplementations.

Compound	Threshold (µg/L)	Concentration (µg/L)					
		DAP		LDH		HDH	
		Mean	Sd	Mean	Sd	Mean	Sd
3-Methyl-1-pentanol	7.5	8.52 a	0.86	9.36 ab	0.53	10.56 b	1.18
Amyl propionate	0.09	0.60	0.83	1.26	0.05	1.28	0.13
Ethyl-2-hexenoate	0.14	3.32 a	0.39	3.70 ab	0.50	4.52 b	0.70
Ethyl butyrate	20	26.06	3.12	26.54	2.41	25.56	2.13
Ethyl heptanoate	2.2	11.54	2.67	9.72	0.99	10.24	0.73
Ethyl hexanoate	62	232.44	14.90	231.26	18.85	246.62	10.42
Ethyl isovalerate	3	11.66	2.21	13.70	1.24	13.82	1.11
Ethyl phthalate	0.33	4.26	3.03	9.96	6.33	10.44	5.93
Isoamyl acetate	30	171.30	20.10	170.94	14.42	196.04	62.98

Different letters indicate significant differences ($p < 0.05$).

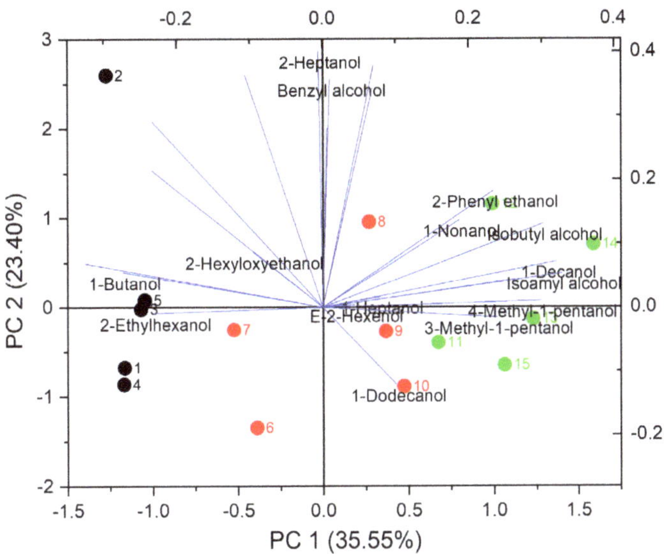

Figure 5. PCA for higher alcohols in Cabernet Sauvignon wines produced by supplemented fermentations with different nitrogen sources. Samples 1 to 5 are DAP (BLACK), 5 to 10 are LDH (red), and 11 to 15 are HDH (green).

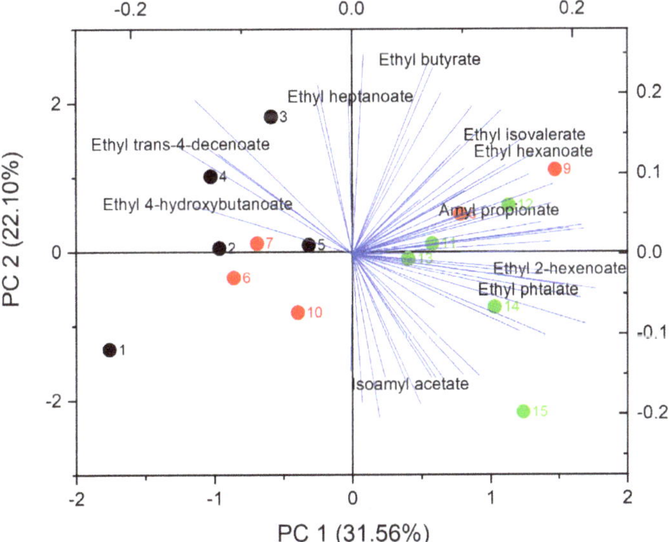

Figure 6. PCA for esters in Cabernet Sauvignon wines produced by supplemented fermentations with different nitrogen sources. Samples 1 to 5 are DAP (black), 5 to 10 are LDH (red), and 11 to 15 are HDH (green).

Other relevant higher alcohols are isoamyl alcohol, isobutanol, and 2-phenylethanol. These are located on the right, together with the HDH samples. The PCA results suggest an influence of HDH treatment on the formation of these compounds. The PCA for esters is plotted in Figure 6. Almost all analyzed esters are located to the right of the plot, where they overlap with the points corresponding to the HDH and LDH samples. Except for citronellyl acetate, all acetate esters are in this zone and are generated by the amino acid metabolism of the yeasts. Regarding other compounds, the variability of isoamyl acetate is mainly involved in PCA 2, which also occurs for ethyl butyrate to a lesser extent. Except for isoamyl acetate, these compounds are ethyl esters generated by fatty acid metabolism. This suggests some correlation between the nitrogen source and the metabolism of these esters.

The evaluation of sensory attributes by the sensorial panel provided information about consumers' preferences and qualifications of attributes of the different wines. The results for ranked sensory attributes are plotted in Figure 7. Significant differences between the wines were detected in aroma and astringency. A significant difference in aroma was detected between both YPHs. Astringency differences were detected between DAP- and both YPH-supplemented wines.

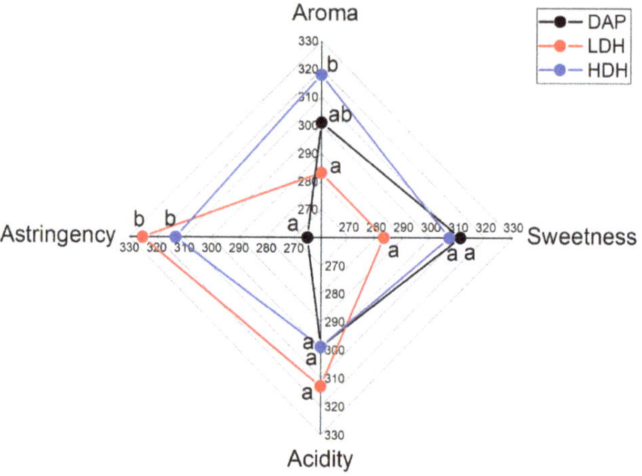

Figure 7. Parameterized results of the sensory attributes evaluation of Cabernet Sauvignon wines produced by supplemented fermentations with different nitrogen sources. Different letters indicate significant differences ($p < 0.05$).

The highest qualification in aroma was assigned to HDH supplementation, while both YPH-supplemented wines were qualified with the highest astringency. The attributes of sweetness and acidity did not present significant differences among the different supplemented wines. The panelists were asked for their preference among the wine samples. The results are presented in Table 3. The preference was significantly different between all treatment comparisons. The panelists' preference for wine samples in decreasing order was DAP-, HDH-, and LDH-supplemented.

Table 3. Ranked sum for the panelists' preference among the different nitrogen-supplemented wines.

N-Supplementation	Ranked Sum
DAP	341 a
LDH	257 b
HDH	306 c

Different letters indicate significant differences ($p < 0.05$).

4. Discussion

Overall, the YPH-supplemented fermentations performed similarly to the DAP-supplemented fermentations. However, differences were noted in the initial fermentation rate, where the HDH-supplemented fermentation was faster than the others. We inferred that shorter-length peptides were more rapidly assimilated by the yeasts. This finding agrees with previous studies on the availability and assimilation of nitrogen sources, especially with Kevvai et al. [19], where a preference for nitrogen assimilation from YPH was detected. For the substrate utilization, residual sugar concentrations were similar among the different nitrogen supplementations and did not present significant differences. Nevertheless, a more interesting and significant impact was observed in the production of higher alcohols and esters. Increased concentrations of higher alcohols and esters compounds were observed in wines produced with YPH-supplementation. The influence of the YPH-supplementations was observed in individual volatile compounds (Table 1) and in the results of PCA (Figures 5 and 6). These findings suggest a metabolic impact by the presence of peptides from organic nitrogen sources. These observations also agreed with previous studies on the effect of organic nitrogen sources on the volatile compounds of wine [13–17]. However, studies on the metabolic effects caused by the rapid-assimilated nitrogen sources are required to evidence the changes in the metabolic fluxes directed to the production of higher alcohols and esters. Some studies reported that higher alcohols are produced by amino acid metabolism [31]. Isoamyl alcohol, isobutyl alcohol, and 2-phenyl ethanol are produced from leucine, valine, and phenyl alanine, respectively. These higher alcohols were detected in the wine samples and presented significant differences between the YPH- and the DAP-supplemented wines. Among the esters, the group formed by ethyl isovalerate, ethyl 2-methyl butyrate, and ethyl isobutyrate display an odorant synergy that could contribute to a red fruit aroma in red wines [32,33]. From the group of volatile compounds presenting higher significant concentrations, the 3-methyl-1-pentanol and the ethyl-2-hexenoate obtained significant OAV. The olfactory descriptors associated with these compounds are fusel, cognac, wine, cocoa, fruit, and pungent for the higher alcohol, and sweet, fruity, and vegetable for the ester. The differences in aroma detected by the panelists could have been caused by the presence of these significant volatile compounds. The preference for the DAP-supplemented wine could be influenced by the higher astringency perceived in the YPH-supplemented wines.

The recycling and utilization of yeast protein hydrolysates as a nitrogen source for wine fermentation is justified by the observed increase in higher alcohols and esters. The impact on wine tasting needs to be evaluated by experts and by a more extended consumers' preference test.

5. Conclusions

Supplementation with yeast protein hydrolysate with a high degree of hydrolysis is an effective treatment to increase the yeast assimilation of amino acids from must. Additionally, it increases the production of esters and higher alcohols, possibly influencing the sensorial perception of wine aroma. Future studies can be directed to modulate the effect of the DH of yeasts proteins and the supplement dose to achieve a compromise between the production of volatile compounds and the consumers' perception.

Supplementary Materials: The following are available online at https://www.mdpi.com/article/10.3390/fermentation7040313/s1, Figure S1: Box charts for acids, sugars and alcohols obtained from FTIR analysis. Figure S2: Calibration curve for polyphenols analysis using gallic acid as standard. Excel file: complete list of all the higher alcohols and esters analyzed by sample.

Author Contributions: Conceptualization, P.R., C.U., A.U. and P.V.; methodology, P.R., D.L., F.I., C.U., A.U. and P.V.; software, P.R.; validation, P.R., D.L., F.I., C.U., W.F., A.U. and P.V.; formal analysis, P.R. and P.V.; investigation, P.R., D.L., F.I., C.U., W.F., A.U. and P.V.; resources, P.R. and P.V.; data curation, P.R. and P.V.; writing—original draft preparation, P.R. and P.V.; writing—review and editing, P.R., D.L., F.I., C.U., W.F., A.U. and P.V.; visualization, P.R. and P.V.; supervision, W.F., A.U. and P.V.; project administration, P.R.; funding acquisition, P.R. and P.V. All authors have read and agreed to the published version of the manuscript.

Funding: This research was funded by ANID FONDEF/CONICYT VIU16P0003 and VIU16E0003.

Institutional Review Board Statement: Not applicable.

Informed Consent Statement: Informed consent was obtained from all subjects involved in the study.

Data Availability Statement: The study did not report any data.

Acknowledgments: Authors thank Concha y Toro, Villaseñor and Indomita Vineyards for supplying grapes and lees used in this research.

Conflicts of Interest: The authors declare no conflict of interest.

References

1. OIV. International Organization of Vine and Wine. *2020 Wine Production—OIV First Estimates*; OIV: Paris, France, 2020.
2. Spigno, G.; Marinoni, L.; Garrido, G.D. 1—State of the Art in Grape Processing By-Products. In *Handbook of Grape Processing By-Products*; Galanakis, C.M., Ed.; Academic Press: Cambridge, MA, USA, 2017; pp. 1–27.
3. Chae, H.J.; Joo, H.; In, M.J. Utilization of brewer's yeast cells for the production of food-grade yeast extract. Part 1: Effects of different enzymatic treatments on solid and protein recovery and flavor characteristics. *Bioresour. Technol.* **2001**, *76*, 253–258. [CrossRef]
4. Pozo-Bayón, M.Á.; Andújar-Ortiz, I.; Moreno-Arribas, M.V. Scientific evidences beyond the application of inactive dry yeast preparations in winemaking. *Food Res. Int.* **2009**, *42*, 754–761. [CrossRef]
5. Kristinsson, H.G.; Rasco, B.A. Fish protein hydrolysates: Production, biochemical, and functional properties. *Crit. Rev. Food Sci. Nutr.* **2000**, *40*, 43–81. [CrossRef] [PubMed]
6. Pérez-Bibbins, B.; Torrado-Agrasar, A.; Salgado, J.M.; Oliveira, R.P.; Domínguez, J.M. Potential of lees from wine, beer and cider manufacturing as a source of economic nutrients: An overview. *Waste Manag.* **2015**, *40*, 72–81. [CrossRef]
7. Crépin, L.; Truong, N.M.; Bloem, A.; Sanchez, I.; Dequin, S.; Camarasa, C. Management of Multiple Nitrogen Sources during Wine Fermentation by Saccharomyces cerevisiae. *Appl. Environ. Microbiol.* **2017**, *83*, 2616–2617. [CrossRef] [PubMed]
8. Waterhouse, A.L.; Sacks, G.L.; Jeffery, D.W. *Understanding Wine Chemistry*, 1st ed.; Wiley: Hoboken, NJ, USA, 2016.
9. Henschke, P.A.; Jiranek, V. Yeasts-metabolism of nitrogen compounds. In *Wine Microbiology and Biotechnology*; Fleet, G.H., Ed.; Harwood Academic Publishers GmbH: Amsterdam, The Netherlands, 1993; pp. 77–164.
10. Crépin, L.; Nidelet, T.; Sanchez, I.; Dequin, S.; Camarasa, C. Sequential use of nitrogen compounds by Saccharomyces cerevisiae during wine fermentation: A model based on kinetic and regulation characteristics of nitrogen permeases. *Appl. Environ. Microbiol.* **2012**, *78*, 8102–8111. [CrossRef] [PubMed]
11. Gobert, A.; Tourdot-Maréchal, R.; Sparrow, C.; Morge, C.; Alexandre, H. Influence of nitrogen status in wine alcoholic fermentation. *Food Microbiol.* **2019**, *83*, 71–85. [CrossRef] [PubMed]
12. Hazelwood, L.A.; Daran, J.-M.; van Maris, A.J.A.; Pronk, J.T.; Dickinson, J.R. The Ehrlich pathway for fusel alcohol production: A century of research on Saccharomyces cerevisiae metabolism. *Appl. Environ. Microbiol.* **2008**, *74*, 2259–2266. [CrossRef] [PubMed]
13. Seguinot, P.; Rollero, S.; Sanchez, I.; Sablayrolles, J.-M.; Ortiz-Julien, A.; Camarasa, C.; Mouret, J.-R. Impact of the timing and the nature of nitrogen additions on the production kinetics of fermentative aromas by Saccharomyces cerevisiae during winemaking fermentation in synthetic media. *Food Microbiol.* **2018**, *76*, 29–39. [CrossRef]
14. Bell, S.J.; Henschke, P.A. Implications of nitrogen nutrition for grapes, fermentation and wine. *Aust. J. Grape Wine Res.* **2005**, *11*, 242–295. [CrossRef]
15. Jiménez-Martí, E.; Aranda, A.; Mendes-Ferreira, A.; Mendes-Faia, A.; del Olmo, M.L. The nature of the nitrogen source added to nitrogen depleted vinifications conducted by a Saccharomyces cerevisiae strain in synthetic must affects gene expression and the levels of several volatile compounds. *Antonie van Leeuwenhoek* **2007**, *92*, 61–75. [CrossRef] [PubMed]
16. Garde-Cerdán, T.; Ancín-Azpilicueta, C. Effect of the addition of different quantities of amino acids to nitrogen-deficient must on the formation of esters, alcohols, and acids during wine alcoholic fermentation. *LWT—Food Sci. Technol.* **2008**, *41*, 501–510. [CrossRef]
17. Torrea, D.; Varela, C.; Ugliano, M.; Ancin-Azpilicueta, C.; Leigh Francis, I.; Henschke, P.A. Comparison of inorganic and organic nitrogen supplementation of grape juice—Effect on volatile composition and aroma profile of a Chardonnay wine fermented with Saccharomyces cerevisiae yeast. *Food Chem.* **2011**, *127*, 1072–1083. [CrossRef] [PubMed]
18. González-Marco, A.; Jiménez-Moreno, N.; Ancín-Azpilicueta, C. Influence of addition of yeast autolysate on the formation of amines in wine. *J. Sci. Food Agric.* **2006**, *86*, 2221–2227. [CrossRef]
19. Kevvai, K.; Kütt, M.-L.; Nisamedtinov, I.; Paalme, T. Simultaneous utilization of ammonia, free amino acids and peptides during fermentative growth of Saccharomyces cerevisiae. *J. Inst. Brew.* **2016**, *122*, 110–115. [CrossRef]
20. Mo, F.; Zhao, H.; Lei, H.; Zhao, M. Effects of nitrogen composition on fermentation performance of brewer's yeast and the absorption of peptides with different molecular weights. *Appl. Biochem. Biotechnol.* **2013**, *171*, 1339–1350. [CrossRef]
21. Zhou, Y.; Yang, H.; Zong, X.; Cui, C.; Mu, L.; Zhao, H. Effects of wheat gluten hydrolysates fractionated by different methods on the growth and fermentation performances of brewer's yeast under high gravity fermentation. *Int. J. Food Sci. Technol.* **2018**, *53*, 812–818. [CrossRef]
22. Yang, H.; Zong, X.; Xu, Y.; Zeng, Y.; Zhao, H. Improvement of Multiple-Stress Tolerance and Ethanol Production in Yeast during Very-High-Gravity Fermentation by Supplementation of Wheat-Gluten Hydrolysates and Their Ultrafiltration Fractions. *J. Agric. Food Chem.* **2018**, *66*, 10233–10241. [CrossRef]

23. Yang, H.; Zong, X.; Cui, C.; Mu, L.; Zhao, H. Wheat gluten hydrolysates separated by macroporous resins enhance the stress tolerance in brewer's yeast. *Food Chem.* **2018**, *268*, 162–170. [CrossRef]
24. Yang, H.; Zong, X.; Xu, Y.; Zeng, Y.; Zhao, H. Wheat gluten hydrolysates and their fractions improve multiple stress tolerance and ethanol fermentation performances of yeast during very high-gravity fermentation. *Ind. Crop. Prod.* **2019**, *128*, 282–289. [CrossRef]
25. Zhao, H.; Wan, C.; Zhao, M.; Lei, H.; Mo, F. Effects of soy protein hydrolysates on the growth and fermentation performances of brewer's yeast. *Int. J. Food Sci. Technol.* **2014**, *49*, 2015–2022. [CrossRef]
26. Li, T.; Wu, C.; Liao, J.; Jiang, T.; Xu, H.; Lei, H. Application of Protein Hydrolysates from Defatted Walnut Meal in High-Gravity Brewing to Improve Fermentation Performance of Lager Yeast. *Appl. Biochem. Biotechnol.* **2020**, *190*, 360–372. [CrossRef]
27. Valencia, P.; Pinto, M.; Almonacid, S. Identification of the key mechanisms involved in the hydrolysis of fish protein by Alcalase. *Process. Biochem.* **2014**, *49*, 258–264. [CrossRef]
28. Gump, B.H.; Zoecklein, B.W.; Fugelsang, K.C.; Whiton, R.S. Comparison of Analytical Methods for Prediction of Prefermentation Nutritional Status of Grape Juice. *Am. J. Enol. Vitic.* **2002**, *53*, 325–329.
29. Blainski, A.; Lopes, G.C.; de Mello, J.C. Application and analysis of the folin ciocalteu method for the determination of the total phenolic content from *Limonium brasiliense* L. *Molecules* **2013**, *18*, 6852–6865. [CrossRef] [PubMed]
30. Meilgaard, M.C.; Civille, G.V.; Carr, B.T. *Sensory Evaluation Techniques*, 4th ed.; CRC Press: Boca Raton, FL, USA, 2007.
31. Waterhouse, A.L.; Sacks, G.L.; Jeffery, D.W. Higher Alcohols. In *Understanding Wine Chemistry*; Waterhouse, A.L., Sacks, G.L., Jeffery, D.W., Eds.; John Wiley & Sons, Ltd.: Hoboken, NJ, USA, 2016; pp. 51–56.
32. Ferreira, V.; Escudero, A.; Campo, E.; Cacho, J. The chemical foundations of wine aroma: A role game aiming at wine quality, personality and varietal expression. In Proceedings of the Thirteenth Australian Wine Industry Technical Conference, Adalaide, Australia, 28 July–2 August 2007; pp. 1–9.
33. Moreno-Arribas, M.V.; Polo, C. *Wine Chemistry and Biochemistry*; Moreno-Arribas, M.V., Polo, C., Eds.; Springer: New York, NY, USA, 2009.

 *fermentation*

Article

Sensory Characteristics of Two Kinds of Alcoholic Beverages Produced with Spent Coffee Grounds Extract Based on Electronic Senses and HS-SPME-GC-MS Analyses

Lu Wang [1,2], Xu Yang [1], Zhuoting Li [1], Xue Lin [1,2,*], Xiaoping Hu [1,2], Sixin Liu [3] and Congfa Li [1,2,*]

1 College of Food Science and Engineering, Hainan University, Haikou 570228, China; lwang@hainanu.edu.cn (L.W.); xu.yang@cansinotech.com (X.Y.); huanyuanjiang@hainanu.edu.cn (Z.L.); huxiaoping03@hainanu.edu.cn (X.H.)
2 Key Laboratory of Food Nutrition and Functional Food of Hainan Province, Haikou 570228, China
3 College of Sciences, Hainan University, Haikou 570228, China; 990495@hainanu.edu.cn
* Correspondence: 993048@hainanu.edu.cn (X.L.); congfa@hainanu.edu.cn (C.L.); Tel./Fax: +86-898-66278326 (X.L. & C.L.)

Abstract: In this work, the hydrothermal extract of spent coffee grounds (SCG) was used to make alcoholic beverages with commercial *S. cerevisiae* strain D254. The sensory characteristics of the SCG alcoholic beverages were analyzed using sensory description, electronic nose, electronic tongue, and gas chromatography-mass spectrometry (GC-MS). The results suggested that the supplement of 0.20% $(NH_4)_2HPO_4$ was effective at improving growth and alcohol fermentation of *Saccharomyces cerevisiae* D254 in SCG extract. SCG fermented beverages (SFB) and SCG distilled spirits (SDS) produced at the optimized fermentation conditions had appropriate physicochemical properties and different sensory characteristics. Fermentation aromas, especially esters, were produced in SFB, increasing the complexity of aroma and lowing the irritating aroma. The combination of original and fermentation components might balance the outstanding sourness, astringency, and saltiness tastes of SFB. The fermentation aroma was partially lost and the sourness, bitterness, astringency, and saltiness tastes were relieved in distillation, leading to the relatively more prominent aroma typicality of coffee and a soft taste. These findings lay a foundation for producing new high-quality coffee-flavored alcoholic beverages or flavoring liquors.

Keywords: spend coffee grounds; fermentation; sensory property; volatile profile

Citation: Wang, L.; Yang, X.; Li, Z.; Lin, X.; Hu, X.; Liu, S.; Li, C. Sensory Characteristics of Two Kinds of Alcoholic Beverages Produced with Spent Coffee Grounds Extract Based on Electronic Senses and HS-SPME-GC-MS Analyses. *Fermentation* **2021**, *7*, 254. https://doi.org/10.3390/fermentation7040254

Academic Editors: Bin Tian and Jicheng Zhan

Received: 10 September 2021
Accepted: 28 October 2021
Published: 1 November 2021

Publisher's Note: MDPI stays neutral with regard to jurisdictional claims in published maps and institutional affiliations.

Copyright: © 2021 by the authors. Licensee MDPI, Basel, Switzerland. This article is an open access article distributed under the terms and conditions of the Creative Commons Attribution (CC BY) license (https://creativecommons.org/licenses/by/4.0/).

1. Introduction

Coffee is one of the most popular beverages worldwide because of its unique flavor and positive effects on health [1]. It has been consumed for over 1000 years and the demand of coffee production is growing steadily [2]. Nevertheless, spent coffee grounds (SCG) are inevitably produced during the processing of instant coffee [3,4]. Producing 1 kg of instant coffee could yield 0.9 kg of by-product SCG simultaneously [4,5]. As the main coffee industry residue, the vast majority of SCG are unconsumed and disposed of into the environment, leading to serious environmental pollution [6]. The reuse of SCG is thus of huge importance from environmental and economic viewpoints.

Resource utilization of SCG is currently of increasing concern. To improve the added value of SCG, some scholars and enterprises try to use SCG in many ways, such as being processed to feed and fertilizer [7], extraction of oil [8], obtaining polyphenols and dietary fiber from SCG [9,10], or using them for fuel production and alcoholic beverages making [11–14]. In South America and Vietnam, people have a tradition of drinking alcoholic beverages of coffee. Coffee-flavored alcoholic beverages, which can be drunk directly or used to prepare cocktails, are generally produced by the mixture of distilled liquor (such as edible alcohols, brandy, rum, and rice wines), coffee, sugar, cream, and other ingredients on the market. This method is relatively convenient for preparing coffee-flavored alcoholic beverages.

However, the fusion of aromas is not satisfactory enough in this type of alcoholic beverage and the flavor is not typical or full.

SCG are rich in high value molecules and organic compounds and might retain the typical flavor and taste of coffee [14,15], making it a source of brewing potential for new coffee-flavored alcoholic beverages or flavoring liquors. Liu et al. [16–18] develops SCG fermented beverages (SFB) using different inoculation strategies and analyzes the changes of strains growth and chemical compositions in fermentation. Sampaio et al. [11] produces SCG distilled spirits (SDS) using SCG extract and analyzes the sensory profile. Machado et al. [14] produces SFB and SDS using SCG extract and also conducts a sensory evaluation. Sensory attributes are one of the most important factors affecting alcoholic beverages' quality and consumer preference. At present, electronic sensing technologies are widely used in the flavor evaluation of fermented food. The amount of research reporting on the combination of electronic nose/electronic tongue, gas chromatography-mass spectrometry (GC-MS), and sensory description to evaluate the sensory characteristics of food is increasing [19–21]. However, the sensory evaluation of SCG alcoholic beverages by integrating the above-mentioned methods has not been reported.

In this study, to further characterize the fermentation potential of SCG extract and enrich the understanding of the sensory quality of the two kinds of SCG extract products SFB and SDS, the sensory characteristics of SFB and SDS produced at the optimized fermentation conditions were evaluated using sensory description, electronic nose, electronic tongue, and headspace solid-phase microextraction (HS-SPME) combined with GC-MS.

2. Materials and Methods

2.1. Raw Material and Yeast Strain

SCG used in this study were the by-products created during the processing of Arabica green coffee beans to instant coffee, in which countercurrent continuous extraction was conducted, and were provided by Hainan Lisun investment holding Co., Ltd. (Haikou, China). SCG were dried at 68 °C to a moisture content of about 10% and refrigerated for later use. The commercial *S. cerevisiae* strain D254 was purchased from LALVIN (Fredericia, Denmark).

2.2. Fermentation Process

The fermentation process of SCG alcoholic beverages is shown in Figure S1. SCG and distilled water were mixed in a proportion of 1:5 (g/mL) and heated at 95 °C for 45 min. After cooling and filtering with 500 mesh filter cloth, the SCG extract was obtained (2.5°Brix, pH 5.26). According to Sampaio et al. [11] and Machado et al. [14], 200 mg/L sodium metabisulfite was used to avoid bacterial contamination.

For the analysis of fermentation blocked in the SCG extract, the following media were used: (1) SCG extracts were mixed with 20% sucrose and pH was adjusted to 4.0 by HCl, obtaining the SCE medium, (2) the modified RAE medium contained 20% sucrose, 1% yeast extract, 1% peptone, 0.3381% $Na_2HPO_4 \cdot 12H_2O$, and 0.151% citric acid and pH was adjusted to 4.0 by HCl, (3) water in the RAE medium was replaced by SCG extract and other compositions remained unchanged, obtaining the RS medium, (4) $Na_2HPO_4 \cdot 12H_2O$ was removed from the RS medium to obtain the RS-P medium, and (5) yeast extract and peptone were removed from the RS medium to obtain the RS-N medium.

To ensure the smooth fermentation of SCG alcoholic beverages, the optimization of nutrition was conducted at the following condition: SCG extract was mixed with 20% sucrose and pH was adjusted to 4.0 by citric acid (equivalent to 2.54 ± 0.09 g/L citric acid in SCG extract) according to Sharma et al. [22] and Hlaing and Joshy [23]. Natively, 2.45 ± 0.03 g/L of citric acid exists in the SCG extract used in this work. Yu et al. [24] reports that citric acid has no obvious effects on the cell count, mortality rate, and alcohol fermentation ability of *Saccharomyces cerevisiae* at a concentration below 19.0 g/L (equivalent to pH 2.51) in apple wine making. Citric acid has no obvious effects on the volume of *S. cerevisiae* cells at a concentration below 5.4 g/L (equivalent to pH 2.97) in the tested conditions. However, citric acid remarkably affects the growth and alcohol fermentation of *S. cerevisiae* at high concentrations (32.8 g/L, 47.1 g/L, and 61.5 g/L, equivalent to pH 2.31, 2.17, and 2.03, respectively). Wang et al. [25] finds similar results using the basic medium YEPD (20 g/L glucose, 20 g/L peptone, and 10 g/L yeast extract). Nitrogen sources were supplemented by NH_4Cl, $(NH_4)_2SO_4$, or $(NH_4)_2HPO_4$ at a concentration of 0.05%, 0.10%, 0.20%, 0.40%, 0.80%, 1.60%, and 2.00%, successively. Then, 4% activated (equivalent to about 5.0 Log CFU/mL cells, hydration in 5% of glucose solution at 37 °C for 30 min) commercial yeast strains were added to the media, and the main fermentation was conducted at 25 °C for 9 days.

After the main fermentation (in which time alcohol fermentation basically finished and the difference of CO_2 production in two consecutive days was below 0.5 g/L) at the optimal condition, the post fermentation (during which time a small amount of residual sugar was used and SCG alcoholic beverages tended to be biologically stable) was conducted at 15 °C for 30 days.

SCG distilled spirits were obtained using the traditional secondary distillation method. The fresh SCG fermented beverages were distilled in a pre-concentration step to obtain the primary distillate product at an ethanol content of about 25%, and the head part (1.5% of the fresh sample) was removed. Then, the primary distillate product was distilled below 85 °C in the second distillation. 1.5% of the distillate was removed. The heart distillate was collected at an ethanol content above 40% and the ethanol content of SCG distilled spirits was adjusted to 50% using ultrapure water. The fermentations were conducted three times.

2.3. Analysis of Yeast Growth

The cell density was measured at 30 °C using a UV spectrophotometer (T6, Persee, Beijing, China). The yeast cells were centrifuged at 1500× *g* for 15 min and washed two times with 4 °C sterile water. Then, yeast cells were dried at 105 °C to constant weight. The final biomass yield was determined by the gram of dry cells per liter fermentation broth.

2.4. Analysis of CO_2 Production

CO_2 weight loss was measured using an analytical balance (EL602, Mettler Toledo, Shanghai, China). The CO_2 weight loss rate was expressed as the gram of CO_2 weight loss production per liter of fermentation broth per fermentation time.

2.5. Analysis of Physicochemical Properties

Residual sugar was determined by titration with methylene blue indicator. An appropriate glucose standard solution (2.5 g/L) mixed with 5 mL Fehling solutionI, 5 mL Fehling solutionII, and 50 mL distilled water. After boiling for 2 min, two drops of methylene blue indicator were added and the glucose standard solution was used to titrate. The total volume of glucose standard solution consumed (V) was recorded when the reaction solution was colorless. The volume of glucose standard solution consumed (V′) was recorded when the appropriate samples replaced the glucose standard solution in the first step of above-mentioned method. The gram of glucose that was equivalent to that in 5 mL Fehling solutionI and II(F) = grams of anhydrous glucose used × V/1000. The residual sugar of sample = F × 1000/(volume of sample used/volume of sample used after dilution) × V′. The results were expressed as glucose. Total acidity was determined by direct titration of organic acids in samples with 0.05 mol/L NaOH solution. A pH of 8.2 was used as the end point of potentiometric titration. The content of total acid of sample = 0.05 × 75 × (volume of NaOH solution for titrating the sample-volume of NaOH solution for titrating the control group)/volume of sample used. The results were expressed as tartaric acid. Ethanol content was obtained by distillation combined with a portable densimeter (DA-130N, KEM, Tokyo, Japan). pH was determined by using a pH meter (SevenEasy S20, METTLER TOLEDO, Zurich, Switzerland). Total esters were determined by titration with 0.1 mol/L sulfuric acid solution. An appropriate 0.1 mol/L NaOH solution was used to neutralize the free acid in 50 mL samples. Two drops of phenolphthalein were used as the reaction indicator. After adding 25 mL of 0.1 mol/L NaOH solution, esters saponification was conducted by heating reflux on a boiling water bath for 30 min. Then, 0.1 mol/L sulfuric acid solution was used to titrate the sample. The content of total ester of sample = 0.01 × 88 × (volume of sulfuric acid solution for titrating the sample-volume of sulfuric acid solution for titrating the control group)/volume of sample used. The results were expressed as ethyl acetate.

2.6. Analysis of Volatile Profile

HS-SPME-GC-MS was used to extract and analyse the volatiles. The SPME fiber (75 μm CAR/PDMS, Supelco, PA, USA) was pretreated in the GC injector at 250 °C for 30 min with 1.0 mL/min flow rate of carrier gas, and 5 g samples were sealed in a 20 mL headspace vial and stirred at 400 r/min. After equilibrating in a 50 °C water bath for 10 min, the aged SPME fiber was suspended 1.6 cm above the sample for 40 min and desorbed in the GC injector at 250 °C for 5 min.

Compound analysis was performed using an Agilent 7890A GC (Agilent, Santa Clara, CA, USA) coupled with an Agilent 5975C mass spectrometer system equipped with a DB-WAX capillary column (60 m × 0.25 mm × 0.25 μm, Agilent, Santa Clara, CA, USA). Helium (purity 99.999%) was used as the carrier gas at 2.0 mL/min. The oven was initially held at 40 °C for 6 min, subsequently raised to 100 °C at a rate of 3 °C/min, and finally raised to 230 °C at a rate of 5 °C/min for 10 min. The electron ionization energy of the mass-selective detector was 70 eV. The chromatogram was recorded by monitoring the total ion currents in the m/z range of 30 to 450 atomic mass units.

The volatile compounds were identified by comparing retention indices and retention times with those obtained for authentic standards, or those of literature data, or with mass spectra in the NIST/WILEY Database. The relative quantification of volatile compounds was performed by peak area method, and 20 μL 2-Methoxy-d3-phenol-3,4,5,6-d4 (410 mg/L) was used as an internal standard.

2.7. Electronic Nose Analysis

The Germany Airsense PEN 3 portable electronic nose system (Schwerin, Mecklenburg-Vorpommern) was used to estimate the aroma profile of SCG alcoholic beverages. The performance of ten metal oxide semiconductor sensors is listed in Table S1.

A 10 mL sample (diluted to 10% vol ethanol with ultrapure water) was sealed in a sample vial for 60 min for data collection. The headspace sampling method was used for electronic nose detection. Detection conditions were set as follows: the sensor cleaning time was 170 s, the auto-zero time was 10 s, the sample preparation time was 5 s, the injection rate of gas and carrier gas speed were 400 mL/min, and the detection time was 120 s. Samples were analyzed using WinMuster statistical analysis software self-contained with PEN 3 and the sensor response values at 98 s were used to establish the radar chart.

2.8. Electronic Tongue Analysis

Electronic tongue analysis was performed on an SA402B system (Insent, Kanagawa, Japan) equipped with seven sensors and reference electrodes, which could analyze bitterness, sourness, saltiness, sweetness, astringency, and bitter anion tastes. The composition of electrode and standard solutions is shown in Table S2. The ethanol concentration of samples was diluted to 10% vol with ultrapure water and the supernatants were used for analysis.

2.9. Sensory Evaluation

SFB and SDS were evaluated by a well-trained panel of 11 members (6 men and 5 women, ranging from 20 to 55 years old). Then, 20 mL of samples was presented to the panelists at 21 ± 1 °C. The sensory scoring criteria and rules were set based on those of dry grape wine, taking into account the sensory characteristics of coffee. The specific sensory scoring rules of SCG fermented beverages were shown in Table S3.

2.10. Statistical Data Analysis

Results were shown as averages of at least three independent experiments ± SD. Statistically significant differences ($p < 0.05$) were determined using one-way ANOVA and Tukey's honestly significant difference (HSD) tests.

3. Results and Discussion

3.1. Analysis of Fermentation Blocked in SCG Extract

In our previous study, the fermentation capacity of four yeast strains, *S. cerevisiae* D254, *S. bayanus* BV818, *S. bayanus* DV10, and *S. bayanus* EC1118, commonly used in fruit wine industry, was evaluated in SCG extract [26]. Surprisingly, the four yeast strains exhibited inferior capacities of producing CO_2 and ethanol in SCE (Table S4, [26]). We speculated that this problem may be related to insufficient nitrogen and phosphorus sources and the existence of a toxic component, such as coffee polyphenols, in the SCG extract. In previous studies, Liu et al. [16–18] comprehensively analyzed the chemical composition of SCG hydrolysates before and after the microbial transformation, and found the insufficient assimilable nitrogen of unfermented SCG hydrolysates. As shown in Table S5, regarding the chemical parameters of unfermented SCG extract used in this work, a similar phenomenon was observed, revealing the necessity of supplementing with a nitrogen source in SCG extract fermentation.

In this study, firstly, the reasons for the weak fermentation of yeast strains in the SCG substrate were analyzed using the "medium components addition/subtraction" method. The strain *S. cerevisiae* D254, which was relatively more suitable for SCG extract fermentation in terms of ethanol content and sensory characteristics, was used in this work. As shown in Figure 1a, compared with that in SCE, the growth of *S. cerevisiae* D254 improved obviously in RAE, RS, and RS-P media. The biomass yield of *S. cerevisiae* D254 achieved or even exceeded 3.0 g/L in RAE, RS, and RS-P media, whereas only 1.4 g/L biomass yield was observed in SCE (Figure 1b). Simultaneously, the residual sugar and ethanol content in RAE, RS, and RS-P were close to the theoretical values (Figure 1c). However, similar slow growth and fermentation to those in SCE were displayed in the RS-N medium.

These results were corresponding to the analysis of chemical parameters of SCG extract and also indicated that an insufficient nitrogen source was the main factor of weak fermentation in SCG substrate, whereas SCG extract and phosphorus source were not the key limiting factors for yeast strains in SCG substrate.

Generally speaking, the lack of assimilatable nitrogen source is a common cause of slow and stagnant fermentation in yeast [27]. The level of nitrogen source could influence the genes expression of yeasts [28]. Yeast cells behave similar as when under carbon starvation at a low concentration of nitrogen, inhibiting their growth and metabolism [29]. In this work, the growth of *S. cerevisiae* D254 was slow with the lack of nitrogen sources; no rapid growth period and few biomass accumulations were observed under the same conditions. This result was consistent with that of Hernandez-Orte et al. [30], which showed that nitrogen source directly affected the biomass of yeast in the rapid growth period at the initial stage of fermentation. Insufficient biomass could lead to the slow fermentation and prolong the fermentation time.

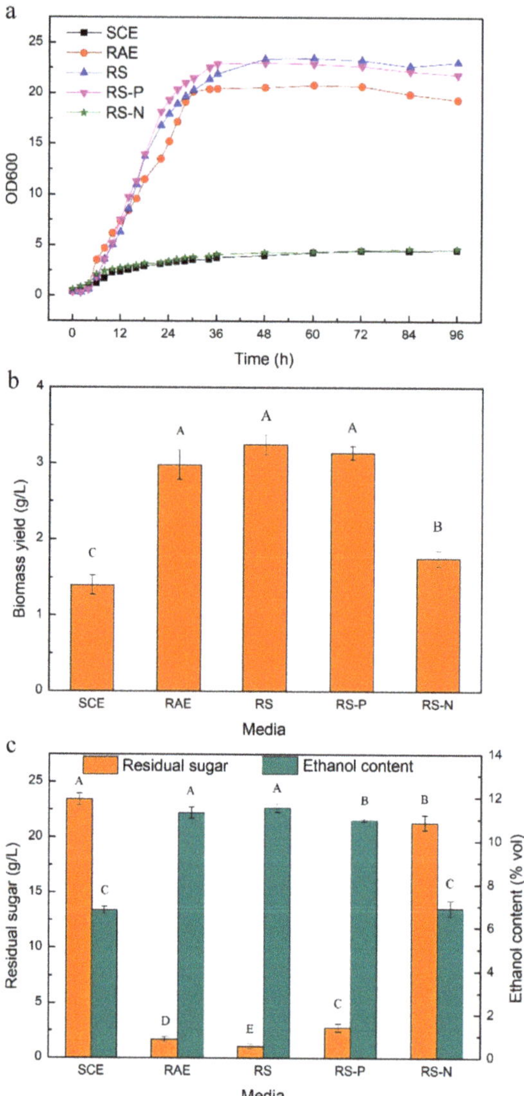

Figure 1. The growth and fermentation properties of *S. cerevisiae* D254 in different substrates. (**a**) The growth of D254 was monitored using OD600 and (**b**) the biomass yield was calculated using cell dry weight per liter medium. (**c**) The residual sugar of fermentation end point was tested. SCE medium: Spent coffee ground (SCG) extracts mixed with 20% sucrose and pH was adjusted to 4.0 by HCl. RAE medium: 20% sucrose, 1% yeast extract, 1% peptone, 0.3381% $Na_2HPO_4 \cdot 12H_2O$, and 0.151% citric acid mixed and pH was adjusted to 4.0 by HCl. RS medium: water in the RAE medium was replaced by SCG extract and other compositions remained unchanged. RS-P medium: $Na_2HPO_4 \cdot 12H_2O$ was removed from the RS medium. RS-N medium: yeast extract and peptone were removed from the RS medium. Values with different capital letters in the columns of the same color are significantly different ($p < 0.05$).

3.2. Optimization of Nitrogen Source for SFB

Considering the high quality, high efficiency, low price, and easy availability of inorganic ammonium salts, different concentrations of NH_4Cl, $(NH_4)_2SO_4$, and $(NH_4)_2HPO_4$ were used to try to improve alcohol fermentation [26]. As shown in Figure 2a, the addition of the three nitrogen sources performed different effects on the metabolism of *S. cerevisiae* D254. Overall, the growth and fermentation capacities enhanced first and then decreased with the increasing concentration of nitrogen sources. The highest biomass yield and ethanol content and the lowest residual sugar were observed at the addition of 0.20% of each nitrogen source, and the biomass yield and ethanol content in the addition of $(NH_4)_2HPO_4$ were higher than those seen with the addition of NH_4Cl and $(NH_4)_2SO_4$ at the same concentration (Figure S2, [26]).

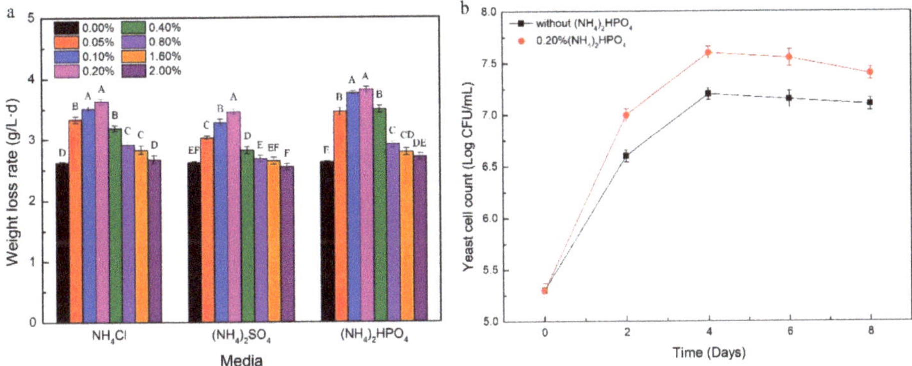

Figure 2. Effects of exogenous nitrogen sources on (**a**) CO_2 weight loss rate of *S. cerevisiae* D254 and (**b**) growth of *S. cerevisiae* D254 in SCG substrate. The YEPD agar plate was used for the viable count of the yeast cells. Values with different capital letters in the columns of the same color are significantly different ($p < 0.05$).

These results reflected that appropriate addition of nitrogen source was promising and essential to the growth and alcohol fermentation of *S. cerevisiae* in SCG extract. The addition of 0.20% $(NH_4)_2HPO_4$ could be used to improve the growth of *S. cerevisiae* D254 (Figure 2b) and solve the slow fermentation in SCG extract.

The addition of 0.20% $NH_4Cl/(NH_4)_2SO_4/(NH_4)_2HPO_4$ was beneficial for cell growth and CO_2 production. The decreased biomass yields at the nitrogen sources' levels above 0.20% might be caused by the negative effect of excess NH^{4+} on the utilization of amino acid in an amino acid starvation condition [31]. The low utilization of sugar under the same conditions might be attributed to the slow growth of yeast cells and the negative effect of excess nitrogen sources on the expression of genes related to the transport of ammonium salts/amino acids, such as *GAP1* and *MEP2* [28]. In addition, excessive nitrogen sources may form carcinogens urethane, biogenic amine, and nitrosamine [32,33], which could negatively affect the quality and safety of wines. $(NH_4)_2HPO_4$ could not only complement the nitrogen source but also provide phosphate ions, further improving the growth and metabolism of yeast cells.

Based on this result, other factors including the ratio of SCG-liquid, inoculation concentration, sugar degree, pH, and fermentation temperature were optimized by single-factor test (Figure S3). Using the content of residual sugar, total acidity, and ethanol and sensory score as evaluation indexes, the optimal main fermentation condition was set as follows: SCG-liquid ratio was 1:4; 0.2% $(NH_4)_2HPO_4$ was added; initial sugar level was 24% sucrose; pH was adjusted to 4.3 by citric acid (equivalent to 2.51 $\pm$ 0.02 g/L citric acid in SCG extract, and the content of citric acid kept at 2.51 $\pm$ 0.10 g/L when the fermentation finished); 6% activated *S. cerevisiae* strain D254 (equivalent to about 5.3 Log CFU/mL cells) was inoculated; and the main fermentation was conducted at 23 °C for 8 days.

3.3. Physicochemical and Sensory Characteristics of SCG Alcoholic Beverages

The contents of residual sugar, total acidity, ethanol, and total esters of SFB and SDS were analyzed at the optimal main fermentation condition. Furthermore, the sensory attributes including colour and lustre, aroma, and taste of SFB and SDS were described. As shown in Table 1, SFB and SDS exhibited suitable content of ethanol. The total acidity of SFB was 9.07 g/L, but that for SDS was only 0.37 g/L. The content of total esters of SDS decreased from 730.42 mg/L of SFB to 426.93 mg/L. For the sensory attributes, compared with SFB, SDS possessed a more typical coffee aroma without fruity aroma, but lacked the multi-level, typical taste of coffee.

Table 1. General physicochemical properties of SFB and SDS.

Indexes	SFB	SDS
Residual sugar (g/L)	6.27 ± 0.51	/
Ethanol (% vol)	12.5 ± 0.1	50.5 ± 0.2
Total acidity (g/L)	9.07 ± 0.43	0.37 ± 0.02
Total esters (mg/L)	730.42 ± 46.65	426.93 ± 44.28
Colour and lustre	Yellowish-brown, clear, and transparent	Light amber, clear, and transparent
Aroma	Typical coffee, sweet fruity, and good wine aromas	A typical and rich coffee aroma with a pure wine aroma
Taste	Full-bodied wine with a harmonious and palatable sour, sweet, and bitter taste	Mellow, soft, and elegant wine

/: not tested. Spent coffee grounds: namely SCG. SCG fermented beverages: namely SFB. SCG distilled spirits: namely SDS.

These results suggested that SFB and SDS had different sensory styles with distinguished physicochemical characteristics. Acids play a role in presenting flavor in wines [34]. The total acidity of SFB was higher than that of most fruit wines, in which the acidity content is commonly below 8 g/L. Meanwhile, the total acidity of most SFB samples possessing a high sensory score was higher than 8 g/L (Figure S3). The high content of acidity could balance the bitter odor of coffee, contributing to a harmonious and palatable taste of SFB. The acids in SFB include dicaffeoylquinic acid, caffeic acid, quinic acid, citric acid, tartaric acid, and acetic acid [35–37]. As the main acids of coffee, the former three ones hardly evaporate in distillation, resulting in the low acidity of SDS. Enough acids could provide more precursors of esters. Esters are one of the most important volatile compositions that contribute to wine flavor [38]. The high content of total esters of SFB could bring about a fruity aroma. Esters were inevitably lost in distillation. The content of total esters of SDS was lower than that of Baijiu, but this level was similar to that of some distilled wines, such as brandy [39]. Therefore, SCG alcoholic beverages have development potential in terms of aroma, flavor, and style.

3.4. Analysis of Electronic Nose and Electronic Tongue

3.4.1. Analysis of Electronic Nose

The electronic nose and electronic tongue were used to relatively objectively evaluate the aroma and taste characteristics of SCG alcoholic beverages, respectively. The response values of SCG extract and coffee liquid were similar in each sensor (Figure 3(a1)). Compared with SCG extract and coffee liquor, the response values of SFB and SDS in sensors S2, S6, S7, S8, and S9 were remarkably increased. PCA was used to study the similarities and differences of SCG extract, coffee, and SCG alcoholic beverages. The locations of SFB and SDS partially overlapped and separated from SCG extract and coffee, and the two groups overlapped in the second quadrant (Figure 3(a2)).

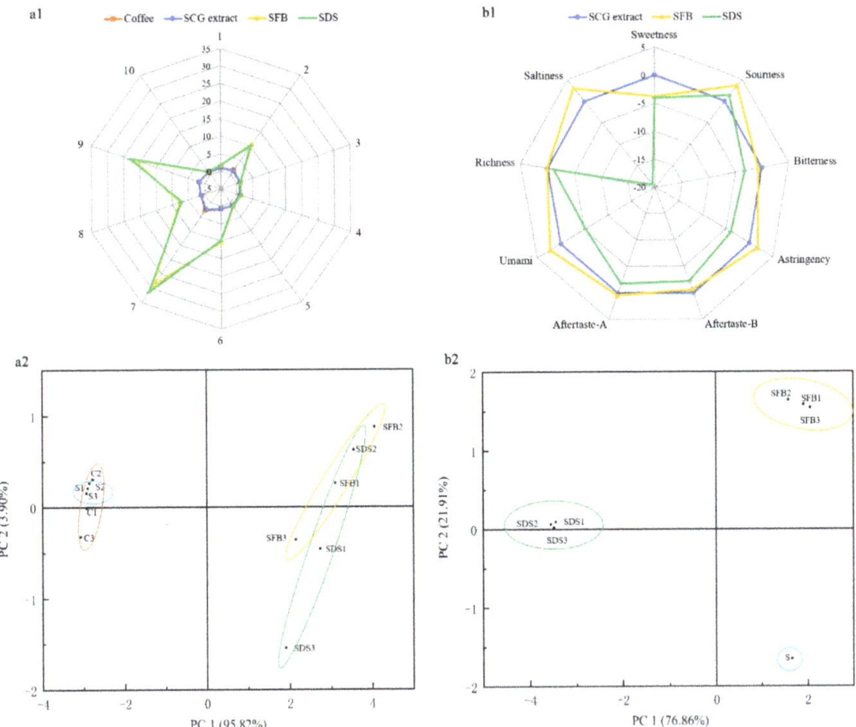

Figure 3. The analysis of the (**a**) electronic nose and (**b**) electronic tongue. (**a1**): Radar graph of samples' aroma, (**b1**): Radar graph of samples' taste. (**a2**,**b2**): PCA of samples. C: Coffee. S: SCG extract. SFB: SCG fermented beverages. SDS: SCG distilled spirits.

These results indicated that the SCG extract well retained the aroma of coffee and had the value for making coffee-flavored products. Fermentation and distillation affected the aroma profile of SCG extract and coffee. The difference between SCG alcoholic beverages and SCG extract/coffee on aromas was probably caused by the increase of nitrogen oxides, sulfides, alcohols, and other substances during fermentation.

3.4.2. Analysis of Electronic Tongue

Compared with SCG extract, SFB exhibited strong sourness, astringency, umami, and saltiness tastes and astringency aftertaste with thin sweetness and bitterness tastes and bitter aftertaste. A similar richness taste to SCG extract was observed in SFB. However, the response to eight taste attributes, excluding sourness, in SDS was lower than those in SCG extract (Figure 3(b1)). SCG extract, SFB, and SDS were located in three different quadrants in the PCA result (Figure 3(b2)).

Compared with SDS, SFB had a prominent coffee taste with strong sourness and astringency feeling. Nevertheless, typical coffee flavor did not represent high consumer acceptance. Some consumers had low acceptance of pure coffee flavor. The final taste characteristic of alcoholic beverages depended on the balance and coordination of all taste attributes.

Sourness, sweetness, bitterness, and saltiness tastes are four main senses of coffee [40]. The sour and astringent tastes of coffee are largely come from a variety of organic acids, such as citric acid, malic acid, tartaric acid, chlorogenic acid, and its degradation product dicaffeoylquinic acid, in the baking process [35,40,41]. The content of organic acids increased during fermentation, and the solubility of dicaffeoylquinic acid in ethanol was higher than that in water [42]. Therefore, sour and astringent substances increased in SFB and the corresponding taste senses enhanced. Sugars of coffee can give a sweet taste, and the caramel compounds produced in the baking process can give sweet and bitter tastes. With the decomposition of sugars in fermentation, the sweetness and bitterness tastes reduced, which in turn made the sourness and saltiness tastes more prominent. Combined with the results of taste evaluation in Table 1, the findings suggested that such changes of taste of coffee in fermentation were favor to the mutual inhibition and balance of sourness, sweetness, bitterness, and saltiness senses in this study. Distillation was a process that selectively extracted the volatile flavor components of fermentation wines. Many compositions with strong water solubility and weak fat solubility could not be brought out with the distillation of ethanol. The prominent sour and astringent tastes in SFB reduced in distillation with less sweetness and saltiness senses, which were suitable for consumer groups with low acceptance of pure coffee. This finding was corresponded to the results of Machado et al. [14], in which SDS exhibited a high global value in quantitative descriptive analysis.

3.5. Volatile Profile of SCG Alcoholic Beverages

A total of 77 volatile compounds were identified in this study, including 15 esters, 8 alcohols, 1 acid, 8 aldehydes, 3 ketones, 4 phenols, 9 pyrazines, 16 furans, 7 terpenes, 3 pyrroles, 1 lactone, and 2 other compounds (Table 2). The most abundant variety but the lowest content of total volatile compositions was observed in SCG extract (Figure 4). The lowest variety of volatile compositions was observed in SDS with the moderate total content (Figure 4). SFB displayed the highest content of total volatiles with the moderate variety (Figure 4). These results suggested that the types and content of volatiles of SCG extract changed in fermentation and distillation. Partial original aromas of SCG extract were lost and converted in alcohol fermentation with the formation of a large amount of fermentation aromas. Both original and fermentation aromas were partially lost in distillation. The types and contents of volatiles differed from those reported in Machado et al. [14] and Liu et al. [16]. This discrepancy could be caused by the difference of extraction and test methods.

More than 800 aromatic components of coffee have been detected in previous studies, of which pyrazines, furans, aldehydes, and sulphur compounds are the key ones and contribute the most to coffee aroma [37,40]. Pyrazines have high contents and low odor thresholds, mainly giving nutty and baking odors [43]. Furans commonly contribute to balsamic, wheat, and baking odors [44]. Aldehydes giving fatty, spicy, sweety, and bread odors could confer a richer, more elegant, and unique aroma to wines [45]. Most sulphur compounds contribute to a negative sulphur aroma. The types and contents of typical aroma of coffee could be changed by different treatment conditions, such as baking temperature, extraction temperature, and extraction method, such as that of the extraction of phenolic/flavonoid compounds using different methods [46,47]. Only one sulphur compound was detected in this work. This result was not consistent to the strong response to sulfides in the electronic nose analysis. The detection method needs to be further optimized in future study. The above-mentioned four kinds of aromatic components accounted for 34.0%, 11.9%, and 32.2% in SCG extract, SFB, and SDS, respectively. A similar proportion of four key aromatic groups to SCG extract in SDS might bring the outstanding typicality of coffee aroma in Table 1 and Figure 4a1, though the fewest types of volatiles were observed in SDS. The proportion of original aroma in SFB decreased with increased fermentation aroma, producing a complex and rich aroma and more flavor levels of SFB.

Table 2. The concentration of volatile compounds identified in SCG extract, SFB, and SDS.

Code	Compounds	Odor Description *	Concentration (μg/L)		
			SCG Extract	SFB	SDS
	Esters				
E1	Ethyl acetate	Pineapple	7105 ± 30^{B}	36690 ± 64^{A}	5094 ± 22^{C}
E2	Ethyl propionate	Fruit	230 ± 4^{B}	936 ± 12^{A}	19 ± 2^{C}
E3	Ethyl 2-methylpropanoate	Sweet, rubber	-	-	33 ± 2
E4	2-Methyl-1-propyl acetate	Fruit, apple, banana	-	33 ± 3	-
E5	Ethyl butanoate	Apple	-	148 ± 5	-
E6	Ethyl 2-methylbutanoate	Apple	-	-	75 ± 3
E7	Ethyl 3-methylbutanoate	Fruit	-	4 ± 0.2	-
E8	Isoamyl acetate	Banana	46 ± 2^{C}	1346 ± 12^{A}	342 ± 5^{B}
E9	Ethyl hexanoate	Apple peel, fruit	63 ± 2^{C}	651 ± 7^{A}	152 ± 4^{B}
E10	Ethyl octanoate	Fruit, fat	172 ± 5^{C}	717 ± 12^{B}	878 ± 16^{A}
E11	Ethyl nonanoate		-	55 ± 3	11 ± 1
E12	Furfuryl acetate		-	2194 ± 22	-
E13	Ethyl decanoate	Grape	62 ± 2^{B}	257 ± 4^{A}	265 ± 4^{A}
E14	Ethyl butyrate	Apple	84 ± 4	-	-
E15	Phenethyl acetate	Rose, honey, tobacco	-	149 ± 2	-
	Alcohols				
A1	2-Methyl-1-butanol	Wine, onion	473 ± 7^{C}	7421 ± 22^{A}	3898 ± 15^{B}
A2	3-Methyl-1-butanol	Whiskey, malt, burnt	810 ± 21^{C}	11804 ± 32^{B}	12565 ± 43^{A}
A3	Isobutanol	Wine, solvent, bitter	78 ± 2^{C}	1480 ± 10^{A}	377 ± 4^{B}
A4	2-Heptanol	Mushroom	35 ± 4	-	-
A5	2,3-Butanediol	Fruit, onion	616 ± 11	-	-
A6	2-Furylmethanol	Burnt	854 ± 22^{B}	1789 ± 14^{A}	-
A7	Benzyl alcohol	Sweet, flower	88 ± 4^{A}	29 ± 1^{B}	-
A8	2-Phenylethyl alcohol	Honey, spice, rose, lilac	435 ± 9^{B}	10183 ± 34^{A}	-
	Acids				
AC1	Acetic acid	Sour	62 ± 4^{B}	1017 ± 23^{A}	58 ± 3^{B}
	Aldehydes				
Al1	Ethanal	Pungent, ether	-	617 ± 10^{B}	1792 ± 14^{A}
Al2	Isovaleraldehyde	Malt	25 ± 2	-	-
Al3	Acetaldehyde diethyl acetal		-	-	6395 ± 21
Al4	(Z)-4-heptenal	Biscuit, cream	42 ± 2	-	-
Al5	Octanal	Fat, soap, lemon, green	34 ± 2	-	-
Al6	Nonanal	Fat, citrus, green	289 ± 5	-	-
Al7	Furfural	Bread, almond, sweet	-	-	1027 ± 22
Al8	Phenylacetaldehyde	Hawthorne, honey, sweet	44 ± 3^{B}	-	1093 ± 13^{A}
	Ketones				
K1	2-Methylpentan-3-one	Mint	29 ± 2^{A}	21 ± 2^{B}	-
K2	2,3,4-Trimethyl-cyclopent-2-enone		137 ± 4	-	-
K3	Cyclopentanone		-	-	19 ± 1
	Phenols				
P1	Phenol	Phenol	322 ± 7^{A}	80 ± 3^{B}	-
P2	4-Ethyl-2-methoxyphenol	Spice, clove	1835 ± 20^{A}	1232 ± 16^{B}	89 ± 4^{C}
P3	2-Ethylphenol		61 ± 3	-	-
P4	4-Allyl-2-methoxyphenol	Clove, honey	249 ± 6^{A}	239 ± 6^{A}	-
	Pyrazines				
Py1	2-Methypyrazine	Popcorn	159 ± 8^{C}	1130 ± 21^{A}	221 ± 5^{B}
Py2	2,5-Dimethyl pyrazine	Cocoa, roasted nut, roast beef, medicine	60 ± 2^{B}	477 ± 13^{A}	-
Py3	2-Eethyl pyrazine	Peanut butter, wood	64 ± 5^{B}	192 ± 4^{A}	24 ± 2^{C}
Py4	2-Ethyl-6-methylpyrazine		-	-	12 ± 1
Py5	2-Ethyl-5-methyl pyrazine	Fruit, sweet	1901 ± 22^{A}	1117 ± 12^{B}	-
Py6	2-Methyl-3-ethylpyrazine	Roast	77 ± 5^{B}	424 ± 8^{A}	-
Py7	2-Ethyl-3,6-dimethylpyrazine	Potato, roast	1761 ± 15^{A}	1538 ± 26^{B}	-
Py8	3,5-Diethyl-2-methylpyrazine	Baked	340 ± 9^{B}	381 ± 8^{A}	-
Py9	2,3-Diethyl-5-methylpyrazine	Potato, meat, roast	313 ± 6^{A}	286 ± 6^{B}	-
	Furans				
F1	2-(2-Propenyl)furan		-	-	52 ± 2
F2	Furfuryl ethyl ether		-	1518 ± 16^{A}	408 ± 7^{B}
F3	2,5-Dimethylfuran		20 ± 1^{B}	11 ± 2^{C}	40 ± 1^{A}
F4	2-(2-Pentenyl)furan		-	-	225 ± 2
F5	5-Methyl-2-acetylfuran		70 ± 2	-	-
F6	2-Vinylfuran		269 ± 7^{A}	39 ± 1^{B}	38 ± 1^{B}
F7	5-Methyl-2-furanaldehyde	Almond, caramel, burnt sugar	-	-	156 ± 3
F8	2-Methyl-Furan		36 ± 2	-	-

Table 2. *Cont.*

Code	Compounds	Odor Description *	Concentration (μg/L)		
			SCG Extract	SFB	SDS
F9	Benzofuran		18 ± 1	-	-
F10	2-Acetylfuran	Balsamic	550 ± 6 ^{B}	2350 ± 21 ^{A}	152 ± 3 ^{C}
F11	5-Methyl-2-propenyl furan		105 ± 2 ^{A}	4 ± 1 ^{C}	27 ± 1 ^{B}
F12	Methyl Furfuryl Disulphide	Smoke	66 ± 3	-	-
F13	2-(2-Furfuryl)furan		768 ± 14 ^{A}	70 ± 3 ^{C}	309 ± 6 ^{B}
F14	5-Methyl-2-propiony furan		252 ± 5	-	-
F15	Difurfuryl ether		1015 ± 10 ^{A}	608 ± 9 ^{B}	-
F16	2-Furfuryl-5-methyl furan		532 ± 11 ^{A}	64 ± 1 ^{C}	82 ± 1 ^{B}
	Terpenes				
T1	β-Myrcene	Balsamic, must, spice	13 ± 1 ^{B}	-	99 ± 3 ^{A}
T2	DL-Limonene		80 ± 2 ^{B}	48 ± 3 ^{B}	396 ± 9 ^{A}
T3	Styrene	Balsamic, gasoline	538 ± 11 ^{B}	-	1030 ± 10 ^{A}
T4	cis-Linaloloxide (2-methyl-2-vinyl-5-(alpha hydroxyisopropyl) tetrahydrofuran)	Flower, wood	308 ± 3 ^{A}	213 ± 6 ^{B}	-
T5	Camphor	Camphor	199 ± 4	-	-
T6	Linalool		273 ± 3 ^{A}	254 ± 6 ^{B}	-
T7	β-Damascenone	Apple. rose, honey	173 ± 2 ^{A}	19 ± 1 ^{B}	-
	Pyrroles				
Pyr1	3-Methylpyrrole		53 ± 4 ^{B}	368 ±7 ^{A}	-
Pyr2	N-methyl-2-acetyl pyrrole		571 ± 14 ^{B}	808 ± 9 ^{A}	-
Pyr3	1-Furfurylpyrrole		893 ± 15	-	-
	Lactones				
L1	γ-Butyrolactone	Caramel, sweet	83 ± 2 ^{B}	167 ± 5 ^{A}	-
	Others				
O1	3,4-Diethylthiophene		-	-	19 ± 2
O2	4-Methylthiazole	Roasted meat	63 ± 4	-	-

Values shown represent the averages of triplicate samples (data are mean ± SD). Values with different superscript letters in the same row are significantly different ($p < 0.05$). -: not detected. Spent coffee grounds: namely SCG. SCG fermented beverages: namely SFB. SCG distilled spirits: namely SDS. * http://www.flavornet.org/flavornet.html (accessed on 22 October 2021).

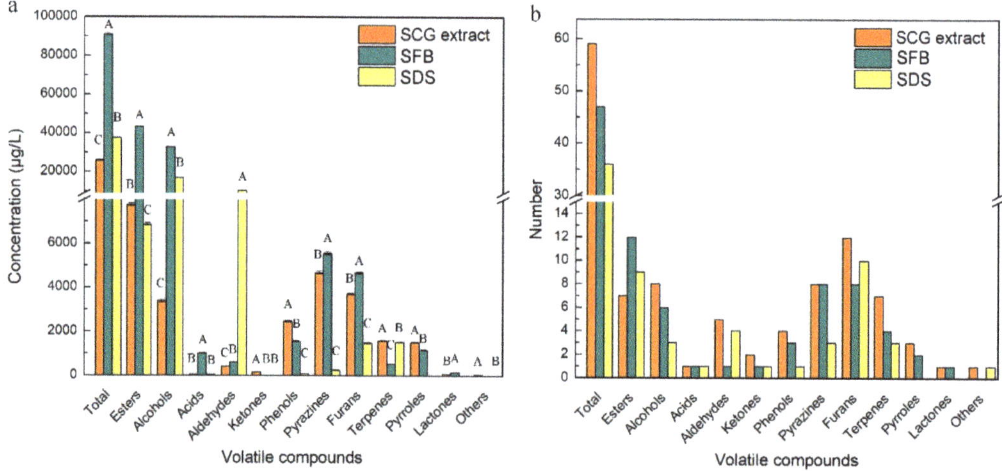

Figure 4. Variation in (**a**) concentration and (**b**) number of volatile compounds in samples. SFB: SCG fermented beverages. SDS: SCG distilled spirits. Values with different capital letters upon the columns of the same item are significantly different ($p < 0.05$).

Compared with SCG extract, 20 volatile compounds including 1 ester, 2 alcohols, 5 aldehydes, 1 ketone, 1 phenol, 5 furans, 3 terpenes, 1 pyrrole, and 1 thiazole lost in SFB, but those for SDS were 36 volatile compounds including 1 ester, 5 alcohols, 4 aldehydes, 2 ketones, 3 phenols, 6 pyrazines, 6 furans, 4 terpenes, 3 pyrroles, 1 lactone, and 1 thiazole.

Simultaneously, 8 volatile compounds, including 6 esters, 1 aldehyde, and 1 furan were added in SFB, compared with SCG extract, and those for SDS were 13 volatile compounds including 3 esters, 3 aldehydes, 1 ketone, 1 pyrazine, 4 furans, and 1 thiophene. These results reflected that the better aroma sense of SDS was not due to the higher retention of original aroma, but might be due to the few fermentation aromas. Two types of aromas balanced and the original aroma was uncovered. Aldehydes, furans, and terpenes were the first three decreasing groups in SFB in terms of the number of compounds, accounting from 1.7% to 0.7%, 14.3% to 5.1%, and 6.1% to 0.6%, respectively. Aldehydes produce in the early stage of alcohol fermentation and are subsequently reduced to alcohols and CO_2. Furans and terpenes can be formed via carbohydrate degradation during alcohol fermentation [48,49]. The loss of several furans and terpenes might convert to other types of volatile compositions, such as increased pyrroles converted from furans. The absence of several aldehydes, furans, and terpenes might affect the typicality of coffee aroma of SFB. However, the stimulating aroma might be decreased in SFB, generating a harmonious overall aroma.

Esters were the most noteworthy group in terms of the variation of number and content of volatile compounds. The number and proportion of esters in SFB increased from 7 to 12 and from 30.0% to 47.4%, respectively. The increased alcohol and acid provided sufficient precursors for the synthesis of esters in SFB. The increase of esters enhanced fruity aroma of SFB, which was consistent with the results of physicochemical and sensory evaluation (Table 1). The rich fruity flavor could alleviate the prominent sour and astringent tastes of SFB to a certain extent.

The odor activity values (OAVs) were used to further investigate the contribution of each compound to the flavor of SFB and SDS (Table 3). The OAVs of 28 volatile compounds were above 1 in this test. Amongst them, 17, 13, and 8 compounds were shown in SCG extract, SFB, and SDS, respectively. The largest number of esters possessing OAV > 1 was observed in SFB, thus contributing the rich fruity odor to SFB. As with SCG extract, several aldehydes of SDS exhibited OAV > 1. However, strong fatty, ether, and sweet odors were negative to wine flavor. Compared with SCG extract, the outstanding alcohols-, ketones-, and pyrroles-induced odors might disappear in SFB and SDS, but phenols- and terpenes-induced odors may remain with OAVs of 4-ethyl-2-methoxyphenol, 4-allyl-2-methoxyphenol, styrene, linalool, and β-damascenone >1. As one of the major groups of the typical flavors of coffee, furans might contribute the most in SFB with the OAV of furfuryl ethyl ether >1, and furans-induced odors might balance with the fruity odor induced by esters. The odor threshold values of many furan compounds did not be retrieved from literature. Therefore, the contribution of furans to SDS could not be ignored.

Table 3. OAVs (>1) for the main volatile compounds in SCG extract, SFB, and SDS.

Code	Compounds	Odor Description *	SCG Extract	SFB	SDS
	Esters				
E1	Ethyl acetate	Pineapple	1421.0	4.9	0.8
E2	Ethyl propionate	Fruit	255.6	20.8	/
E3	Ethyl 2-methylpropanoate	Sweet, rubber	-	-	7.3
E5	Ethyl butyrate	Apple	560.0	-	-
E7	Ethyl 3-methylbutanoate	Fruit	-	1.3	-
E8	Isoamyl acetate	Banana	0.2	44.9	6.2
E9	Ethyl hexanoate	Apple peel, fruit	12.6	46.5	0
E10	Ethyl octanoate	Fruit, fat	8.9	35.9	6.0
E13	Ethyl decanoate	Grape	12.4	1.3	0.2
E15	Phenethyl acetate	Rose, honey, tobacco	-	0.6	-
	Alcohols				
A1	2-Methyl-1-butanol	Wine, onion	29.7	0.2	0.1
A2	3-Methyl-1-butanol	Whiskey, malt, burnt	/	0.4	0.7
A4	2-Heptanol	Mushroom	0.5	-	-
	Aldehydes				
Al1	Ethanal	Pungent, ether	-	0.5	1629.1

Table 3. *Cont.*

Code	Compounds	Odor Description *	SCG Extract	SFB	SDS
Al3	Acetaldehyde diethyl acetal		-	-	6.4
Al4	(Z)-4-heptenal	Biscuit, cream	12.4	-	-
Al5	Octanal	Fat, soap, lemon, green	1.5	-	-
Al6	Nonanal	Fat, citrus, green	15.9	-	-
Al8	Phenylethanal	Hawthorne, honey, sweet	55.0	-	1.0
	Ketones				
K1	2-Methylpentan-3-one	Mint	5.8	/	-
	Phenols				
P2	4-Ethyl-2-methoxyphenol	Spice, clove	26.4	37.3	1.8
P4	4-Allyl-2-methoxyphenol	Clove, honey	1464.7	39.8	-
	Furans				
F2	Furfuryl ethyl ether		-	138.0	0.9
	Terpenes				
T3	Styrene	Balsamic, gasoline	/	-	4.1
T6	Linalool		2.7	10.2	-
T7	β-Damascenone	Apple. rose, honey	/	380.0	-
	Pyrroles				
Pyr3	1-Furfurylpyrrole		8.9	-	-
	Lactones				
L1	γ-Butyrolactone	Caramel, sweet	2.4	4.8	-

-: not detected./not calculated due to unreferenced odor threshold values. * http://www.flavornet.org/flavornet.html (accessed on 22 October 2021). OAVs were calculated by the ratio of the concentration of a compound to its odor threshold value. To calculate the OAVs of compounds in SCG extract, SFB, and SDS, odor threshold values in water, low content of alcohol (<15%), and high content of alcohol (>40%) were referred from literature, respectively.

4. Conclusions

The supplement of 0.20% $(NH_4)_2HPO_4$ was effective for improving growth and alcohol fermentation of *S. cerevisiae* D254 in SCG extract. SFB and SDS produced at the optimized fermentation conditions (SCG-liquid ratio 1:4; 24% sucrose; pH 4.3; 6% activated *S. cerevisiae* strain D254; 23 °C for 8 days) had appropriate physicochemical properties (12.5% and 50.5% ethanol, respectively; 9.07 g/L and 0.37 g/L total acidity, respectively) and different sensory characteristics, including aroma and taste senses. Fermentation aromas, especially esters, produced in SFB, increased the complexity of aroma and lowered the irritating aromas, such as nitrogen oxides-induced ones. The combination of original and fermentation aromatic components could balance the outstanding sourness, astringency, and saltiness of SFB. Fermentation aroma was partially lost (e.g., the total contents of esters decreased by 84%, compared with SFB) and the sourness, bitterness, astringency, and saltiness tastes were relieved in distillation according to the electronic tongue system, leading to the relatively more prominent aroma typicality of coffee and a soft taste.

Supplementary Materials: The following are available online at https://www.mdpi.com/article/10.3390/fermentation7040254/s1, Figure S1: The brewing process of SCG beverages, Figure S2: Effects of types and concentrations of nitrogen source on **a** biomass yield of D254, **b** residual sugar, and c ethanol content in SCG substrate, Figure S3: Effects of **a** the ratio of SCG **b** the inoculation amount of *S. cerevisiae* D254 **c** the initial content of sucrose **d** pH **e** fermentation temperature on the physicochemical and sensory properties of SFB, Table S1: Chemical sensors used in electronic nose corresponding to different types of volatile substances, Table S2: Standard solution for electronic tongue analysis, Table S3: Sensory scoring rules for the SCG fermented beverages, Table S4: Performance of different yeast strains in SCG extract. Table S5: Chemical parameters of unfermented SCG extract.

Author Contributions: Conceptualization, C.L. and S.L.; Methodology, Z.L.; Investigation, Z.L.; Resources, C.L. and S.L.; Data Curation, X.Y.; Writing—Original Draft Preparation, L.W., X.Y. and X.L.; Writing—Review & Editing, X.L. and X.H.; Supervision, C.L. All authors have read and agreed to the published version of the manuscript.

Funding: This research was funded by [the Hainan Provincial Natural Science Foundation of China] grant number [2019RC083], [the National Natural Science Foundation of China] grant number [31760456], [the Education Department of Hainan Province] grant number [Hnky2019-7], and [the Scientific Research Foundation of Hainan University] grant number [KYQD1660]. The APC was funded by [the National Natural Science Foundation of China].

Institutional Review Board Statement: Not applicable.

Informed Consent Statement: Not applicable.

Data Availability Statement: The original contributions presented in the study are included in the article/Supplementary Material, further inquiries can be directed to the corresponding author.

Acknowledgments: This research was funded by the Hainan Provincial Natural Science Foundation of China (grant number 2019RC083), the National Natural Science Foundation of China, grant number 31760456, the Education Department of Hainan Province, grant number Hnky2019-7, and the Scientific Research Foundation of Hainan University, grant number KYQD1660.

Conflicts of Interest: The authors declare no conflict of interest.

References

1. Açıkalın, B.; Sanlier, N. Coffee and its effects on the immune system. *Trends Food Sci. Technol.* **2021**, *114*, 625–632. [CrossRef]
2. Ruta, L.L.; Farcasanu, I.C. Coffee and yeasts: From flavor to biotechnology. *Fermentation* **2021**, *7*, 9. [CrossRef]
3. Campos-Vega, R.; Loarca-Piña, G.; Vergara-Castañeda, H.A.; Oomah, B.D. Spent coffee grounds: A review on current research and future prospects. *Trends Food Sci. Technol.* **2015**, *45*, 24–36. [CrossRef]
4. Mussatto, S.J.; Machado, E.M.S.; Martins, S.; Teixeira, J.A. Production, composition and application of coffee and its industrial residues. *Food Bioprocess Technol.* **2011**, *4*, 661–672. [CrossRef]
5. Silva, M.; Nebra, S.A.; Machado Silva, M.J.; Sanchez, C.G. The use of biomass residues in the Brazilian soluble coffee industry. *Biomass Bioenerg.* **1998**, *14*, 457–467. [CrossRef]
6. da Silveira, J.S.; Durand, N.; Lacour, S.; Belleville, M.P.; Perez, A.; Loiseau, G.; Dornier, M. Solid-state fermentation as a sustainable method for coffee pulp treatment and production of an extract rich in chlorogenic acids. *Food Bioprod. Process.* **2019**, *115*, 175–184. [CrossRef]
7. Simões, G.; Demétrio, G.B.; de Paula, G.F.; Ladeira, D.C.; Matsumoto, L.S. Influence of spent coffee grounds on soil microbiological attributes and maize crop. *Res. Soc. Dev.* **2020**, *9*, e818986400. [CrossRef]
8. Araújo, M.N.; Azevedo, A.Q.P.L.; Hamerski, F.; Voll, F.A.P.; Corazza, M.L. Enhanced extraction of spent coffee grounds oil using high-pressure CO_2 plus ethanol solvents. *Ind. Crop Prod.* **2019**, *141*, 111723. [CrossRef]
9. Martinez-Saez, N.; García, A.T.; Pérez, I.D.; Rebollo-Hernanz, M.; Mesías, M.; Morales, F.J.; Martín-Cabrejas, M.A.; del Castillo, M.D. Use of spent coffee grounds as food ingredient in bakery products. *Food Chem.* **2017**, *216*, 114–122. [CrossRef]
10. Panusa, A.; Zuorro, A.; Lavecchia, R.; Marrosu, G.; Petrucci, R. Recovery of natural antioxidants from spent coffee grounds. *J. Agric. Food Chem.* **2013**, *61*, 4162–4168. [CrossRef]
11. Sampaio, A.; Dragone, G.; Vilanova, M.; Oliveira, J.M.; Teixeira, J.A.; Mussatto, S.I. Production, chemical characterization, and sensory profile of a novel spirit elaborated from spent coffee ground. *LWT-Food. Sci. Technol.* **2013**, *54*, 557–563. [CrossRef]
12. Gouvea, B.M., Torres, C.; Franca, A.S.; Oliveira, L.S.; Oliveira, E.S. Feasibility of ethanol production from coffee husks. *Biotechnol. Lett.* **2009**, *31*, 1315–1319. [CrossRef]
13. Hughes, S.R.; López-Núñez, J.C.; Jones, M.A.; Moser, B.R.; Cox, E.J.; Lindquist, M.; Galindo-Leva, L.A.; Riaño-Herrera, N.M.; Rodriguez-Valencia, N.; Gast, F.; et al. Sustainable conversion of coffee and other crop wastes to biofuels and bioproducts using coupled biochemical and thermochemical processes in a multi-stage biorefinery concept. *Appl. Microbiol. Biotechnol.* **2014**, *98*, 8413–8431. [CrossRef]
14. Machado, E.; Mussatto, S.; Teixeira, J.; Vilanova, M.; Oliveira, J. Increasing the sustainability of the coffee agro-industry: Spent coffee grounds as a source of new beverages. *Beverages* **2018**, *4*, 105. [CrossRef]
15. Garcia, C.V.; Kim, Y.T. Spent coffee grounds and coffee silverskin as potential materials for packaging: A review. *J. Polym. Environ.* **2021**, *29*, 2372–2384. [CrossRef]
16. Liu, Y.J.; Yuan, W.Q.; Lu, Y.Y.; Liu, S.Q. Biotransformation of spent coffee grounds by fermentation with monocultures of *Saccharomyces cerevisiae* and *Lachancea thermotolerans* aided by yeast extracts. *LWT-Food Sci. Technol.* **2021**, *138*, 110751. [CrossRef]
17. Liu, Y.J.; Seah, R.H.; Abdul Rahaman, M.S.; Lu, Y.Y.; Liu, S.Q. Concurrent inoculations of *Oenococcus oeni* and *Lachancea thermotolerans*: Impacts on non-volatile and volatile components of spent coffee grounds hydrolysates. *LWT-Food Sci. Technol.* **2021**, *148*, 111795. [CrossRef]
18. Liu, Y.J.; Lu, Y.Y.; Liu, S.Q. The potential of spent coffee grounds hydrolysates fermented with *Torulaspora delbrueckii* and *Pichia kluyveri* for developing an alcoholic beverage: The yeasts growth and chemical compounds modulation by yeast extracts. *Curr. Res. Food Sci.* **2021**, *4*, 489–498. [CrossRef]

19. Buratti, S.; Ballabio, D.; Benedetti, S.; Cosio, M.S. Prediction of Italian red wine sensorial descriptors from electronic nose, electronic tongue and spectrophotometric measurements by means of Genetic Algorithm regression models. *Food Chem.* **2007**, *100*, 211–218. [CrossRef]
20. Jo, Y.; Gu, S.Y.; Chung, N.; Gao, Y.; Kim, H.J.; Jeong, M.H.; Jeong, Y.J.; Kwon, J.H. Comparative analysis of sensory profiles of commercial cider vinegars from Korea, China, Japan, and US by SPME/GC-MS, E-nose, and E-tongue October Korean. *J. Food Sci. Technol.* **2016**, *48*, 430–436.
21. Zhao, G.Z.; Feng, Y.X.; Hadiatullah, H.; Zheng, F.P.; Yao, Y.P. Chemical characteristics of three kinds of Japanese soy sauce based on electronic senses and GC-MS analyses. *Front. Microbiol.* **2021**, *11*, 3222. [CrossRef]
22. Sharma, S.; Mahant, K.; Sharma, S.; Thakur, A.D. Effect of nitrogen source and citric acid addition on wine preparation from Japanese persimmon. *J. I. Brewing* **2017**, *123*, 144–150. [CrossRef]
23. Hlaing, M.T.; Joshy, C.G. Optimization of fermentation process for the preparation of coffee wine by response surface methodology. *Dagon Univ. Commem. 25th Anniv. Silver Jubil. Res. J.* **2019**, *9*, 2.
24. Yu, L.; Du, J.H.; Wang, X.J.; Yin, Q. Study on resistance of *Saccharomyces cerevisiae* sp. to citric acid. *Sci. Technol. Food Ind.* **2008**, *29*, 148–151.
25. Wang, B.S.; Li, L.B.; Wu, Z.W.; Zhang, L.; Chen, F.; Chen, L.; Zhang, M.X. Inhibiting effects of high concentration of citric acid on the growth of *Saccharomyces cerevisiae*. *China Brew.* **2018**, *37*, 56–60.
26. Li, Z.T.; Li, C.F.; Jia, Y.Y.; Wang, Q.K.; Lu, Y.S.; Wang, M.Y.; Huang, Q.; Liu, S.X. Screening of yeast for coffee-grounds wine fermentation and research on nitrogen source for nutritional condition optimization. *China Brew.* **2016**, *35*, 61–64.
27. Gutiérrez, A.; Chiva, R.; Sancho, M.; Beltran, G.; Arroyo-López, F.N.; Guillamon, J.M. Nitrogen requirements of commercial wine yeast strains during fermentation of a synthetic grape must. *Food Microbiol.* **2012**, *31*, 25–32. [CrossRef]
28. Beltran, G.; Novo, M.; Rozès, N.; Mas, A.; Guillamón, J.M. Nitrogen catabolite repression in *Saccharomyces cerevisiae* during wine fermentations. *FEMS Yeast Res.* **2004**, *4*, 625–632. [CrossRef]
29. Lucero, P.; Moreno, E.; Lagunas, R. Catabolite inactivation of the sugar transporters in *Saccharomyces cerevisiae* is inhibited by the presence of a nitrogen source. *FEMS Yeast Res.* **2002**, *1*, 307–314.
30. Hernandez-Orte, P.; Ibarz, M.J.; Cacho, J.; Ferreira, V. Effect of the addition of ammonium and amino acids to musts of Airen variety on aromatic composition and sensory properties of the obtained wine. *Food Chem.* **2005**, *89*, 163–174. [CrossRef]
31. Santos, J.; Leitão-Correia, F.; Sousa, M.J.; Leão, C. Ammonium is a key determinant on the dietary restriction of yeast chronological aging in culture medium. *Oncotarget* **2015**, *6*, 6511–6523. [CrossRef] [PubMed]
32. Forkert, P.G. Mechanisms of lung tumorigenesis by ethyl carbamate and vinyl carbamate. *Drug Metab. Rev.* **2010**, *42*, 355–378. [CrossRef] [PubMed]
33. Silla, S.M.H. Biogenic amines: Their importance in foods. *Int. J. Food Microbiol.* **1996**, *29*, 213–231.
34. Hernanz, D.; Gallo, V.; Recamales, A.F.; Melendez-Martinez, A.J.; Gonzalez-Miret, M.L.; Heredia, F.J. Effect of storage on the phenolic content, volatile composition and colour of white wines from the varieties zalema and colombard. *Food Chem.* **2009**, *113*, 530–537. [CrossRef]
35. Bressani, A.P.P.; Martinez, S.J.; Sarmento, A.B.I.; Borém, F.M.; Schwan, R.F. Organic acids produced during fermentation and sensory perception in specialty coffee using yeast starter culture. *Food Res. Int.* **2020**, *128*, 108773. [CrossRef]
36. De Bruyn, F.; Zhang, S.J.; Pothakos, V.; Torres, J.; Lambot, C.; Moroni, A.V.; Callanan, M.; Sybesma, W.; Weckx, S.; De Vuyst, L. Exploring the impacts of postharvest processing on the microbiota and metabolite profiles during green coffee bean production. *Appl. Environ. Microbiol.* **2016**, *83*, e02398-16. [CrossRef]
37. Evangelista, S.R.; da Cruz Pedrozo Miguel, M.G.; de Souza Cordeiro, C.; Silva, C.F.; Pinheiro, A.C.M.; Schwan, R.F. Inoculation of starter cultures in a semi-dry coffee (Coffea arabica) fermentation process. *Food Microbiol.* **2014**, *44*, 87–95. [CrossRef]
38. Dzialo, M.C.; Park, R.; Steensels, J.; Lievens, B.; Verstrepen, K.J. Physiology, ecology and industrial applications of aroma formation in yeast. *FEMS Microbiol. Rev.* **2017**, *41*, S95–S128. [CrossRef]
39. Ma, P.X.; Zhang, B.C.; Zhang, Y. Research on the difference between Chinese quality brandy and French quality brandy. *Liquor-Mak. Sci. Technol.* **2011**, *1*, 47–51.
40. Sunarharum, W.B.; Williams, D.J.; Smyth, H.E. Complexity of coffee flavor: A compositional and sensory perspective. *Food Res. Int.* **2014**, *62*, 315–325. [CrossRef]
41. Lee, S.J.; Kim, M.K.; Lee, K.G. Effect of reversed coffee grinding and roasting process on physicochemical properties including volatile compound profiles. *Innov. Food Sci. Emerg. Technol.* **2017**, *44*, 97–102. [CrossRef]
42. Ribeiro, L.S.; Miguel, M.G.; Evangelista, S.R.; Machado Martins, P.M.; van Mullem, J.; Belizario, M.H.; Schwan, R.F. Behavior of yeast inoculated during semi-dry coffee fermentation and the effect on chemical and sensorial properties of the final beverage. *Food Res. Int.* **2017**, *92*, 26–32. [CrossRef]
43. Czerny, M.; Christlbauer, M.; Christlbauer, M.; Fischer, A.; Granvogl, M.; Hammer, M.; Hartl, C.; Hernandez, N.M.; Schieberle, P. Re-investigation on odour thresholds of key food aroma compounds and development of an aroma language based on odour qualities of defined aqueous odorant solutions. *Eur. Food Res. Technol.* **2008**, *228*, 265–273. [CrossRef]
44. Lin, X.; Wang, Q.K.; Wu, W.Y.; Zhang, Y.X.; Liu, S.X.; Li, C.F. Evaluation of different *Saccharomyces cerevisiae* strains on the profile of volatile compounds in pineapple wine. *J. Food Sci. Technol.* **2018**, *55*, 4119–4130. [CrossRef]
45. Nyanga, L.K.; Nout, M.J.; Smid, E.J.; Boekhout, T.; Zwietering, M.H. Fermentation characteristics of yeasts isolated from traditionally fermented masau (*Ziziphus mauritiana*) fruits. *Int. J. Food Microbiol.* **2013**, *166*, 426–432. [CrossRef]

46. Lin, X.; Wu, L.F.; Wang, X.; Yao, L.L.; Wang, L. Ultrasonic-assisted extraction for flavonoid compounds content and antioxidant activities of India *Moringa oleifera* L. leaves: Simultaneous optimization, HPLC characterization and comparison with other methods. *J. Appl. Res. Med. Aromat. Plants* **2021**, *20*, 100284.
47. Wu, L.F.; Li, L.; Chen, S.J.; Wang, L.; Lin, X. Deep eutectic solvent-based ultrasonic-assisted extraction of phenolic compounds from *Moringa oleifera* L. leaves: Optimization, comparison and antioxidant activity. *Sep. Purif. Technol.* **2020**, *274*, 117014. [CrossRef]
48. Lin, X.; Jia, Y.; Li, K.; Hu, X.P.; Li, C.F.; Liu, S.X. Effect of the inoculation strategies of selected *Metschnikowia agaves* and *Saccharomyces cerevisiae* on the volatile profile of pineapple wine in mixed fermentation. *J. Food Sci. Technol.* **2021**. [CrossRef]
49. Sun, S.Y.; Jiang, W.G.; Zhao, Y.P. Evaluation of different *Saccharomyces cerevisiae* strains on the profile of volatile compounds and polyphenols in cherry wines. *Food Chem.* **2011**, *2*, 547–555. [CrossRef]

Article

Increase in Fruity Ester Production during Spine Grape Wine Fermentation by Goal-Directed Amino Acid Supplementation

Zijian Zhu [1], Kai Hu [1], Siyu Chen [1], Sirui Xiong [1] and Yongsheng Tao [1,2,*]

[1] College of Enology, Northwest A&F University, Yangling 712100, China; zhuzijian@nwafu.edu.cn (Z.Z.); kh@nwafu.edu.cn (K.H.); csy18302937517@163.com (S.C.); Xsrxn1805@163.com (S.X.)

[2] Shaanxi Engineering Research Center for Viti-Viniculture, Yangling 712100, China

[*] Correspondence: taoyongsheng@nwsuaf.edu.cn; Tel.: +86-8709-2233

Abstract: The aim of this work was to enhance the levels of fruity esters in spine grape (*Vitis davidii* Foëx) wine by goal-directed amino acid supplementation during fermentation. HPLC and GC-MS monitored the amino acids and fruity esters, respectively, during alcoholic fermentation of spine grape and Cabernet Sauvignon grape. HPLC was also used to determine the extracellular metabolites and precursors involved in the synthesis of fruity esters. Alanine, phenylalanine, and isoleucine levels in spine grape were less than those in Cabernet Sauvignon. Pearson correlation between amino acid profile and fruity ester content in the two systems indicated that deficiencies in alanine, phenylalanine, and isoleucine levels might have limited fruity ester production in spine grape wine. Supplementation of these three amino acids based on their levels in Cabernet Sauvignon significantly increased fruity ester content in spine grape wine. Interestingly, goal-directed amino acid supplementation might have led to changes in the distribution of carbon fluxes, which contributed to the increase in fruity ester production.

Keywords: amino acid; fruity ester; wine aroma; nitrogen management; Pearson correlation analysis; carbon metabolism; *Vitis davidii* Foëx

Citation: Zhu, Z.; Hu, K.; Chen, S.; Xiong, S.; Tao, Y. Increase in Fruity Ester Production during Spine Grape Wine Fermentation by Goal-Directed Amino Acid Supplementation. *Fermentation* **2021**, 7, 231. https://doi.org/10.3390/fermentation7040231

Academic Editor: Bin Tian

Received: 12 September 2021
Accepted: 4 October 2021
Published: 16 October 2021

Publisher's Note: MDPI stays neutral with regard to jurisdictional claims in published maps and institutional affiliations.

Copyright: © 2021 by the authors. Licensee MDPI, Basel, Switzerland. This article is an open access article distributed under the terms and conditions of the Creative Commons Attribution (CC BY) license (https://creativecommons.org/licenses/by/4.0/).

1. Introduction

Yeast assimilable nitrogen (YAN), composed of organic (amino acids) and inorganic (ammonium ions) forms, is an important macronutrient involved in yeast metabolism [1]. Their concentrations vary depending on variety, geographical location, climate, and viticulture practices [2,3]. During alcoholic fermentation, YAN is used to sustain growth and produce biomass as well as synthesize fermentative aroma compounds [4]. Esters are the major fermentative aroma compounds, and they can impart aroma even at concentrations below their odor threshold [5]. Fruity esters, including acetate esters and fatty acid ethyl esters (FAEEs), are considered target components to enhance wine aroma [6].

Studies on the correlation between amino acids and aroma compounds have revealed an association between amino acids and fruity ester production [7–9]. Amino acid supplementation in grape must enhanced the production of fermentation-derived aroma components, such as acetate esters and ethyl esters [10–12] and improved the scores for the descriptors "confectionary", "red fruit", and "dark fruit" [13]. These findings suggest that amino acid supplementation potentially improves wine aroma quality. The synthesis of fermentation-derived aroma compounds as well as their association with nitrogen and sugar metabolism in *Saccharomyces cerevisiae* has been extensively studied. α-keto acids have been considered as important intermediaries in the formation of higher alcohols and volatile acids. Either amino acid catabolism via the Ehrlich pathway [14] or glucose/fructose metabolism via central carbon metabolism [15] produces these precursors. Nevertheless, mechanisms underlying the increase in fruity esters under amino acid supplementation are still not clearly understood. Recent studies have shown that only a small fraction of higher alcohols is produced by catabolism of consumed amino acids [16,17]. The interaction between central carbon metabolism and nitrogen metabolism may significantly influence the production of volatile compounds in *S. cerevisiae* [2]. Traditionally, nitrogen was supplemented without the determination of amino acid content of grape must. Injudicious supplementation of nitrogen is wasteful and has negative impacts on wine aroma due to high acetic acid and ethyl acetate production [4]. Therefore, several studies have highlighted the need to determine the nutritional status of grape juice before nitrogen supplementation to increase aromatic compound production by yeast [11,18]. Further studies are necessary to evaluate the increase in wine aroma and explore the possible mechanisms of increase in fruity esters under goal-directed amino acid modulation.

Spine grape is a wild species widely distributed in southern China. Spine grape wine has high anthocyanin and polyphenol content [19] with poor fruity aroma. A study has shown that spine grape wine has weak aroma characteristics, which may be related to low levels of varietal aroma precursors and fermentative aroma compounds. In particular, fruity esters were positively correlated with wine fruity aroma [20]. On the contrary, Cabernet Sauvignon, a well-known *Vitis vinifera* grape, has elegant fruity aroma with high fruity ester content. Variations in the ester profiles of Shiraz and Cabernet Sauvignon wines were related to specific grape-juice nitrogen composition [21], and our previous study also indicated the significant correlation between amino acid composition and fruity ester production during fermentation of spine grapes [22]. Therefore, modulation based on the amino acid profile of Cabernet Sauvignon can be an effective approach to increasing the fruity ester content of spine grape wine.

In this study, we attempted to enhance the fruity ester content in spine grape wine by goal-directed amino acid supplementation. Pearson's correlation was used to describe the association between amino acid profile and fruity ester content in spine grape and Cabernet Sauvignon grape during fermentation. Subsequently, we supplemented these amino acids before spine grape winemaking and evaluated the fruity ester content at different concentrations of the nitrogen supplement. We also analyzed the extracellular metabolites and precursors involved in fruity ester synthesis. These findings can provide novel insights into the mechanisms underlying the increase in fruity esters under goal-directed amino acid supplementation.

2. Materials and Methods

2.1. Grape Material

Cabernet Sauvignon grapes were harvested from Heyang, Shaanxi Province, China on 28 August 2019. Grape berries were disease-free, pest-free, and healthy and had 207.5 g/L reducing sugars, 5.0 g/L titratable acidity (tartaric acid), and a pH value of 3.34. Spine grape 'Xiang Pearl' was harvested from Huaihua, Hunan Province, China on 25 August 2019. Grape berries had 172.5 g/L reducing sugars, 4.3 g/L titratable acidity (tartaric acid), and a pH value of 3.57.

2.2. Chemicals, Standards, and Microorganisms

The analytical reagents glucose, sodium hydroxide (NaOH), sodium chloride (NaCl), potassium phosphate monobasic (KH_2PO_4), sodium carbonate (Na_2CO_3), sodium bicarbonate ($NaHCO_3$), sodium acetate (NaAc), 2,4-dinitrofluorobenzene (DNFB), and diammonium phosphate (DAP) were purchased from Kermel (Tianjin, China).

Non-volatile standards (purity $\geq$ 98.0%), included L-aspartic acid, L-tryptophan, L-isoleucine, L-alanine, L-arginine, γ-aminobutyric acid (GABA), L-methionine, L-phenylalanine, L-tyrosine, and L-proline (Yuanye Shanghai, China); L-glutamic acid, L-valine, and L-leucine (Solarbio, Beijing, China); and L-malic acid, citric acid, acetic acid, succinic acid, and glycerol (DrE, Ratzeburg, Germany) (Table S1). Methanol and acetonitrile were obtained from Kermel (Tianjin, China). Water was obtained from a Milli-Q purification system (Millipore, Bedford, MA, USA).

Volatile chemical standards (purity $\geq$ 97.0%), including ethyl acetate, isobutyl acetate, isoamyl acetate, phenylethyl acetate, ethyl butyrate, ethyl hexanoate, ethyl octanoate, ethyl decanoate, isobutyl alcohol, isoamyl alcohol, 2-phenylethanol, 2-octanol, hexanoic acid, octanoic acid, and decanoic acid, were obtained from Sigma-Aldrich (Shanghai, China) (Table S2).

S. cerevisiae strain Angel RV171 (Angel Yeast, Yichang, China), a neutral wine yeast strain for dry wine fermentation, was used in this study.

2.3. Winemaking and Nitrogen Supplementation

Alcoholic fermentation was carried out as described by Kong et al. [20] Mature and healthy grape berries were manually selected, destemmed, and crushed. Sulfur dioxide (SO_2; 60 mg/L) was added to the grape must to avoid oxidation. Subsequently, cold maceration was conducted at 4 °C for 24 h, and the must was divided into triplicates. Sucrose was added to the must to achieve a final alcohol concentration of 12% (w). Fermentation was performed in 20 L bottles after yeast inoculation. Fermentation temperature was maintained between 20 °C and 22 °C, and cap management to push the floating caps into the juice was done thrice daily. For the analysis of the nitrogen compounds, aroma compounds and other metabolites, fermented juice was sampled every 24 h. Each sample was centrifuged for 10 min at 8000 rpm to remove yeast cells, and 20 mL of the supernatant was collected and stored in a freezer at −20 °C until further analysis. After fermentation (reducing sugar level < 2 g/L), clear wine was separated from skin and lees, followed by sulfur dioxide addition (60 mg/L).

To determine the Pearson's correlation coefficient between amino acids and fruity esters during fermentation, spine grape must was divided into six groups for different nitrogen supplements. Samples were analyzed before and after nitrogen addition to ensure similar YAN concentrations in spine grape must and Cabernet Sauvignon grape must (Table S3). Six different nitrogen supplements were used in the study, including IN, AA-Ala, AA-Phe, AA-Ile, MAA, and IN+MAA (Table 1). The treatment IN+MAA was used in an attempt to modulate the concentration of alanine, phenylalanine, and isoleucine in spine grape must to the same level as that in Cabernet Sauvignon must, and the difference in YAN was adjusted by DAP addition. MAA treatment was used to add the three key amino acids multiple times to attain a final proportion similar to IN+MAA without DAP. All supplements contained approximately 50 mg N/L YAN. After nitrogen supplementation, YAN in spine grape and Cabernet Sauvignon must was between 185.3 and 187.9 mg N/L. Similar alcoholic fermentation was carried out after nitrogen addition. All fermentation experiments were performed in triplicate. Wine samples were stored at −20 °C until further analysis.

Table 1. Composition of each nitrogen supplement.

Supplement	Composition
IN	DAP, 236 mg/L
AA-Ala	Alanine, 318 mg/L
AA-Phe	Phenylalanine, 589 mg/L
AA-Ile	Isoleucine, 468 mg/L
MAA	Alanine, 163 mg/L; phenylalanine, 71 mg/L; isoleucine, 170 mg/L
IN+MAA	Alanine, 39 mg/L; phenylalanine, 17 mg/L; isoleucine, 39 mg/L; DAP, 178 mg/L

2.4. Amino Acid Analysis

Amino acids were analyzed by precolumn derivatization using DNFB as described by Li et al. [23] with some modifications. For precolumn derivatization, 100 μL of clarified wine or standard was mixed with 100 μL of 0.05 M $NaHCO_3$ and 40 μL of DNFB solution (5 mg/mL; w). The mixture was incubated in a water bath at 60 °C for 60 min in the dark. After returning to room temperature, 760 μL of KH_2PO_4 buffer (0.01 M; pH 7.0) was added to the tube, vortexed, and kept in the dark for 15 min.

Amino acids were analyzed in a Waters Alliance 2695 HPLC system (Milford, MA, USA) using Agilent ZORBAX SB-C18 column (Analytical 4.6 × 250 mm; 5 micron). The solvent system consisted of mobile phase A (acetonitrile), mobile phase B (ultrapure water), and mobile phase C (NaAc buffer; pH = 6.4). Samples were filtered through a nylon filter (0.22 μm), injected (20 μL) onto the column, and eluted at 33 °C at a flow rate of 1 mL/min according to the following gradient: initial—8% A, 8% B, 84% C; 0–2 min/8–15% A, 8–15% B, 84–70% C; 2–4 min/15–7% A, 15–17% B, 70–66% C; 4–8 min/17–20% A, 17–20% B, 66–60% C; 8–14 min/20–21% A, 20–22% B, 60–57% C; 14–24 min/21–28% A, 22–27% B, 57–45% C; 24–27 min/28–28% A, 27–27%B, 45–45% C; 27–36 min/28–49% A; 27–49% B, 45–2% C; and 36–40 min/49–8% A, 49–8% B, 2–84% C. The separated amino acid derivatives were detected using Waters 2996 photodiode array detector at 360 nm. The external standard method was used to qualitatively and quantitatively analyze the separated amino acids.

2.5. *Organic Acid and Glycerol Analyses*

Malic acid, acetic acid, citric acid, and succinic acid were determined in a Waters Alliance 2695 HPLC system (Milford, MA, USA) with a Welch Ultimate AQ-C18 column (Analytical 4.6 mm × 250 mm; 5 μm) at 30 °C. The column was eluted with 98% 0.02 mM KH_2PO_4:2% methanol at 1 mL/min. The separated organic acids were detected using a Waters 2996 photodiode array detector at 210 nm. Glycerol was measured with a Welch Ultimate XB-NH_2 column (Analytical 4.6 mm × 250 mm; 5 μm) at 30 °C. The column was eluted with 85% acetonitrile at 1 mL/min. Glycerol was detected using a Waters 2410 refractive index detector. Glycerol and organic acids were quantitated using the calibration curves.

2.6. *Yeast Assimilable Nitrogen Analysis*

Yeast assimilable nitrogen (YAN) was calculated using the following Equation (1):

$$\text{YAN} = \text{TAN} + N_{\text{ammonium}} - N_{\text{proline}} \tag{1}$$

where TAN indicates total amino nitrogen concentration, N_{ammonium} indicates contribution of nitrogen corresponding to ammonium, and N_{proline} indicates contribution of nitrogen corresponding to proline.

The formaldehyde titration method was used to analyze the concentration of TAN and N_{ammonium} [24]. Briefly, 10 mL of filtered juice was titrated with NaOH solution until neutralized (pH = 8.0), and then 2 mL of neutral formaldehyde (pH = 8.0) was added. The mixture was titrated with NaOH solution until pH = 8.0. TAN and N_{ammonium} were calculated using the following Equation (2):

$$\text{TAN} + N_{\text{ammonium}} = (V_1 \times C \times 14 \times D \times 1000)/V_2 \tag{2}$$

where V_1 indicates the volume of NaOH solution used in titration, C indicates the concentration of NaOH titrant, D indicates dilution factor of juice sample, and V_2 indicates the volume of juice sample.

2.7. *Quantification of Volatiles*

The volatiles were analyzed using headspace solid-phase microextraction (HS-SPME) coupled with GC–MS as described by Hu et al. [25]. A 50/30 μm DVB/CAR/PDMS fiber (Supelco, Bellefonte, PA, USA) was used to extract the volatiles. In a 20 mL gas-tight vial, 2 g of NaCl, 2 mL of wine, 6 mL of pure water, and 20 μL of internal standard (16 mg/L, 2-octanol) were added and incubated in a 40 °C water bath with stirring at 600 rpm for 15 min, extracted for 30 min, and desorbed in the GC injection port (230 °C) for 5 min using a Shimadzu QP2020 GC–MS (Shimadzu Corporation, Kyoto, Japan) and a DB-WAX column (60 m × 0.25 mm × 0.25 μm; Agilent Corporation, Santa Clara, CA, USA). The carrier gas used was helium (99.999%) at a flow rate of 1.5 mL/min. The GC program was as follows: 40 °C for 3 min, raised to 160 °C at 4 °C/min, followed by an increase to 220 °C at 7 °C/min, and hold for 8 min. MS transfer line and ion source temperatures were set to 220 °C and 200 °C, respectively. Electron ionization (EI) mass spectrometric data from m/z 35 to 350 were scanned at 0.2 s intervals. Esters were identified by comparing their retention time and mass spectra with those of pure standards using the NIST 17 mass spectral library. Target compound concentration was calculated by interpolation of relative areas in the calibration graphs obtained with the pure reference compound (2-octanol).

2.8. Statistical Analysis

All experiments were performed at least in duplicate. Data were expressed as mean ± standard deviation. One-way analysis of variance (ANOVA) was used to compare the means with Duncan's test. Data correlation was evaluated using bivariate (two-tailed) Pearson correlation coefficient (r) in SPSS 23.0 (IBM Corp., Armonk, IL, USA).

3. Results

3.1. Evolution of Amino Acids and Fruity Esters during Alcoholic Fermentation

We investigated the dynamic evolution of 13 amino acids in spine grape and Cabernet Sauvignon during alcoholic fermentation (Table S4). Total amino acid content (except proline) of spine grape must was more than that of Cabernet Sauvignon must. In spine grape must, aspartic acid, valine, glutamic acid, GABA, and arginine levels were significantly ($p < 0.01$) more than those in Cabernet Sauvignon grape must, while proline, alanine, phenylalanine, and isoleucine levels were significantly less than those in Cabernet Sauvignon grape must (Figure 1). With progress in alcoholic fermentation, almost all amino acids gradually decreased, except proline and tyrosine. Proline was not assimilated and finally produced by yeast, and tyrosine content decreased slightly during Cabernet Sauvignon fermentation and increased during spine grape fermentation.

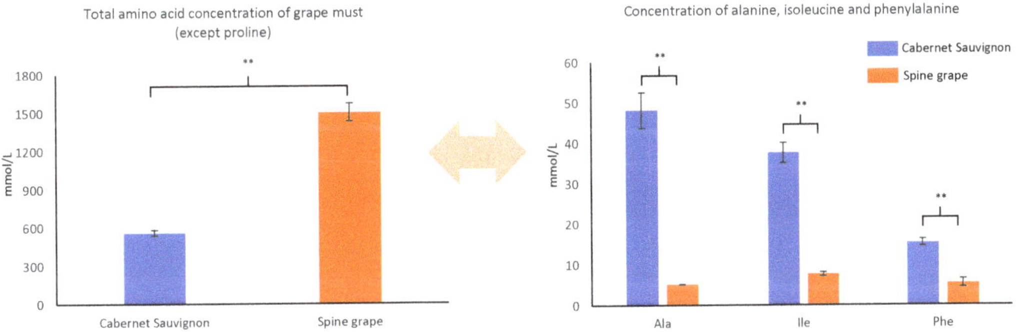

Figure 1. Concentrations of total amino acid (except proline), alanine, isoleucine, and phenylalanine in Cabernet Sauvignon and spine grape must. Difference significant at 99% (**) confidence level.

Evolution of fruity esters during alcoholic fermentation was monitored by HPME-GC-MS (Figure 2, Tables S5 and S6). Acetates and ethyl esters rapidly increased once fermentation started, and the highest ethyl acetate and short-chain fatty acid ethyl ester (SCFAEE) levels were detected at the end of fermentation. In contrast, acetates of higher alcohols (AHAs) and medium-chain fatty acid ethyl esters (MCFAEEs) decreased during the later stage of fermentation (day 7 to day 12). The level of ethyl acetate in spine grape wine was 58% of that in Cabernet Sauvignon wine, AHA was 5%, SCFAEE was 89%, and MCFAEE was 67%.

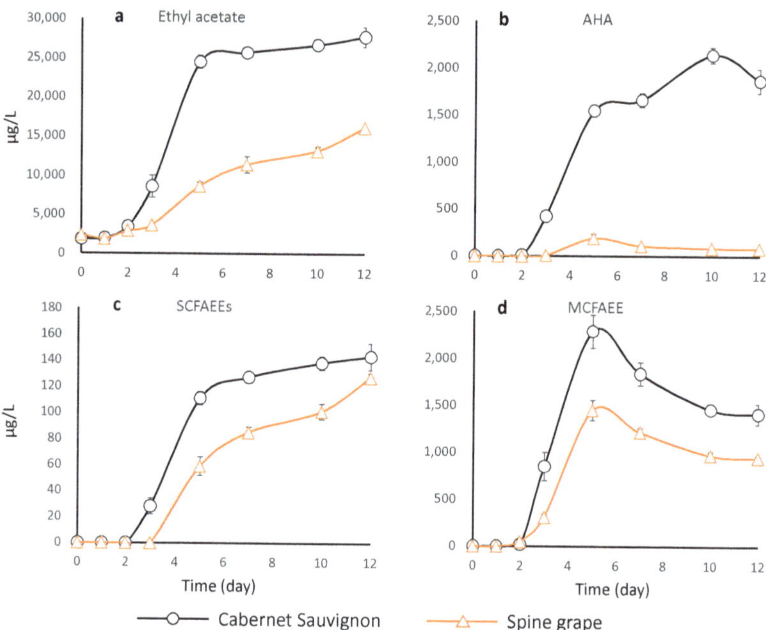

Figure 2. Evolution of fruity esters during Cabernet Sauvignon and spine grape alcoholic fermentation: (**a**) Ethyl acetate, (**b**) Acetate of higher alcohol, AHA, (**c**) Short-chain fatty acid ethyl ester, SCFAEE, (**d**) medium-chain fatty acid ethyl ester, MCFAEE. AHAs included isobutyl acetate, isoamyl acetate, hexyl acetate, and β-Phenylethyl acetate; SCFAEEs included ethyl butyrate; and MCFAEEs included ethyl hexanoate, ethyl octanoate, and ethyl decanoate.

3.2. Correlation Analysis and Identification of Goal-Directed Amino Acid

To estimate the goal-directed amino acid, Pearson correlation analysis was used to establish a fitting model between the amino acid profile and fruity ester levels in spine grape and Cabernet Sauvignon grape during fermentation (Table S7), and the differences in these Pearson's correlation coefficients (r) were further assessed (Table 2). During fermentation, almost all amino acids were rapidly consumed and the fruity ester content gradually increased, thus all r values were negative. Correlation coefficients between these amino acids and AHAs were slightly less than those during Cabernet Sauvignon fermentation, nevertheless, a correlation coefficient similar to or higher than that of Cabernet Sauvignon fermentation was observed between the amino acids arginine, aspartic acid, glutamic acid, GABA, and valine and the fruity esters SCFAEEs and MCFAEEs during spine grape fermentation. In contrast, the correlation coefficients between the amino acids alanine, phenylalanine, and isoleucine, and the fruity esters AHAs, SCFAEEs, and MCFAEEs during fermentation of spine grape were less than those during fermentation of Cabernet Sauvignon. As mentioned in Section 3.1, arginine, aspartic acid, glutamic acid, GABA, and valine levels in spine grape must were more than those in Cabernet Sauvignon grape must, while alanine, phenylalanine, and isoleucine levels were less than those in Cabernet Sauvignon grape must. These findings indicate that the lesser abundance of alanine, phenylalanine, and isoleucine may have led to lower fruity ester production during spine grape fermentation. Therefore, alanine, phenylalanine, and isoleucine were used for nitrogen supplementation.

Table 2. Difference in Pearson's correlation coefficients (r) during Cabernet Sauvignon and spine grape fermentation. Data were calculated by $r_{\text{cabernet sauvignon}} - r_{\text{spine grape}}$, where $r_{\text{cabernet sauvignon}}$ indicates the correlation coefficient of Cabernet Sauvignon fermentation and $r_{\text{spine grape}}$ indicates the correlation coefficient of spine grape fermentation. Red cell means the correlation coefficient during spine grape fermentation was less than that during Cabernet Sauvignon fermentation. The darker the red, the weaker the correlation between the amino acids and fruity ester during spine grape fermentation. AHA, acetate of higher alcohol; SCFAEE, short-chain fatty acid ethyl ester; MCFAEE, medium-chain fatty acid ethyl ester.

Amino Acid	AHAs	SCFAEEs	MCFAEEs
Total AAs	−0.274	−0.011	−0.075
Arg	−0.185	−0.010	0.031
Asp	−0.186	0.080	−0.002
Glu	0.038	0.123	0.143
GABA	−0.164	0.149	−0.044
Leu	−0.143	0.011	−0.002
Met	−0.101	0.095	0.072
Trp	−0.123	0.004	0.024
Val	−0.136	−0.064	0.045
Ala	−0.251	−0.156	−0.022
Ile	−0.099	−0.020	−0.157
Phe	−0.406	−0.266	−0.237

3.3. Modification of Fruity Ester Production under Nitrogen Supplementation

Based on the results described in Section 3.2, different nitrogen sources were supplemented to the six groups of spine grape must with Cabernet Sauvignon control (CK-CS) and spine grape control (CK-SP) (Table S3). Fruity esters, especially ethyl acetate, AHAs, and MCFAEEs, in MAA and IN+MAA treatments were considerably more than that in CK-SP and their levels reached or even exceeded the ester levels in CK-CS (Table 3). A significant increase in ethyl acetate concentration was observed in MAA, which was 2.1-fold and 1.5-fold more than CK-SP and CK-CS, respectively, and even 1.6-fold more than that in the IN+MAA treatment. Meanwhile, the content of FAEEs was 1.2-fold less than that in IN+MAA treatment. These findings show that a comprehensive supplementation of alanine, phenylalanine, and isoleucine significantly influences the production of fruity esters.

With single nitrogen supplements, no significant difference was observed in the acetate and FAEE levels between inorganic nitrogen treatment (DAP addition, IN) and CK-SP. AA-Ala led to significantly higher levels of ethyl acetate, SCFAEEs, and MCFAEEs compared with other single amino acid supplements. AA-Phe and AA-Ile showed no significant increase in ethyl acetate content, while AA-Ile reduced the content. However, AHA production with AA-Ala was less than with AA-Phe and AA-Ile. AA-Phe significantly increased phenethyl acetate content, and AA-Ile significantly increased isobutyl acetate and isoamyl acetate levels. AA-Phe and AA-Ile increased FAEE content compared to CK-SP; no significant difference was observed compared with IN+MAA treatment.

Table 3. Concentration of fruity esters and their precursors in spine grape wines obtained from different nitrogen nutrient treatments and controls. Data are mean ± standard deviation. Values displaying different letters within each row are significantly different according to the Duncan test at 95% confidence level. AHA, acetate of higher alcohol; SCFAEE, short-chain fatty acid ethyl ester; MCFAEE, medium-chain fatty acid ethyl ester; CK-CS, Cabernet Sauvignon control; CK-SP, spine grape control; IN, DAP supplement; AA-Ala, single alanine supplement; AA-Phe, single phenylalanine supplement; AA-Ile, single isoleucine supplement; MAA, mixed nitrogen supplement with alanine, phenylalanine, and isoleucine; IN+MAA, mixed nitrogen supplement with DAP, alanine, phenylalanine, and isoleucine.

Compounds	CK-CS	CK-SP	IN	AA-Ala	AA-Phe	AA-Ile	MAA	IN+MAA
Ethyl acetate (μg/L)	30,905 ± 1083b	21,973 ± 436cd	23,043 ± 1397cd	25,454 ± 5621bc	20,857 ± 2039cd	18,409 ± 2093d	45,946 ± 899a	28,962 ± 2052b
AHAs (μg/L)	1573 ± 14a	745 ± 99cde	574 ± 41e	698 ± 62de	1346 ± 97b	1161 ± 83b	937 ± 44c	847 ± 4cd
Isobutyl acetate	41 ± 2c	29 ± 2c	76 ± 4b	112 ± 18a	74 ± 5b	110 ± 12a	119 ± 15a	86 ± 4b
Isoamyl acetate	1346 ± 7a	584 ± 87cd	411 ± 31e	505 ± 39de	508 ± 55de	964 ± 66b	689 ± 22c	634 ± 0de
Hexyl acetate	54 ± 1a	36 ± 7bcd	33 ± 4cd	32 ± 1cd	38 ± 1bcd	31 ± 0d	39 ± 1bc	42 ± 0b
β-Phenylethyl acetate	132 ± 4b	96 ± 3c	54 ± 2d	49 ± 4d	726 ± 36a	56 ± 5d	90 ± 6c	85 ± 0c
SCFAEEs (μg/L)	140 ± 3b	54 ± 1c	38 ± 30c	250 ± 57a	210 ± 8a	212 ± 39a	106 ± 9bc	232 ± 26a
Ethyl butyrate	140 ± 3	54 ± 1	38 ± 30	250 ± 57	210 ± 8	212 ± 39	106 ± 9	232 ± 26
MCFAEEs (μg/L)	2101 ± 43b	1867 ± 63c	1858 ± 75c	2290 ± 242ab	2423 ± 220a	1954 ± 97bc	2125 ± 39b	2485 ± 65a
Ethyl hexanoate	817 ± 2a	576 ± 24d	575 ± 39d	695 ± 77bc	739 ± 63ab	668 ± 34bcd	691 ± 22bc	762 ± 42ab
Ethyl octanoate	754 ± 1c	751 ± 27c	753 ± 35c	934 ± 88b	1056 ± 79a	847 ± 34bc	910 ± 17b	1050 ± 21a
Ethyl decanoate	530 ± 40ab	540 ± 12ab	530 ± 1ab	661 ± 77a	628 ± 78a	439 ± 29b	524 ± 0ab	673 ± 2a
Higher alcohols (mg/L)	972.15 ± 26.45a	597.29 ± 54.96cd	471.12 ± 29.12d	604.04 ± 93.42cd	796.24 ± 53.59b	994.98 ± 64.11a	707.40 ± 19.11bc	558.89 ± 19.84cd
Isobutanol	39.15 ± 2.52a	27.10 ± 2.87bc	20.48 ± 1.57c	42.75 ± 7.03a	23.20 ± 1.19bc	41.03 ± 2.52a	43.75 ± 1.31a	30.14 ± 2.09b
Isopentanol	800.05 ± 19.86a	502.08 ± 48.40bc	398.67 ± 24.57c	500.95 ± 78.78bc	436.37 ± 34.42c	890.35 ± 56.28a	577.80 ± 12.87b	477.30 ± 16.68bc
1-hexanol	4.91 ± 0.15a	2.26 ± 0.30d	3.03 ± 0.15c	2.99 ± 0.43c	3.02 ± 0.22c	1.66 ± 0.09e	3.09 ± 0.13c	3.63 ± 0.19b
2-phenylethanol	128.04 ± 3.92b	65.85 ± 3.39cde	48.94 ± 2.83e	57.35 ± 7.18de	333.65 ± 17.76a	61.94 ± 5.22de	82.76 ± 4.80c	47.82 ± 0.88cd
Volatile acids (mg/L)	7.70 ± 0.53d	7.74 ± 0.58d	8.08 ± 0.50cd	13.37 ± 0.97a	11.75 ± 0.47b	8.39 ± 1.02cd	9.28 ± 0.37c	11.22 ± 0.30b
Hexanoic acid	4.24 ± 0.26d	4.15 ± 0.27d	4.51 ± 0.34cd	7.43 ± 0.70a	5.76 ± 0.28b	4.64 ± 0.22cd	5.29 ± 0.26bc	6.14 ± 0.09b
Octanoic acid	3.27 ± 0.36c	3.44 ± 0.28c	3.41 ± 0.15c	5.70 ± 0.26a	5.74 ± 0.18a	3.58 ± 0.78c	3.82 ± 0.10c	4.86 ± 0.20b
Decanoic acid	0.19 ± 0.01bc	0.18 ± 0.03bc	0.16 ± 0.01c	0.24 ± 0.01a	0.25 ± 0.01a	0.17 ± 0.02c	0.17 ± 0.01bc	0.22 ± 0.01ab

3.4. Analysis of Extracellular Metabolites and Precursors Involved in Fruity Ester Synthesis

Fruity ester precursors and corresponding extracellular metabolites, including higher alcohols (isobutanol, isopentanol, 1-hexanol, and 2-phenylethanol), volatile linear fatty acids (hexanoic acid, octanoic acid, and decanoic acid), and carbon-metabolism-derived extracellular metabolites (L-malic acid, acetic acid, citric acid, succinic acid, and glycerol), under different treatments were analyzed (Tables 3 and 4). Nitrogen supplementation significantly increased the levels of higher alcohols and fatty acids. AA-Phe and AA-Ile dramatically increased the higher alcohol level; AA-Phe increased 2-phenylethanol content (5-fold compared with CK-SP); and AA-Ile increased the isoamyl alcohol level (1.8-fold compared with CK-SP), whereas AA-Ala and IN+MAA did not contribute to higher alcohol production. In contrast, AA-Ala, AA-Phe, and IN+MAA treatments significantly enhanced the production of medium-chain fatty acids (MCFAs). Analysis of carbon-metabolism-derived extracellular metabolites indicated that CK-CS had the highest organic acid content while AA-Ile produced the highest amount of glycerol. All nitrogen treatments reduced citric acid content in spine grape wine. Compared with CK-SP, the treatments AA-Ala, AA-Phe, AA-Ile, and MAA led to a significant increase in glycerol content in spine grape wine while IN+MAA led to a significant decrease in glycerol, malic acid, and citric acid levels. Other nitrogen treatments, except AA-Phe, increased acetic acid content. In addition, IN+MAA produced the highest level of extracellular succinic acid, while other treatments had no significant effect on succinic acid production.

Table 4. Concentrations of extracellular metabolites in wines obtained from different nitrogen nutrient treatments (g/L). Data are mean ± standard deviation. Values displaying different letters within each row are significantly different according to the Duncan test at 95% confidence level. The Duncan test was carried out in columns CK-SP, IN, MAA, and IN+MAA. CK-CS, Cabernet Sauvignon control; CK-SP, spine grape control; IN, DAP supplement; AA-Ala, single alanine supplement; AA-Phe, single phenylalanine supplement; AA-Ile, single isoleucine supplement; MAA, mixed nitrogen supplement with alanine, phenylalanine, and isoleucine; IN+MAA, mixed nitrogen supplement with DAP, alanine, phenylalanine, and isoleucine.

Compounds	CK-CS	CK-SP	IN	AA-Ala	AA-Phe	AA-Ile	MAA	IN+MAA
Glycerol	7.47 ± 0.12	7.26 ± 0.22b	7.36 ± 0.28ab	7.65 ± 0.24	7.63 ± 0.19	7.76 ± 0.26	7.73 ± 0.30a	6.98 ± 0.15c
Malic acid	2.51 ± 0.14	0.78 ± 0.04a	0.72 ± 0.01a	0.65 ± 0.01	0.71 ± 0.01	0.67 ± 0.01	0.73 ± 0.02a	0.64 ± 0.01b
Acetic acid	1.41 ± 0.01	0.71 ± 0.04b	0.67 ± 0.03b	0.77 ± 0.07	0.66 ± 0.03	0.74 ± 0.01	0.85 ± 0.03a	0.81 ± 0.01a
Citric acid	0.70 ± 0.06	0.33 ± 0.08a	0.33 ± 0.04a	0.25 ± 0.01	0.22 ± 0.01	0.25 ± 0.01	0.22 ± 0.01ab	0.19 ± 0.01b
Succinic acid	4.61 ± 0.10	1.01 ± 0.02b	1.06 ± 0.02b	0.92 ± 0.01	1.11 ± 0.01	0.92 ± 0.04	1.03 ± 0.01b	1.22 ± 0.04a

Furthermore, the differences in yeast extracellular metabolites involved in MCFA and acetate synthesis pathways are illustrated in Figure 3. Only MAA and IN+MAA treatments, which resulted in significant modification of fruity ester profile, were selected in Figure 3 while CK-SP and IN were used as controls. Compared with CK-SP and IN, IN+MAA treatment decreased glycerol and acetic acid levels and increased citric acid, succinic acid, ethyl octanoate, and ethyl decanoate levels. In contrast, IN had no significant influence compared with CK-SP.

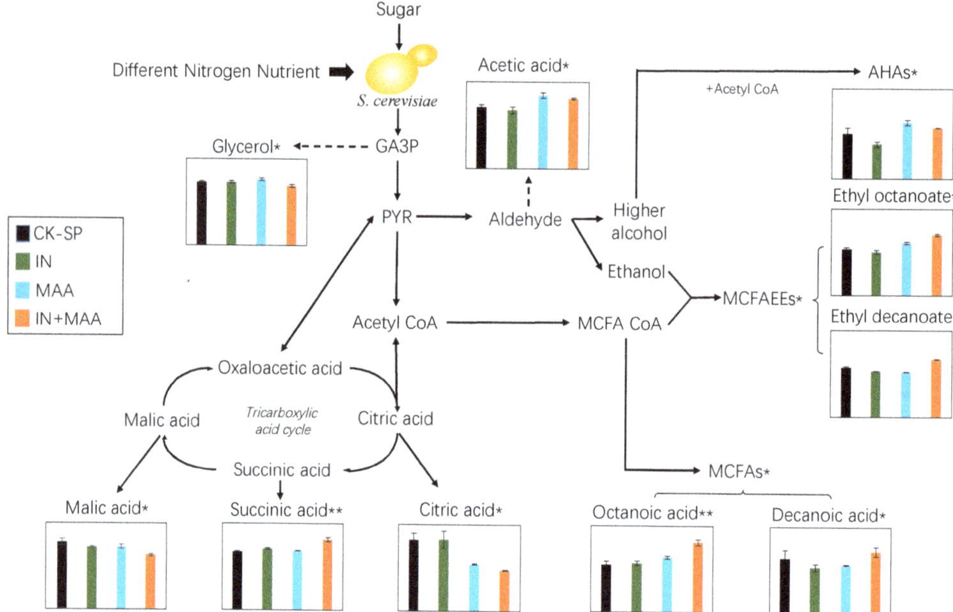

Figure 3. Redistribution of carbon fluxes in the central metabolic network by different nitrogen treatments. PYR, pyruvate; GA3P, glyceraldehyde 3-phosphate; ACTA, acetaldehyde. Data are mean ± standard deviation. Differences are significant at 95% (*), 99% (**)confidence level. AHA, acetate of higher alcohol; MCFA, medium-chain fatty acid; MCFAEE, medium-chain fatty acid ethyl ester; CK-SP, spine grape control; IN, DAP supplement; MAA, mixed nitrogen supplement with alanine, phenylalanine, and isoleucine; IN+MAA, mixed nitrogen supplement with DAP, alanine, phenylalanine, and isoleucine.

4. Discussion

Amino acids are important factors involved in the synthesis of fermentation-derived aroma compounds. Studies have correlated amino acid profiles with grape varieties [26]. In this study, we show differences in amino acid composition and consumption dynamics during fermentation between spine grape and Cabernet Sauvignon. *S. cerevisiae* has a nitrogen-catabolite repression (NCR) system that suppresses alanine assimilation until ammonium gets exhausted from the media. Therefore, alanine is always consumed late [27]. Here, the start of alanine consumption was delayed by more than 3 days during Cabernet Sauvignon fermentation, while it was consumed rapidly to depletion during spine grape fermentation. Meanwhile, tyrosine was not consumed in large amounts in either system. This observation is consistent with the findings of Gao et al. [28] who reported that tyrosine is not fully consumed during wine alcoholic fermentation due to the presence of aromatic amino acid biosynthetic pathway in certain *S. cerevisiae* [29]. However, Garde-Cerdán et al. [10] showed complete consumption of tyrosine during alcoholic fermentation of four different *V. vinifera* varieties. These contradictory results indicate that tyrosine evolution might be highly dependent on fermentation conditions. The levels of leucine, tryptophan, and methionine were not different between the two grape varieties. Alanine, isoleucine, and phenylalanine were lowest in spine grape must while their levels were still within the rational concentration range [1]. Moreover, yeast cannot assimilate proline [30], and therefore, its evolution and correlation with esters are not discussed.

Fruity esters are the main aroma compounds in wine [31]. Only a few studies have so far focused on aroma compounds in spine grape wine. Meng et al. analyzed the characteristic aroma components in spine grape berries instead of spine grape wine [32]. Kong et al. analyzed ester content in spine grape wine without comparing it with other types of wine and found that the fruity ester content in spine grape wine was significantly less than that in Cabernet Sauvignon wine [20]. Fruity esters, which are the fermentative volatiles, contributed significantly to the aroma of spine grape wine. This indicates that a low level of fruity esters is responsible for the weak fruity aroma of spine grape wine.

The amino acid profile of grapes and wine varied between varieties [33]. Different methods have been used to establish a model to determine the correlation between amino acids and aroma compounds. Procopio et al. [8] established a fingerprint of the amino acid and aroma compound profiles by partial least-squares regression (PLS) analysis, and Fairbairn et al. [7] established a linear model between amino acids and volatile aroma components of synthetic juice. In this study, Pearson's correlation was used to determine the association between amino acid profile and fruity ester content during spine grape and Cabernet Sauvignon grape fermentation. The analysis revealed that alanine, phenylalanine, and isoleucine at low levels and with low correlation coefficients may have led to limited fruity ester production during spine grape fermentation, which is in accordance with our previous study that the low content of specific amino acids could negatively affect ester productions during alcoholic fermentation [22]. Accordingly, we speculate that the modification of amino acid profile in spine grape must by supplementing alanine, phenylalanine, and isoleucine has the potential to increase fruity ester production.

Studies have shown that amino acid addition affects wine ester concentration [7,12,34]. Generally, nitrogen management during winemaking is based on the catabolism of the amino acids combined with de novo synthesis of proteinogenic amino acids [16,17]. However, there are limited studies on amino acid supplementation based on the amino acid profile of grape must and on the correlation between amino acids and aroma compounds. We found that goal-directed supplementation of amino acids led to significant increase in the production of AHAs and MCFAEEs. In contrast, IN had no significant effect on fruity ester content. Earlier, Hernández-Orte et al. reported no effect of DAP supplementation on the ester content of synthetic juice [35]. Meanwhile, Vilanova et al. [4] showed an increase in ethyl ester content with DAP supplementation. These differences in findings may be due to the differences in nutrient availability and yeast strains [2]. Amino acid utilization via Ehrlich pathway had pronounced effects on the balance between higher alcohols and corresponding acetate esters. In the present study, AA-Phe and AA-Ile caused a higher production of β-phenylethyl acetate and isoamyl acetate due to increase in 2-phenylethanol and isoamyl alcohol precursors. However, AA-Ile had no marked effect on total MCFAEE content and, similar to AA-Phe, had reduced ethyl acetate production. In contrast to AA-Phe and AA-Ile, AA-Ala significantly enhanced ethyl acetate production. This may be because alanine is converted to pyruvate [36], a main precursor of acetyl-CoA, and the accumulation of acetyl-CoA could have led to increase in ethyl acetate and fatty acid biosynthesis [15,37]. To conclude, AA-Ala enhanced fruity ester production; however, AHA and MCFAEE levels in AA-Ala were less than that in IN+MAA.

Despite similar YAN content, differences in amino acid composition significantly influenced the ester content in wine. The production of fruity esters was considerably higher under both MAA and IN+MAA treatments; however, a higher (2.1-fold) increase in ethyl acetate content with MAA treatment may lead to off-flavor and finally decreased wine quality [38]. Meanwhile, IN did not contribute to fruity ester production. Single amino acid supplements led to an imbalance in the fruity ester profile. Therefore, IN+MAA, a balanced and goal-directed amino acid supplement, can be used to enhance fermentative aroma quality of spine grape wine.

In addition to the aroma compounds produced during fermentation, changes in metabolic intermediates also greatly affect wine ester profile. Amino acids act as precursors of fermentation-derived aroma compounds principally via the Ehrlich pathway [14]. Research has indicated that transamination of amino acids is essential for the subsequent redistribution of nitrogen for the de novo synthesis of proteinogenic amino acids [16], which consequently releases α-keto acid intermediates [39]. Crepin et al. [16] showed that the carbon skeletons of consumed amino acids contribute less to the production of volatile compounds. This is in contrast to the findings of Fairbairn et al. who reported that even minor changes in the amino acid profile of synthetic grape must significantly influence volatile production [7]. These contradicting results indicated an interaction between carbon and nitrogen metabolism for the production of aroma compounds. Accordingly, we analyzed fruity ester precursors and corresponding extracellular metabolites to illustrate the mechanism involved in fruity ester production. CK-SP, CK-CS, IN, and AA-Ile showed lower production of MCFAs, which are markers for limited acetyl-CoA levels [40]. Saerens et al. [41] showed that the level of fatty acids is the most limiting factor in ethyl ester production. An increase in MCFA production enhanced MCFAEE production [25]. Studies have shown that nitrogen management positively influences fatty acid biosynthesis via intensified glycolysis, impaired TCA cycle, and enhanced metabolic fluxes channeling pyruvate and acetyl-CoA [42]. Similarly, IN+MAA, the goal-directed supplementation of specific amino acids in moderate amounts based on the levels in Cabernet Sauvignon, seemingly triggered carbon-flux redistribution combined with an increase in α-keto acid precursors via the central carbon metabolism. These mechanisms led to higher acetyl-CoA and MCFA-CoA levels and enhanced MCFA production, which finally improved the contents of corresponding ethyl esters in spine grape wine. To conclude, IN+MAA may redistribute carbon flux and favor fruity ester production.

Other treatments also have led to an increase in MCFA and MCFAEE, which supports that amino acid supplements favor fruity ester production. AA-Ala improved MCFA production, while it resulted in lesser AHA production compared with IN+MAA. Acetate esters are synthesized from alcohols and acetyl-CoA by alcohol acetyltransferases (AATases) [43]. Amino acid profiles in grape juice influence the redox potential [44], which can modulate the enzymatic activity or gene expression [45] thereby affecting wine ester production. The variation in the formation of AHA derivatives from their precursors may be due to an imbalance in the amino acid profile that affects the intracellular redox potential. This might have led to limited activity or expression of AATases [45] under excessive supplementation of specific amino acids (MAA and AA-Ala). Besides, MAA treatment led to an increase in glycerol, acetic acid, citric acid, succinic acid, and ethyl acetate levels. This may be due to the increase in total carbon fluxes derived from the carbon skeletons of these specific amino acids and resultant increase in metabolites of carbon metabolism pathway [17]. However, in-depth studies are needed to reveal the transcriptomic and metabolomic differences under goal-directed amino acid supplementation.

5. Conclusions

This work demonstrates a change in fruity ester production with goal-directed amino acid supplementation during alcoholic fermentation of spine grape. Alanine, phenylalanine, and isoleucine may be the defect in the amino acid profile of spine grape must that lowered fruity ester production. Supplementation of these three amino acids based on the amino acid profile of Cabernet Sauvignon grape must dramatically enhanced the fruity ester content in spine grape wine. In addition, modification of the amino acid profile triggered carbon-flux redistribution, promoted MCFA production, and, finally, enhanced MCFAEE content. However, further work is needed to reveal the molecular mechanisms of fruity ester biosynthesis under nitrogen intervention.

Supplementary Materials: The following are available online at https://www.mdpi.com/article/10.3390/fermentation7040231/s1, Table S1: Qualitative and quantitative information of chromatographically pure standards in HPLC analysis, Table S2: Qualitative and quantitative information of chromatographically pure standards in SPME-GC-MS analysis, Table S3: Concentrations of yeast assimilable nitrogen in different nitrogen treatments and controls, Table S4: Evolution of amino acid content during alcoholic fermentation of spine grape and Cabernet Sauvignon grape (μmol/L), Table S5: Concentrations of wine fruity esters during alcoholic fermentation of Cabernet Sauvignon grape (μg/L), Table S6: Concentration of wine fruity esters during alcoholic fermentation of spine grape (μg/L), Table S7: Pearson's correlation coefficients (*r*) between amino acids and wine fruity esters during alcoholic fermentation, Table S8 Physiochemical indices of wine samples.

Author Contributions: Conceptualization, Z.Z.; methodology, Z.Z. and K.H.; validation, K.H.; formal analysis, Z.Z., S.C., and S.X.; investigation, Z.Z., S.C., and S.X.; writing—original draft preparation, Z.Z.; writing—review and editing, Z.Z., K.H., and Y.T.; visualization, Z.Z.; supervision, Y.T.; project administration, Y.T.; funding acquisition, Y.T. All authors have read and agreed to the published version of the manuscript.

Funding: This research was funded by National Natural Science Foundation of China (31972199), the Shaanxi Science Fund for Distinguished Young Scholars (2020JC-22), and the Scientific and Technological Key Projects of Xinjiang production and Construction Corps (2019AB025).

Institutional Review Board Statement: Not applicable.

Informed Consent Statement: Not applicable.

Data Availability Statement: Data can be found within the manuscript and Supplementary Materials.

Acknowledgments: The authors would like to thank Aihua Li from the College of Food Science and Engineering, Northwest A&F University for the excellent technical assistance of instrumental analysis.

Conflicts of Interest: The authors declare no conflict of interest.

References

1. Bell, S.J.; Henschke, P.A. Implications of nitrogen nutrition for grapes, fermentation and wine. *Aust. J. Grape Wine Res.* **2005**, *11*, 242–295. [CrossRef]
2. Gobert, A.; Tourdot-Marechal, R.; Sparrow, C.; Morge, C.; Alexandre, H. Influence of nitrogen status in wine alcoholic fermentation. *Food Microbiol.* **2019**, *83*, 71–85. [CrossRef]
3. Gutierrez-Gamboa, G.; Carrasco-Quiroz, M.; Martinez-Gil, A.M.; Perez-Alvarez, E.P.; Garde-Cerdan, T.; Moreno-Simunovic, Y. Grape and wine amino acid composition from Carignan noir grapevines growing under rainfed conditions in the Maule Valley, Chile: Effects of location and rootstock. *Food Res. Int.* **2018**, *105*, 344–352. [CrossRef] [PubMed]
4. Vilanova, M.; Ugliano, M.; Varela, C.; Siebert, T.; Pretorius, I.S.; Henschke, P.A. Assimilable nitrogen utilisation and production of volatile and non-volatile compounds in chemically defined medium by *Saccharomyces cerevisiae* wine yeasts. *Appl. Microbiol. Biotechnol* **2007**, *77*, 145–157. [CrossRef]
5. Lytra, G.; Tempere, S.; Le Floch, A.; de Revel, G.; Barbe, J.C. Study of sensory interactions among red wine fruity esters in a model solution. *J. Agric. Food Chem.* **2013**, *61*, 8504–8513. [CrossRef] [PubMed]
6. Hu, K.; Jin, G.J.; Mei, W.C.; Li, T.; Tao, Y.S. Increase of medium-chain fatty acid ethyl ester content in mixed *H. uvarum/S. cerevisiae* fermentation leads to wine fruity aroma enhancement. *Food Chem.* **2018**, *239*, 495–501. [CrossRef]
7. Fairbairn, S.; McKinnon, A.; Musarurwa, H.T.; Ferreira, A.C.; Bauer, F.F. The Impact of Single Amino Acids on Growth and Volatile Aroma Production by *Saccharomyces cerevisiae* Strains. *Front. Microbiol.* **2017**, *8*, 2554. [CrossRef] [PubMed]
8. Procopio, S.; Krause, D.; Hofmann, T.; Becker, T. Significant amino acids in aroma compound profiling during yeast fermentation analyzed by PLS regression. *LWT-Food Sci. Technol.* **2013**, *51*, 423–432. [CrossRef]
9. Hernández-Orte, P.; Cacho, J.F.; Ferreira, V. Relationship between varietal amino acid profile of grapes and wine aromatic composition. Experiments with model solutions and chemometric study. *J. Agric. Food Chem.* **2002**, *50*, 2891–2899. [CrossRef]
10. Garde Cerdán, T.; Martínez-Gil, A.M.; Lorenzo, C.; Lara, J.F.; Pardo, F.; Salinas, M.R. Implications of nitrogen compounds during alcoholic fermentation from some grape varieties at different maturation stages and cultivation systems. *Food Chem.* **2011**, *124*, 106–116. [CrossRef]
11. Torrea, D.; Varela, C.; Ugliano, M.; Ancin-Azpilicueta, C.; Leigh Francis, I.; Henschke, P.A. Comparison of inorganic and organic nitrogen supplementation of grape juice—Effect on volatile composition and aroma profile of a Chardonnay wine fermented with *Saccharomyces cerevisiae* yeast. *Food Chem.* **2011**, *127*, 1072–1083. [CrossRef] [PubMed]
12. Chen, D.; Chia, J.Y.; Liu, S.Q. Impact of addition of aromatic amino acids on non-volatile and volatile compounds in lychee wine fermented with *Saccharomyces cerevisiae* MERIT.ferm. *Int. J. Food Microbiol.* **2014**, *170*, 12–20. [CrossRef]
13. Ugliano, M.; Travis, B.; Francis, I.L.; Henschke, P.A. Volatile composition and sensory properties of Shiraz wines as affected by nitrogen supplementation and yeast species: Rationalizing nitrogen modulation of wine aroma. *J. Agric. Food Chem.* **2010**, *58*, 12417–12425. [CrossRef] [PubMed]
14. Hazelwood, L.A.; Daran, J.M.; van Maris, A.J.; Pronk, J.T.; Dickinson, J.R. The Ehrlich pathway for fusel alcohol production: A century of research on *Saccharomyces cerevisiae* metabolism. *Appl. Environ. Microbiol.* **2008**, *74*, 2259–2266. [CrossRef] [PubMed]
15. Malcorps, P.; Dufour, J.P. Short-chain and medium-chain aliphatic-ester synthesis in *Saccharomyces cerevisiae*. *Eur. J. Biochem.* **1992**, *210*, 1015–1022. [CrossRef]
16. Crepin, L.; Truong, N.M.; Bloem, A.; Sanchez, I.; Dequin, S.; Camarasa, C. Management of Multiple Nitrogen Sources during Wine Fermentation by *Saccharomyces cerevisiae*. *Appl. Environ. Microbiol.* **2017**, *83*. [CrossRef]
17. Rollero, S.; Mouret, J.R.; Bloem, A.; Sanchez, I.; Ortiz-Julien, A.; Sablayrolles, J.M.; Dequin, S.; Camarasa, C. Quantitative (13) C-isotope labelling-based analysis to elucidate the influence of environmental parameters on the production of fermentative aromas during wine fermentation. *Microb. Biotechnol.* **2017**, *10*, 1649–1662. [CrossRef]

18. Liu, P.; Wang, Y.; Ye, D.; Duan, L.; Duan, C.; Yan, G. Effect of the addition of branched-chain amino acids to non-limited nitrogen synthetic grape must on volatile compounds and global gene expression during alcoholic fermentation. *Aust. J. Grape Wine Res.* **2018**, *24*, 197–205. [CrossRef]
19. Meng, J.; Xu, T.; Qin, M.; Zhuang, X.; Fang, Y.; Zhang, Z. Phenolic characterization of young wines made from spine grape (*Vitis davidii* Foex) grown in Chongyi County (China). *Food Res. Int.* **2012**, *49*, 664–671. [CrossRef]
20. Kong, C.; Li, A.; Su, J.; Wang, X.; Chen, C.; Tao, Y. Flavor modification of dry red wine from Chinese spine grape by mixed fermentation with *Pichia fermentans* and *S. cerevisiae*. *LWT-Food Sci. Technol.* **2019**, *109*, 83–92. [CrossRef]
21. Antalick, G.; Suklje, K.; Blackman, J.W.; Meeks, C.; Deloire, A.; Schmidtke, L.M. Influence of Grape Composition on Red Wine Ester Profile: Comparison between Cabernet Sauvignon and Shiraz Cultivars from Australian Warm Climate. *J. Agric. Food Chem.* **2015**, *63*, 4664–4672. [CrossRef]
22. Zhu, Z.; Chen, S.; Su, J.; Tao, Y. Correlation Analysis between Amino Acids and Fruity Esters during Spine Grape Fermentation. *Sci. Agric. Sin.* **2020**, *53*, 2272–2284. [CrossRef]
23. Li, N.; Liu, Y.; Zhao, Y.; Zheng, X.; Lu, J.; Liang, Y. Simultaneous HPLC determination of amino acids in tea infusion coupled to pre-column derivatization with 2, 4-dinitrofluorobenzene. *Food Anal. Methods* **2016**, *9*, 1307–1314. [CrossRef]
24. Zoecklein, B.; Fugelsang, K.; Gump, B.; Nury, F. *Wine Analysis and Production*; Chapman & Hall. Inc.: New York, NY, USA, 1995.
25. Hu, K.; Jin, G.J.; Xu, Y.H.; Xue, S.J.; Qiao, S.J.; Teng, Y.X.; Tao, Y.S. Enhancing wine ester biosynthesis in mixed *Hanseniaspora uvarum*/*Saccharomyces cerevisiae* fermentation by nitrogen nutrient addition. *Food Res. Int.* **2019**, *123*, 559–566. [CrossRef] [PubMed]
26. Huang, Z.; Ough, C.S. Amino acid profiles of commercial grape juices and wines. *Am. J. Enol. Vitic.* **1991**, *42*, 261–267.
27. Roca-Mesa, H.; Sendra, S.; Mas, A.; Beltran, G.; Torija, M.J. Nitrogen Preferences during Alcoholic Fermentation of Different Non-Saccharomyces Yeasts of Oenological Interest. *Microorganisms* **2020**, *8*, 157. [CrossRef] [PubMed]
28. Gao, N.; Li, L.; Deng, X.; Song, L. Changes of free amino acids during wine fermentation. *China Brew.* **2011**, *1*, 28–33. [CrossRef]
29. Braus, G.H. Aromatic amino acid biosynthesis in the yeast *Saccharomyces cerevisiae*: A model system for the regulation of a eukaryotic biosynthetic pathway. *Microbiol. Rev.* **1991**, *55*, 349–370. [CrossRef]
30. Beltran, G.; Esteve-Zarzoso, B.; Rozès, N.; Mas, A.; Guillamón, J.M. Influence of the timing of nitrogen additions during synthetic grape must fermentations on fermentation kinetics and nitrogen consumption. *J. Agric. Food Chem.* **2005**, *53*, 996–1002. [CrossRef]
31. Ebeler, S.E.; Thorngate, J.H. Wine chemistry and flavor: Looking into the crystal glass. *J. Agric. Food Chem.* **2009**, *57*, 8098–8108. [CrossRef]
32. Meng, J.; Xu, T.; Song, C.; Li, X.; Yue, T.; Qin, M.; Fang, Y.; Zhang, Z.; Xi, Z. Characteristic free aromatic components of nine clones of spine grape (Vitis davidii Foex) from Zhongfang County (China). *Food Res. Int.* **2013**, *54*, 1795–1800. [CrossRef]
33. Martínez-Pinilla, O.; Guadalupe, Z.; Hernández, Z.; Ayestarán, B. Amino acids and biogenic amines in red varietal wines: The role of grape variety, malolactic fermentation and vintage. *Eur. Food Res. Technol.* **2013**, *237*, 887–895. [CrossRef]
34. Garde-Cerdán, T.; Ancín-Azpilicueta, C. Effect of the addition of different quantities of amino acids to nitrogen-deficient must on the formation of esters, alcohols, and acids during wine alcoholic fermentation. *LWT-Food Sci. Technol.* **2008**, *41*, 501–510. [CrossRef]
35. Hernández-Orte, P.; Ibarz, M.; Cacho, J.; Ferreira, V. Effect of the addition of ammonium and amino acids to musts of Airen variety on aromatic composition and sensory properties of the obtained wine. *Food Chem.* **2005**, *89*, 163–174. [CrossRef]
36. Frick, O.; Wittmann, C. Characterization of the metabolic shift between oxidative and fermentative growth in *Saccharomyces cerevisiae* by comparative 13 C flux analysis. *Microb. Cell Fact.* **2005**, *4*, 1–16. [CrossRef]
37. Saerens, S.M.; Delvaux, F.R.; Verstrepen, K.J.; Thevelein, J.M. Production and biological function of volatile esters in *Saccharomyces cerevisiae*. *Microb. Biotechnol.* **2010**, *3*, 165–177. [CrossRef]
38. Boutou, S.; Chatonnet, P. Rapid headspace solid-phase microextraction/gas chromatographic/mass spectrometric assay for the quantitative determination of some of the main odorants causing off-flavours in wine. *J. Chromatogr. A* **2007**, *1141*, 1–9. [CrossRef] [PubMed]
39. Magasanik, B.; Kaiser, C.A. Nitrogen regulation in *Saccharomyces cerevisiae*. *Gene* **2002**, *290*, 1–18. [CrossRef]
40. Bloem, A.; Sanchez, I.; Dequin, S.; Camarasa, C. Metabolic Impact of Redox Cofactor Perturbations on the Formation of Aroma Compounds in *Saccharomyces cerevisiae*. *Appl. Environ. Microbiol.* **2016**, *82*, 174–183. [CrossRef] [PubMed]
41. Saerens, S.; Delvaux, F.; Verstrepen, K.; Van Dijck, P.; Thevelein, J.; Delvaux, F. Parameters affecting ethyl ester production by *Saccharomyces cerevisiae* during fermentation. *Appl. Environ. Microbiol.* **2008**, *74*, 454–461. [CrossRef]
42. Zhu, Q.; Wang, X.; Xu, Y. Palladium-Catalyzed Tandem Reactions of β-(2-Bromophenyl)-α,β-Unsaturated Carbonyl Compounds with 2-Hydroxyphenylboronic Acid: A New Route to Benzo[c]chromenes. *Eur. J. Org. Chem.* **2012**, *2012*, 1112–1114. [CrossRef]
43. Mason, A.B.; Dufour, J.P. Alcohol acetyltransferases and the significance of ester synthesis in yeast. *Yeast* **2000**, *16*, 1287–1298. [CrossRef]
44. Mauricio, J.C.; Valero, E.; Millán, C.; Ortega, J.M. Changes in nitrogen compounds in must and wine during fermentation and biological aging by flor yeasts. *J. Agric. Food Chem.* **2001**, *49*, 3310–3315. [CrossRef] [PubMed]
45. Farina, L.; Medina, K.; Urruty, M.; Boido, E.; Dellacassa, E.; Carrau, F. Redox effect on volatile compound formation in wine during fermentation by *Saccharomyces cerevisiae*. *Food Chem.* **2012**, *134*, 933–939. [CrossRef]

 *fermentation*

MDPI

Article

Chemical and Sensory Characterization of Vidal Icewines Fermented with Different Yeast Strains

Ke Tang [1,2], Yulu Sun [1,2], Xiaoqian Zhang [1,2], Jiming Li [3] and Yan Xu [1,2,*]

1 Key Laboratory of Industrial Biotechnology of Ministry of Education, Jiangnan University, 1800 Lihu Avenue, Wuxi 214122, China; tandy81@jiangnan.edu.cn (K.T.); sunyulu1120@163.com (Y.S.); z18861823809@163.com (X.Z.)
2 Lab of Brewing Microbiology and Applied Enzymology, School of Biotechnology, Jiangnan University, 1800 Lihu Avenue, Wuxi 214122, China
3 Center of Science and Technology, Changyu Group Company Ltd., Yantai 264000, China; zyljm@163.com
* Correspondence: yxu@jiangnan.edu.cn

Abstract: The aim of this study is to comprehensively investigate the aroma composition and sensory attributes of Vidal icewine fermented with four yeast strains (ST, K1, EC1118, and R2). A total of 485 kinds of volatile components were identified by comprehensive two-dimensional gas chromatography-time of flight mass spectrometry, among which 347 kinds of volatile compounds were the same in four kinds of sample. The heat map was conducted with 156 volatile compounds, which have aroma contributions, and the analysis results identified the characteristics of the aroma composition of icewine fermented with different yeasts. Quantitative descriptive analysis was performed with a trained panel to obtain the sensory profiles. The aroma attributes of honey and nut of the icewine fermented by R2 were much higher than others. Partial least squares discriminant analysis further provided 40 compounds that were mainly responsible for the differences of the aroma characteristics of the icewines fermented by four yeasts. This study provides more data on the current status of Vidal icewines by main commercial yeasts.

Keywords: icewine; Vidal; yeast; aroma compounds; sensory analysis

Citation: Tang, K.; Sun, Y.; Zhang, X.; Li, J.; Xu, Y. Chemical and Sensory Characterization of Vidal Icewines Fermented with Different Yeast Strains. *Fermentation* **2021**, 7, 211. https://doi.org/10.3390/fermentation7040211

Academic Editor: Bin Tian

Received: 31 August 2021
Accepted: 27 September 2021
Published: 29 September 2021

Publisher's Note: MDPI stays neutral with regard to jurisdictional claims in published maps and institutional affiliations.

Copyright: © 2021 by the authors. Licensee MDPI, Basel, Switzerland. This article is an open access article distributed under the terms and conditions of the Creative Commons Attribution (CC BY) license (https://creativecommons.org/licenses/by/4.0/).

1. Introduction

Icewine is a unique sweet wine, made by delaying grape harvest, freezing, and air-drying the fruits hanging on the branches at −7 °C (EU regulations) and −8 °C (Canada, Vintners Quality Alliance 1999), pressing and low-temperature sugar-preserving fermentation in the freezing state. The ripening condition of icewine grapes at low temperatures is quite different from that of ordinary wine grapes. In this stage, the grapes are dried and shriveled, and the flavor substances such as sugar and aroma are continuously "concentrated". Therefore, icewine has more unique flavor characteristics than ordinary wine [1]. At present, only a few countries in the world (Canada, Germany, Austria, Switzerland, China, etc.) can produce icewine in a few regions [2]. In addition, the special post-freezing ripening process requires the ice grape berries to have thick skins, easy to preserve, and the ice grape to resist cold. Vidal is a hardy hybrid grape that is grown mainly in Canada and the northeastern United States. It has thick skin, strong resistance to cold, and high natural acidity, thus good for icewine [1,3].

Similar to regular wines, the aroma is also a core factor in determining the quality of icewine [4]. In general, the aroma of icewine is mainly influenced by grape raw material and vinification, and among the vinification, the yeast is one of the important factors to determine the aroma characters of icewine [5]. During fermentation, yeast converts sweet and low-aroma grape juice into high-aroma wine through the glycolytic pathway. In this process, the fructose and glucose are converted to ethanol, carbon dioxide, and volatile metabolites [6–8]. Furthermore, many volatile metabolites are also released from non-volatile grape-derived precursors by yeast enzymes [9]. Unlike the regular grapes, the grapes for making icewine have to be harvested in their naturally frozen state and pressed while still frozen under a higher-than-normal pressure. Normally, the sugar concentration of grape juice for making icewine reach a minimum 35° Brix, and sometimes it can be as high as 50° Brix [10]. However, the more noteworthy are the challenges of making icewine that is caused by the high reducing sugar and viscosity of ice grape juice. The high sugar fermentation environment means that it takes longer for icewine to reach the expected alcohol content, sometimes even last for several months. Besides, high sugar also can cause high osmotic stress, which leads to the higher volatile acid content of icewine than ordinary wine [11]. For all of this, special measures should be taken to ensure the safety of long-term fermentation, the excellent quality, the harmony of taste, and the volatile acid content of icewine. The main methods are selecting appropriate yeast and controlling fermentation temperature, especially the yeast species that have significant effects on the formation of acetic acid and glycerol, the fermentation speed, and the sensory properties of icewine [1].

Due to the ability to ferment under high-sugar conditions, be alcohol-tolerant, and produce relatively lower volatile acidity, V1116 (*Saccharomyces cerevisiae*), ST (*S. cerevisiae*), N96 (*S. cerevisiae*), EC1118 (*S. bayanus*), R2 (*S. bayanus*), VL1 (*S. cerevisiae*), and AWRI 1572 (a hybrid between *S. cerevisiae* and *S. bayanus*) are normally utilized to ferment icewine [12–15]. Erasmus et al. (2004) [13] found that N96 and EC1118 could be used to produce high-quality icewine with a strong fruit aroma and low sulfur-like aroma level by studying the differences of seven commercial yeast strains (ST, N96, VIN13, VIN7, EC1118, 71B, V1116) in Riesling icewine production. Crandles et al. (2015) [12] researched the aroma compounds of icewine fermented by different yeasts (V1116, VL1, EC1118, natural fermentation) and found that yeast strain impacted odor-active compounds in Riesling and Vidal icewines, but yeast effects depended upon cultivar and vintage. Synos et al. (2015) [16] used three kinds of yeast (V1116, EC1118, VL1) to ferment and compare with natural fermentation in Cabernet Franc icewine. It was found that the content of aroma active substances in EC1118 and naturally fermented icewine was the highest, but there were great differences between them in the kinds of aroma active substances. Pigeau et al. (2007) [17] studied the effects of ice grape juice with a sugar content over 40 Brix on fermentation, and the results showed that with the increase of ice grape juice sugar content, more acetic acid and glycerol were produced during fermentation, the opposite was that the yeast growth rate and ethanol production were reduced. However, studies assessing yeast effects on icewine aroma compounds are still uncommon, especially in Vidal icewine.

In this work, volatile compounds and sensory were evaluated in Vidal icewine fermented by four different commercial yeasts using comprehensive two-dimensional gas chromatography-time of flight mass spectrometry (GC×GC-TOFMS) and quantitative descriptive analysis (QDA), respectively. Furthermore, relationships between the aroma compounds and sensory attributes were analyzed by multivariate statistical analysis to compare the differences and characteristics of volatile components in icewine fermented by different yeasts. The aim of this study was to comprehensively investigate aroma characteristics in Vidal icewine and to provide more data on the current status of Vidal icewine by main commercial yeasts.

2. Materials and Methods

2.1. Icewine Samples

Experimental Vidal icewines were available from Vidal grapes harvested from ChangYu Winery in Huanren-on-the-Huanlong Lake, LiaoNing province, China, in 2018. The total sugar and total acid (in tartaric acid) were 380.5 g/L and 9.1 g/L for the icewine juice, respectively. The icewine samples were fermented by 4 different commercial yeasts (ST, K1, EC1118, R2) at 0.2 g/L—in three replicates for each one ($n = 3$). The alcoholic fermentation temperature was 13 ± 1 °C in 10 L carboys with 100 mg/L of $K_2S_2O_5$. Fermentation proceeded until sugar consumption by yeast stopped, after which carboys were moved to a -2 °C chamber for cold stabilization. The icewine samples were then racked off the gross lees and bottled for analysis. At bottling, 50 mg/L $K_2S_2O_5$ were added. Details of the icewine samples are provided in Table 1 (the physico-chemical method of wine based on GB/T 15037-2006).

Table 1. Summary of icewine samples by four different commercial yeast strains.

Treatment	Residual Sugar (g/L)	Ethanol (%, *v*/*v*)	Titratable Acidity (g/L)	Acetic Acid (g/L)	pH
EC1118	183.45 ± 1.22	10.5 ± 0.1	11.2 ± 0.5	1.11 ± 0.02	3.51
R2	170.21 ± 1.05	11.4 ± 0.2	12.8 ± 0.4	1.40 ± 0.02	3.45
K1	181.33 ± 2.11	10.6 ± 0.1	12.1 ± 0.4	1.22 ± 0.03	3.52
ST	178.92 ± 1.57	10.8 ± 0.1	12.2 ± 0.3	0.92 ± 0.11	3.52

2.2. Chemicals and Reagents

All chemical standards and internal standards (IS) were of the highest available purity (GC-grade). The analytical standards employed for the positive identification of the aroma compounds were obtained from Sigma-Aldrich (St. Louis, MO, USA) with at least 97% purity. n-hexyl- d13-alcohol (≥98, IS1), 2-methoxy-d3-phenol (≥98, IS2), menthyl acetate (≥98, IS3) were purchased from ANPEL Scientific Instrument Co., Ltd. (Shanghai, China). Ethanol (99.9%, HPLC grade) was purchased from Sigma-Aldrich (Shanghai, China). Sodium chloride (NaCl) was purchased from China National Pharmaceutical Group Corp. Ultrapure water was obtained from a Milli-Q purification system (Millipore, Bedford, MA, USA). Four different *S. cerevisiae* (ST, K1, EC1118, R2) were purchased from France Lallemand Co., Ltd., Paris, France.

2.3. Descriptive Sensory Analysis (DA)

2.3.1. Panel

Panel candidates were recruited among students and employees of the school of Biotechnology in Jiangnan University. The candidates were selected based on interest, health status, availability, and familiarity with wine using an initial recruitment questionnaire. Thereafter, the candidates were required to complete sensory ability tests for aroma identification, ranking, and response scales (10 cm unstructured scale, ranging from "none" on the left end to "strong" on the right end) [18]. A total of 30 candidates were selected who had achieved at least 70% acuity and were available during the designated time. The general training schedule consisted of 16 h (2 h/week) for introduction to sensory analysis, aroma description and identification, ranking, and triangle tests [19]. Commercial icewines were provided for descriptive tests after 2 to 3 months of general training. The consistency, discernibility, and repeatability of panelists were then evaluated. A total of 12 assessors (5 females and 7 males, aged between 21 and 31 years old) with good sensory performance according to the evaluation were finally selected to participate a further training of descriptive analysis.

2.3.2. Training

The training process was the same as the previous method [20] and took 3 months (once a week). Firstly, 6 descriptors (nut, tropical fruit, apricot, honey, caramel, and rose) of Vidal icewines were obtained by a panel discussing to describe the icewine flavor. Then, assessors were trained for the identification and intensive evaluation of the selected descriptors with reference standards. The reference standards were prepared by adding corresponding aroma standards from Le nez du vin (Jean Lenoir, Provence, France) to 10% *v*/*v* aqueous ethanol (pH 3.4) and diluted in series. The sensory reference standards of the 6 descriptors are shown in Table S1. The assessors' performance was assessed by PanelCheck in terms of their ability in consistency, stability, and repeatability for giving scores before sample evaluation.

2.3.3. Sample Evaluation

Samples were provided to assessors in standard wine glasses covered, coded with random 3-digit numbers, and containing 30 mL icewine (8–12 °C) per glass [21]. Each assessor scored the icewines for each attribute with unipolar 10 cm line scale [22], anchored on the left end with 0 (none) means low intensity, and on the right end with 10 (extreme). Scores were converted to scores from 0 to 10 and exported to an Excel spreadsheet. Between samples, the panelists were asked to take a short break and smell the water to minimize any carry-over effect from the previous sample. All the testing took place in isolated booths illuminated with standard yellow light to eliminate color differences at 20 °C. All sample evaluation was performed in duplicate.

2.4. Analysis of Volatile Aroma Compounds

Headspace solid-phase microextraction coupled with comprehensive 2-dimensional gas chromatography and time-of-flight mass spectrometry (HS-SPME-GC×GC-TOFMS) was employed to determine the volatile profile of the icewine samples, and each sample was analyzed in 3 replicates. Based on previously described methods with slight modifications [23], 5 mL of icewine were placed into 20 mL glass with a silicon septum and saturated with 1.5 g of NaCl. Internal standard mixture (10 uL; IS1: n-hexyl- d13-alcohol, 403.76 μg/L; IS2: 2-methoxy-d3-phenol, 197.6 μg/L; IS3: menthyl acetate, 20.08 μg/L) was added as an internal standard used for the semi-quantification of aroma compounds. A MultiPurpose autosampler (Gerstel GmbH and Co. KG, Mülheim an der Ruhr, Germany) with a 50/30 μm divinylbenzene/Carboxen/polydimethylsiloxane (DVB/CAR/PDMS) fiber (2 cm; Supelco Inc., Bellefonte, PA, USA) was used to extract the volatile compounds from the headspace of the sample vial based on the previous work with slight modifications. The sample was equilibrated at 50 °C for 5 min and then extracted for 45 min under stirring (250 rpm). After extraction, the fiber was inserted into the gas chromatograph injection port (250 °C) and desorbed for 5 min (splitless) to GC × GC-TOFMS analysis.

A LECO Pegasus 4D GC × GC-TOFMS instrument (LECO Corporation, St. Joseph, MI, USA) equipped with an Agilent 7890B gas chromatograph (Agilent Technologies, Palo Alto, CA, USA) was used in GC × GC-TOFMS analysis. A polar/moderately polar column set was optimized for the GC × GC separation. The first dimension (1D) column was DB-FFAP column (60 m × 0.25 mm i.d. and 0.25 μm film thickness, Agilent Technologies Inc., Santa Clara, CA, USA), and the second dimension (2D) column was Rxi-17Sil MS Cap. column (1.5 m × 0.25 mm i.d.and 0.25 μm film thickness, Restek Technologies Inc, Bellefonte, PA, USA). The oven temperature of the first column was held at 45 °C for 2 min, ramped at 4 °C/min to 230 °C, held for 15 min. The temperature of the second column was held at 40 °C for 2 min, ramped at 5 °C/min to 250 °C, held for 5 min. A quad-jet, dual-stage thermal modulator was equipped between the 1D and 2D columns. The modulation period as set for 4 s with a 1 s hot pulse time. Helium (99.9995% purity) was used as the carrier gas at a constant flow of 1.0 mL/min. For the TOFMS system, the temperatures of the transfer line and ionization sources were 280 and 230 °C, respectively. Electron impact mass spectra were recorded at 70 eV, and the acquisitions were performed over an m/z scan range of 35–400 amu.

LECO ChromaTOFTM Workstation (version 4.44) was used for all acquisition control and data processing. Automated peak detection and spectral deconvolution were employed. The minimum value for the signal to noise (S/N) ratio necessary to record a chromatographic peak was set as 50 in GC × GC. The positive identification of the compounds was achieved by comparing the retention data and mass spectra of the standard compounds. For unavailable standards, tentative identification of the compounds was achieved by comparing their experimental retention indices (I) with the retention indices reported in the scientific literature (IT). A compound was considered to be tentatively identified if the similarity between mass spectrometric information of each chromatographic peak and the National Institute of Standards and Technology (NIST) mass spectra library was at least 75%, and the difference between I and IT did not exceed 30 units. Semi-quantitative analysis was performed in triplicate, and average values of concentration were used in further data elaboration.

2.5. Statistical Analysis

Panelcheck (version 1.4.2, Nofima Mat and DTU-Informatics and Mathematical Modelling, Norway) was used to assess the consistency, discernibility, and repeatability of the results from panelists. Partial least squares (PLS) regression analysis was conducted using XLSTAT software (version 2014; Addinsoft, Paris, France), employed to investigate the relationships between the compositional variables (x data) and the sensory attributes (y data) in icewine samples fermented by 4 different yeast strains. Heatmap were visualized with Gephi (Version 0.9.1) using classic FruchtermanReingold algorithm.

3. Results and Discussion

3.1. Samples of Icewines

The four commercial yeast strains produced icewines with small differences in residual sugar values (170.21–183.45 g/L), pH (3.45–3.52), ethanol values (10.5–11.4%), and TA (11.2–12.8 g/L) (Table 1). These results are consistent with Fuleki (1994) [14], who also found that little differences in Riesling icewines fermented by five commercial yeasts. Furthermore, R2 icewines contained the lowest residual sugar, had the lowest pH, and highest ethanol.

Acetic acid is a key indicator during the wine-making progress, and its content determines the quality of wine directly. Equally, acetic acid is important to the production of icewine. The yeast will be exposed to a high osmotic pressure due to the high sugar concentration of ice grape juice, which results in yeast metabolism abnormality. On the one hand, achieving the desired alcohol content needs to take more time. On the other hand, high osmotic pressure will lead to the formation of a large number of volatile acids that are dominated by acetic acid [13,24]. Appropriate acetic acid will not have a bad effect on icewine. Under the catalysis of ethanol acetyl transferase, acetic acid can react with ethanol to produce ethyl acetate, which will bring a fruit flavor for icewine. In Canada, VQA (1999) limits that the volatile acid content in icewine must be below 2.1 g/L. However, Cliff and Pickering (2006) [25] found that the threshold of acetic acid in icewine was 3.185 g/L, meaning the acetic acid content in most icewine would not be above this value. A previous study of Canadian icewine reported that there was a wide range of acetic acid in Canadian icewine, from 0.49 to 2.29 g/L. The average level was around 1.3 g/L, which was still far below the sensory threshold and will not have a great change on the flavor of icewine [3]. Table 1 shows that the concentration of acetic acid was between 0.92 and 1.40 g/L in the four yeast-fermented icewine samples, also far below the 3.185 g/L. The content of acetic acid in icewine fermented by ST yeast was the lowest, which was only 0.92 g/L, and EC1118 yeast also had a good effect on the control of acetic acid, which was 1.11 g/L. The results were the same as Erasmus et al. (2004) [13], who compared the differences of acetic acid in icewines that fermented by seven different commercial yeasts and found that ST yeast produced the least acetic acid, followed by N96 and EC1118.

3.2. Sensory Analysis

In order to describe the sensory differences of Vidal icewine fermented by different yeasts, QDA was used to analyze the flavor of icewine samples. Based on the average strength of six aroma descriptors (nut, tropical fruit, apricot, honey, caramel, and rose) in icewine [4,20] to plot a spider web (Figure 1). It shows that the aroma intensity of honey and nut was the highest in icewine fermented by R2 yeast, especially the honey aroma, was much higher than others, and the aroma intensity of rose and tropical fruit was ranked only second to EC1118 yeast-fermented wine. For the icewine fermented with EC1118 yeast, it has a stronger aroma intensity on tropical fruits and caramel than others. However, the aroma intensity in icewine fermented by K1 was much lower than others except for the apricot aroma.

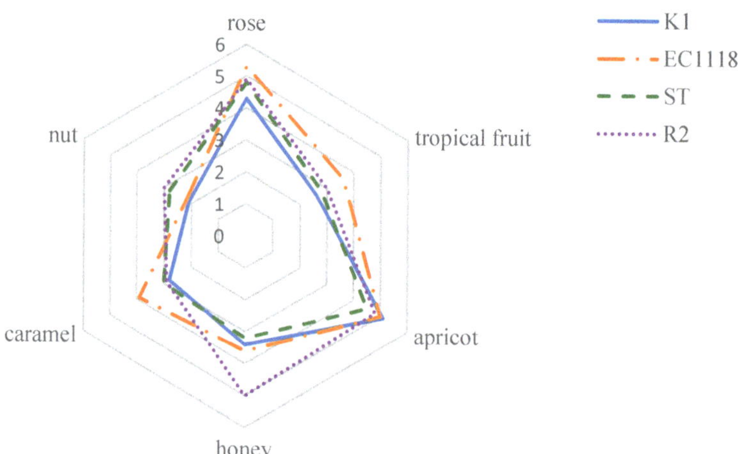

Figure 1. Sensory profiles of Vidal icewines fermented by different yeasts.

3.3. Chemical Analysis

3.3.1. Volatile Profiling of Icewines Fermented by Different Yeasts

Due to the complexity of wine aroma composition and matrix, there were some problems in the separation and detection of wine by one-dimensional gas chromatography, such as insufficient peak capacity, co-outflow, low sensitivity, and so on. Therefore, this method cannot satisfy our needs for the study of wine flavor chemistry. However, the emergence of GC×GC-TOFMS provides a useful tool for better separating and identifying these complex components [26]. In this study, the volatile compounds of four different Vidal icewine fermented by four yeasts (ST, K1, EC1118, R2) were analyzed by GC×GC-TOFMS. A total of 3784 chromatographic peaks were detected. Figure S1A shows the chromatogram of the volatile components in R2 fermented icewine. It can be seen from Figure S1B that the three compounds, 4-methyl-1-pentanol, 1-butanol, and ethyl valerate, co-flowed in one dimension and were difficult to identify and quantify, but they were well separated in the two-dimensional chromatography. Consequently, it can be proved that GC×GC-TOFMS can effectively solve the co-outflow phenomenon and improve the accuracy of qualitative analysis.

Finally, 485 volatile compounds were identified in four icewine samples, including 123 esters, 26 terpenes, 105 alcohols, 115 carbonyl compounds, 23 furans, 21 nitrogen-containing compounds, 19 volatile phenols, 7 sulfides, 35 aromatic compounds, and 11 lactones. A total of 391, 399, 364, and 457 volatile compounds were detected in four icewines, which fermented by K1, EC1118, ST, and R2 yeast, respectively (Table 2). It can be seen that the number of volatile compounds in icewine fermented by R2 was much higher than the other. In addition, there were 347 compounds in common, accounting for 72% of the total volatile compounds. It can also be seen from Table 2 that the icewines fermented by different yeasts differed greatly in yeast fermentation products such as esters, alcohols, carbonyls (aldehydes and ketones), and nitrogen-containing compounds. In previous studies, the aroma compounds were analyzed using GC-MS in Riesling, Vidal, and Cabernet Franc icewines, which was fermented by different yeasts, the differences of aroma compounds mainly in alcohol and ester compounds, and the sum of these two classes of compounds accounted for more than 50% of the total [12,16]. The results of this study were basically consistent with the above research.

Table 2. Volatile compounds in Vidal icewine fermented by different yeasts.

	K1	EC1118	ST	R2
Esters	97	91	93	117
Alcohols	79	81	73	105
Ketones	50	49	46	55
Aromatic compounds	34	35	32	31
Terpenes	25	24	24	26
Aldehydes	18	23	14	26
Furans	18	19	19	23
Acids	19	21	19	22
Nitrogen-containing compounds	17	21	10	18
Phenols	17	18	17	17
Lactones	11	11	11	11
Sulfides	6	6	6	6
Total	391	399	364	457

3.3.2. Chemical Characteristic of Icewines Fermented by Different Yeasts

Among the 485 volatile compounds identified in all icewine samples, not all of them have aroma contributions, and the study of flavor pays more attention to the compounds with aroma contributions [27]. Therefore, based on the qualitative results and matched it with the aroma substance database (Flavornet, http://www.flavornet.org/flavornet.html, accessed on 11 December 2018), 156 volatile compounds were selected to further analyze, details can be seen in Table S2.

The semi-quantitative results were obtained by the internal standard method and then drawn on a heat map (Figure 2). In the figure, the rows represent the compounds, and the columns represent the yeasts used for fermentation. The distance measure used in clustering rows (compounds) was the Euclidean method, and the data were standardized before cluster analysis. According to the results of heat map analysis, the 156 compounds can be divided into three classes, the compounds of class A were the highest content in R2 yeast fermented icewine, class B substances were highest in ST yeast fermented icewine, and class C aroma components had the highest content in K1 and EC1118 yeast fermented icewine.

In class A, most of the compounds were esters, alcohols, terpenes, and carbonyl compounds. In this group, ethyl 2-methylbutyrate, ethyl isobutyrate, ethyl octanoate, and ethyl caproate were all the major aroma contributors in Vidal icewine, which smells similar to the apple, pineapples, and so on. In addition, the 1-octen-3-ol and 1-octen-3-one showed a mushroom characteristic, 1-hexanol showed a resin characteristic, β-damascenone, linalool, and geraniol with the aroma characteristics of honey, lavender, and rose, respectively. α-Terpinene, γ-terpinene, and 1,4-cineole were the aroma characteristics of pine. Benzaldehyde showed the aroma characteristics of almonds. These compounds were all key aroma compounds reported in Vidal icewine [4,20], especially β-damascenone, which has been reported in many studies that is the strongest aroma in Vidal icewine [4,20,28]. In red wines, β-damascenone has been found to enhance fruit aroma while can also inhibit the plant aroma produced by methoxypyrazine [29]. Similar phenomena have also been found in the study of icewine, β-damascenone not only affects the perception of honey but also affects the perception of other aromas such as apricot peach [4].

Class B is mainly composed of aromatic compounds and ester compounds, including seven aromatic compounds, six ester, four alcohol, six aldehydes and ketones, two nitrogen compounds, and one volatile phenol, of which phenethyl acetate (flowers), homofuraneol (sweet), ethyl valerate (fruit) were reported as the important aroma compounds in icewine [4]. Class C compounds gather all lactones, which usually have aroma characteristics such as apricot [4]. From the above analysis, we can see that some key aromas, which has reported on Vidal icewine were significantly higher in R2 than other samples, such as 1-octen-3-ol, 1-hexanol, β-damascenone, linalool, geraniol, 1-octen-3-one, and so on, and the alcohol, ester, and carbonyl compounds in the wine also had obvious advantages.

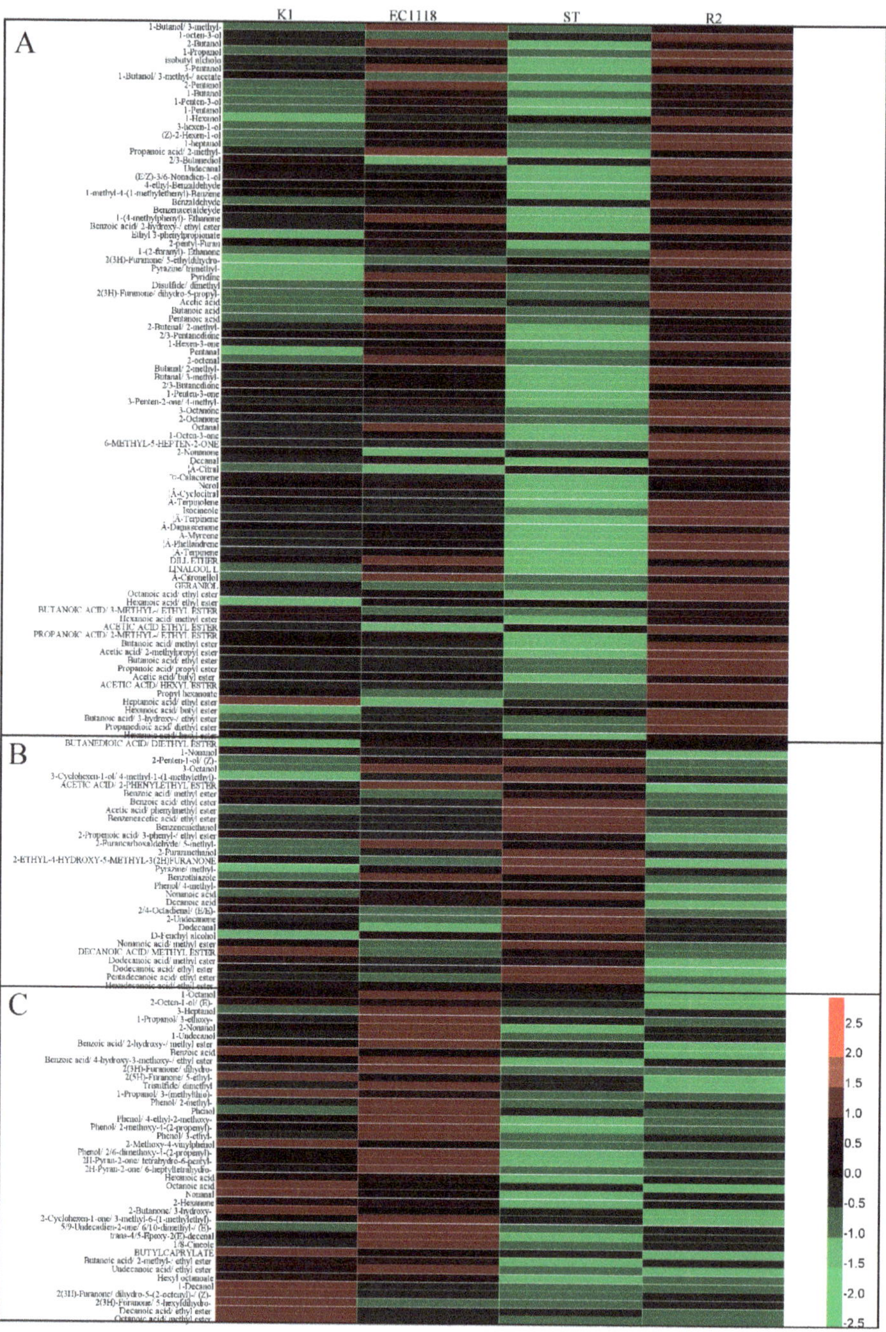

Figure 2. Heat map of 156 compounds with different aroma characteristics of four different yeast fermented icewines. In the figure, the rows represent the compounds, and the columns represent the yeasts used for fermentation. Distance measure used in clustering rows (compounds) was Euclidean method, and the data were standardized before cluster analysis. (**A**), the compounds were higher in R2 yeast fermented icewine; (**B**), the compounds were higher in ST yeast fermented icewine; (**C**), the compounds were higher in K1 and EC1118 yeast fermented icewine.

3.4. Relationships between the Sensory Attributes and Aroma Compounds

To further understand the chemical origin of the sensory descriptors, PLS regression analysis was employed to determine the associations between the aroma attributes (y variables, $n = 6$) and the significantly different aroma compounds (x variables, $n = 65$) in four different icewine samples. The result showed that differences existing in Vidal icewines fermented by different yeasts could be revealed by the chemometric methods. The aroma compounds plotted in the vicinity of the sensory descriptors were positively associated with those attributes (Figure 3).

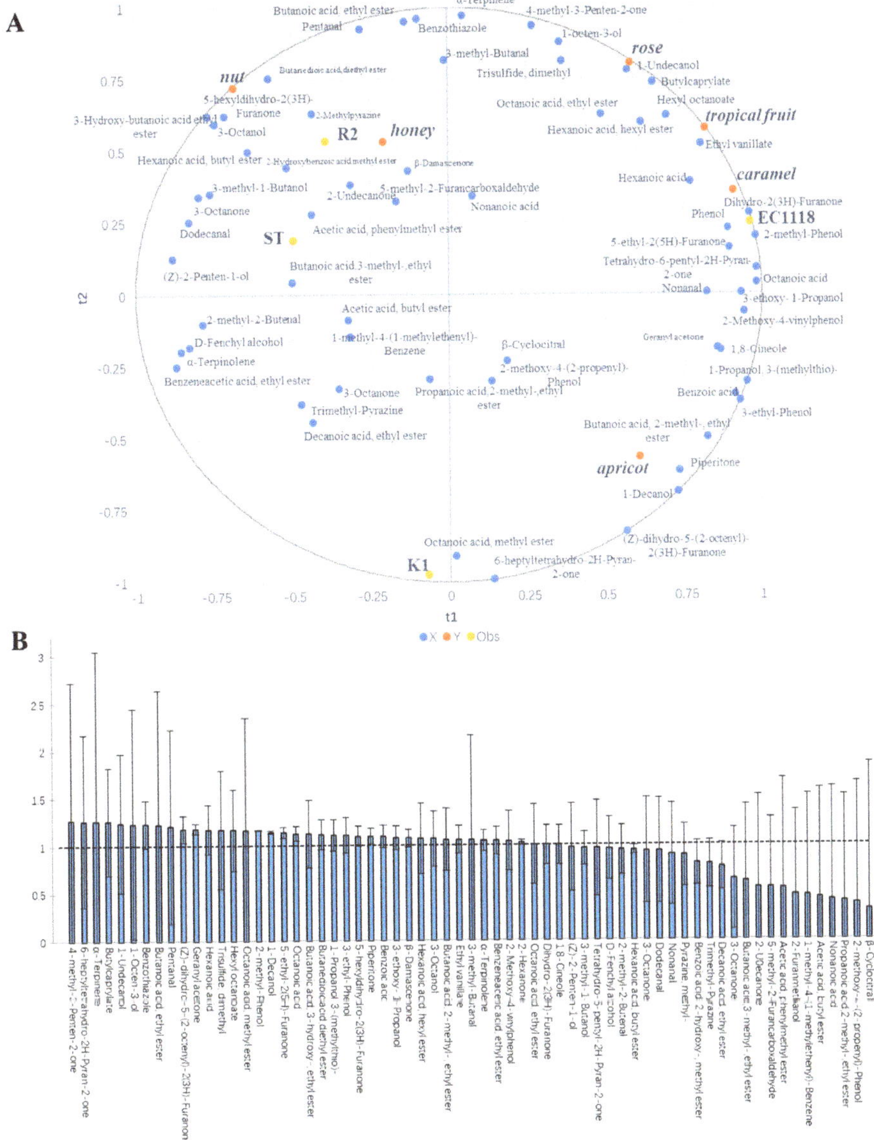

Figure 3. Projection on the PLS regression between sensory attributes and aroma compounds ($p < 0.05$) of four different yeast fermented icewines (**A**); the VIP values were calculated from the PLS regression model (**B**).

As we can see from Figure 3A, the aroma attributes of honey and nut can be used to distinguish the aroma of the icewine fermented by R2 and other icewines. At the same time, R2 was also well related to β-damascenone, 2-hydroxybenzoic acid methyl ester, 2-methylpyrazine, Butanedioic acid diethyl ester, 5-hexyldihydro-2(3H)-furanone, 3-hydroxy-butanoic acid ethyl ester, and 3-octanol. The icewines fermented by EC1118 had high correlation with some sweet aromas such as caramel and tropical fruit. The caramel was well related to dihydro-2(3H)-furanone and hexanoic acid, which showed the aroma of caramel and sweet. The esters, such as ethyl vanillate, hexanoic acid hexyl ester, octanoic acid ethyl ester, hexyl octanoate together with butylcaprylate were associated with the "tropical fruit" aroma. The esters generally contributed to the fruity aroma of the wine, which conforms to the PLS regression results. In previous studies, Huang et al. (2018) [20] find the sensory profiles of icewines from China were characterized as nut and honey aromas, while icewines from Canada expressed caramel and rose aromas. Bowen and Reynolds (2015) [30] stress the importance of time of harvest, whereby early-harvested fruit (e.g., following a "cold snap" in late November) typically produces wines that were fruit-forward (citrus, peach, etc.), whereas fruit harvested after several freeze-thaw cycles normally produce wines that were more caramel, nutty, and sherry-like. According to the results of this study, the possible reason for the different aroma characteristics of Chinese and Canadian Vidal icewine was not only the harvest time but also the selection of different yeast during the fermentation.

In addition, the variable importance for projection (VIP) values was obtained from the PLS regression model (Figure 3B). The VIP scores for the 40 aroma compounds were >1, including 11 esters (butylcaprylate, butanoic acid ethyl ester, hexyl octanoate, octanoic acid methyl ester, butanoic acid 2-methyl- ethyl ester, ethyl vanillate, benzeneacetic acid ethyl ester, hexanoic acid hexyl ester, octanoic acid ethyl ester, butanoic acid 3-hydroxy- ethyl ester, butanedioic acid diethyl ester), five alcohols (3-octanol, 1-undecanol, 1-octen-3-ol, 1-decanol, 3-ethoxy-1-propanol), five terpenes (α-terpinene, piperitone, geranyl acetone, α-terpinolene, 1,8-cineole), five aldehydes and ketones (pentanal, 3-methyl-butanal, 4-methyl-3-penten-2-one, 6-heptyltetrahydro-2H-pyran-2-one, 2-hexanone), four furans (5-ethyl-2(5H)-furanone, 5-hexyldihydro-2(3H)-furanone, dihydro-2(3H)-furanone, (Z)-dihydro-5-(2-octenyl)-2(3H)-furanone), four phenols (3-ethyl-phenol, 2-methoxy-4-vinylphenol, phenol, 2-methyl-phenol), three acids (hexanoic acid, octanoic acid, henzoic acid), and three sulfides (trisulfide dimethyl, 1-propanol 3-(methylthio)-, benzothiazole). These aroma compounds can be considered as being mainly responsible for the differences in the aroma characteristics between the icewines fermented by four yeasts.

4. Conclusions

In this study, four different yeast fermented icewines were further analyzed by evaluating the chemical and sensory profiles and then the potential associations between them. A total of 485 volatile components were identified by HS-SPME-GC×GC-TOFMS, among which 347 kinds of volatile compounds were the same in four kinds of the sample. Matching it with aroma substance database (http://www.flavornet.org/flavornet.html, accessed on 22 September 2021), finally, 156 aromas were selected to draw the heat map, and the analysis results identified the characteristics of aroma composition of icewines fermented by different yeasts. The sensory differences of different Vidal icewine samples were distinguished by QDA. PLS-DA also provided some candidate compounds that were correlated with particular sensory characteristics in different yeast fermented icewine samples. This study not only comprehensively investigates aroma characteristics in icewine fermented by different yeasts but also provides a theoretical reference for the production of icewine.

Supplementary Materials: The following are available online at https://www.mdpi.com/article/10.3390/fermentation7040211/s1. Figure S1: Analytical ion chromatogram contour plot analysis of volatile components in icewine fermented R2 yeast (A) and modulated peaks of 3 compounds found in icewine (B) and deconvoluted mass spectra of 3 compounds(C). Table S1: Aroma descriptor reference. Table S2: Information on 156 compounds with aroma characteristics ($\mu g \cdot L^{-1}$).

Author Contributions: Conceptualization and funding acquisition, K.T. and Y.X.; methodology, formal analysis and project administration, K.T. and X.Z.; investigation and data curation, X.Z. and J.L.; writing—original draft preparation and writing-review and editing, K.T. and Y.S. All authors have read and agreed to the published version of the manuscript.

Funding: This research was funded by the National Natural Science Foundation of China (31501470), and National Key R&D Program (2016YFD0400504).

Institutional Review Board Statement: Not applicable.

Informed Consent Statement: Not applicable.

Data Availability Statement: Not applicable.

Conflicts of Interest: The authors declare no conflict of interest.

References

1. Cliff, M.A.; Yuksel, D.; Girard, B.; King, M. Characterization of Canadian ice wines by sensory and compositional analyses. *Am. J. Enol. Viticult.* **2002**, *53*, 46–53.
2. Tang, K.; Li, J.M.; Wang, B.; Ma, L.; Xu, Y. Evaluation of nonvolatile flavor compounds in Vidal icewine from China. *Am. J. Enol. Viticult.* **2013**, *64*, 110–117. [CrossRef]
3. Nurgel, C.; Pickering, G.J.; Inglis, D. Sensory and chemical characteristics of Canadian ice wines. *J. Sci. Food Agric.* **2004**, *84*, 1675–1684. [CrossRef]
4. Ma, Y.; Tang, K.; Xu, Y.; Li, J.M. Characterization of the key aroma compounds in Chinese Vidal icewine by gas chromatography-olfactometry, quantitative measurements, aroma recombination, and omission tests. *J. Agric. Food Chem.* **2017**, *65*, 394–401. [CrossRef]
5. Ma, Y.; Xu, Y.; Tang, K. The Aroma of Icewine: A Review on How Environmental, Viticultural and Oenological Factors Affect the Aroma of Icewine. *J. Agric. Food Chem.* **2021**, *69*, 6943–6957. [CrossRef]
6. Belda, I.; Ruiz, J.; Esteban-Fernández, A.; Navascués, E.; Marquina, D.; Santos, A.; Moreno-Arribas, M.V. Microbial contribution to wine aroma and its intended use for wine quality improvement. *Molecules* **2017**, *22*, 189. [CrossRef]
7. Borren, E.; Tian, B. The Important Contribution of Non-Saccharomyces Yeasts to the Aroma Complexity of Wine: A Review. *Foods* **2021**, *10*, 13. [CrossRef]
8. Molina, A.M.; Guadalupe, V.; Varela, C.; Swiegers, J.H.; Pretorius, I.S.; Agosin, E. Differential synthesis of fermentative aroma compounds of two related commercial wine yeast strains. *Food Chem.* **2009**, *117*, 189–195. [CrossRef]
9. Swiegers, J.H.; Bartowsky, E.J.; Henschke, P.A.; Pretorius, I.S. Yeast and bacterial modulation of wine aroma and flavour. *Aust. J. Grape Wine Res.* **2005**, *11*, 139–173. [CrossRef]
10. Pigeau, G.M. Icewine Fermentation by Saccharomyces Cervisiae: Fundamental Stress Responses and Comparative Fermentation Dynamics. Ph.D. Thesis, Brock University, St. Catharines, ON, Canada, 2006.
11. Kontkanen, D.; Inglis, D.; Pickering, G.; Reynolds, A. Effect of yeast inoculation rate, acclimatization, and nutrient addition on Icewine fermentation. *Am. J. Enol. Viticult.* **2004**, *55*, 363–370.
12. Crandles, M.; Reynolds, A.G.; Khairallah, R.; Bowen, A. The effect of yeast strain on odor active compounds in Riesling and Vidal blanc icewines. *LWT-Food Sci. Technol.* **2015**, *64*, 243–258. [CrossRef]
13. Erasmus, D.J.; Cliff, M.; van Vuuren, H.J.J. Impact of yeast strain on the production of acetic acid, glycerol, and the sensory attributes of icewine. *Am. J. Enol. Viticult.* **2004**, *55*, 371–378.
14. Fuleki, T. Performance of five yeast strains in icewine fermentation. *Am. Soc. Enol. Vitic.* **1994**, *45*, 359A.
15. Yang, F. Study of New Yeast Strains as Novel Starter Cultures for Riesling Icewine Production. Master's Thesis, Brock University, St. Catharines, ON, Canada, 2011.
16. Synos, K.; Reynolds, A.G.; Bowen, A.J. Effect of yeast strain on aroma compounds in Cabernet franc icewines. *LWT-Food Sci. Technol.* **2015**, *64*, 227–235. [CrossRef]
17. Pigeau, G.M.; Bozza, E.; Kaiser, K.; Inglis, D.L. Concentration effect of Riesling Icewine juice on yeast performance and wine acidity. *J. Appl. Microbiol.* **2007**, *103*, 1691–1698. [CrossRef]
18. ISO-8586-1. *Sensory Analysis: General Guidance for the Selection, Training and Monitoring of Assessors, Selected Assessors*; International Organization for Standardization: Geneva, Switzerland, 1993.
19. ISO-8586-2. *Sensory Analysis: General Guidance for the Selection, Training and Monitoring of Assessors, Expert Sensory Assessors*; International Organization for Standardization: Geneva, Switzerland, 2008.
20. Huang, L.; Ma, Y.; Tian, X.; Li, J.M.; Li, L.X.; Tang, K.; Xu, Y. Chemosensory characteristics of regional Vidal icewines from China and Canada. *Food Chem.* **2018**, *261*, 66–74. [CrossRef]
21. Albert, A.; Varela, P.; Salvador, A.; Hough, G.; Fiszman, S. Overcoming the issues in the sensory description of hot served food with a complex texture. Application of QDA®, flash profiling and projective mapping using panels with different degrees of training. *Food Qual. Prefer.* **2011**, *22*, 463–473. [CrossRef]

22. ISO-4121. *Sensory Analysis-Guidelines for the Use of Quantitative Response Scales*; International Organization for Standardization: Geneva, Switzerland, 2003.
23. Wang, L.L.; Fan, S.S.; Yan, Y.; Yang, L.; Chen, S.; Xu, Y. Characterization of potent odorants causing a pickle-like off-odor in moutai-aroma type baijiu by comparative aroma extract dilution analysis, quantitative measurements, aroma addition, and omission studies. *J. Agric. Food Chem.* **2020**, *68*, 1666–1677. [CrossRef]
24. Erasmus, D.J.; Van der Merwe, G.K.; van Vuuren, H.J.J. Genome-wide expression analyses: Metabolic adaptation of Saccharomyces cerevisiae to high sugar stress. *FEMS Yeast Res.* **2003**, *3*, 375–399. [CrossRef]
25. Cliff, M.A.; Pickering, G.J. Determination of odour detection thresholds for acetic acid and ethyl acetate in ice wine. *J. Wine Res.* **2006**, *17*, 45–52. [CrossRef]
26. He, Y.X.; Liu, Z.P.; Qian, M.; Yu, X.W.; Xu, Y.; Chen, S. Unraveling the chemosensory characteristics of strong-aroma type Baijiu from different regions using comprehensive two-dimensional gas chromatography-time-of-flight mass spectrometry and descriptive sensory analysis. *Food Chem.* **2020**, *331*, 127335. [CrossRef]
27. Dunkel, A.; Steinhaus, M.; Kotthoff, M.; Nowak, B.; Krautwurst, D.; Schieberle, P.; Hofmann, T. Nature's chemical signatures in human olfaction: A foodborne perspective for future biotechnology. *Angew. Chem. Int. Edit.* **2014**, *53*, 7124–7143. [CrossRef]
28. Bowen, A.J.; Reynolds, A.G. Odor potency of aroma compounds in Riesling and Vidal blanc table wines and icewines by gas chromatography-olfactometry-mass spectrometry. *J. Agric. Food Chem.* **2012**, *60*, 2874–2883. [CrossRef]
29. Pineau, B.; Barbe, J.C.; Van Leeuwen, C.; Dubourdieu, D. Which impact for β-damascenone on red wines aroma? *J. Agric. Food Chem.* **2007**, *55*, 4103–4108. [CrossRef]
30. Bowen, A.J.; Reynolds, A.G. Aroma compounds in Ontario Vidal and Riesling icewines. I. Effects of harvest date. *Food Res. Int.* **2015**, *76*, 540–549. [CrossRef]

Article

Increased Varietal Aroma Diversity of Marselan Wine by Mixed Fermentation with Indigenous *Non-Saccharomyces* Yeasts

Xiaomin Xi, Aili Xin, Yilin You, Weidong Huang and Jicheng Zhan *

Beijing Key Laboratory of Viticulture and Enology, College of Food Science and Nutritional Engineering, China Agricultural University, Beijing 100083, China; xxm1113@cau.edu.cn (X.X.); xaljy68751807@163.com (A.X.); yilinyou@cau.edu.cn (Y.Y.); huanggwd@263.net (W.H.)
* Correspondence: zhanjicheng@cau.edu.cn

Abstract: The common use of commercial yeasts usually leads to dull wine with similar aromas and tastes. Therefore, screening for novel indigenous yeasts to practice is a promising method. In this research, aroma discrepancies among six wine groups fermentated with indigenous yeasts were analyzed. Three *Saccharomyces* yeasts (FS36, HL12, YT28) and three matched *non-Saccharomyces* yeasts (FS31, HL9, YT2) were selected from typical Chinese vineyards. The basic oenological parameters, aroma compounds, and sensory evaluation were analyzed. The results showed that each indigenous *Saccharomyces* yeast had excellent fermentation capacity, and mixed-strain fermentation groups produced more glycerol, contributing to sweeter and rounder taste. The results from GC-MS, principal components analysis (PCA), and sensory evaluation highlighted that the HL mixed group kept the most content of Marselan varietal flavors such as calamenene and β-damascone hereby ameliorated the whole aroma quality. Our study also implied that the indigenous yeast from the same region as the grape variety seems more conducive to preserve the natural variety characteristics of grapes.

Keywords: marselan wine; aroma compounds; indigenous yeast strains; *Saccharomyces*; *non-Saccharomyces*

Citation: Xi, X.; Xin, A.; You, Y.; Huang, W.; Zhan, J. Increased Varietal Aroma Diversity of Marselan Wine by Mixed Fermentation with Indigenous *Non-Saccharomyces* Yeasts. *Fermentation* **2021**, *7*, 133. https://doi.org/10.3390/fermentation7030133

Academic Editor: Rosanna Tofalo

Received: 24 June 2021
Accepted: 22 July 2021
Published: 27 July 2021

Publisher's Note: MDPI stays neutral with regard to jurisdictional claims in published maps and institutional affiliations.

Copyright: © 2021 by the authors. Licensee MDPI, Basel, Switzerland. This article is an open access article distributed under the terms and conditions of the Creative Commons Attribution (CC BY) license (https://creativecommons.org/licenses/by/4.0/).

1. Introduction

A balanced and complex aroma is one of the most attractive factors that determine consumers' preference for wine, which also reflects the wine quality and value [1]. Yeasts are one of the most significant factors that impact the wine aroma profile by producing a large array of volatile compounds [2]. However, with the wide application of commercial *Saccharomyces* strains, inherent limitation gradually appears due to reducing the uniqueness and particularity of regional wine bouquets [3]. The current set of commercial *S. cerevisiae* strains and its derived hybrids is insufficient to provide novel properties to wine, stressing the need for new and improved strains for the industry [4]. Moreover, different yeast species and even different genotypes of yeasts displayed distinct wine aroma profiles [5]. The use of standard starter cultures, mainly applied to reduce the risk of spoilage and unpredictable changes of wine flavor, can ensure a balanced wine flavor, but it also causes a loss of characteristic aroma and flavor determinants. Therefore, this awareness opened new requirements to meet wine makers' demand for "special yeasts for special traits". Recently, the role of indigenous *Saccharomyces* yeast strains has gained much more attraction as a novel tool to convey regional characters to wine itself. Indeed, the application of a "location-specific" starter culture highlighted the association between the profile of wine aromas and the geographical origin of the yeast used for the fermentation process [3,6–8]. There are plenty of yeast resources distributed widely in Chinese vineyards, requiring deeper research and proper application urgently. Therefore, using autochthonous *Saccharomyces* yeasts instead of commercial ones has been greatly encouraged in current winemaking due to their excellent adaptability for the local climate, soil, and micro-conditions and improve the aroma quality of wine by generating unique regional characters [9].

Usually, wine fermentations constitute complex microbial ecosystems, including *Saccharomyces* yeasts and *non-Saccharomyces* yeasts. The term "*non-Saccharomyces* yeasts" was used in the past to refer to a group of species with secondary relevance during the fermentation of grape musts to wine, considered even as spoilage organisms [10]. More recently, they are recognized for their beneficial role in improving the wine quality by producing some distinctive metabolites and are selected and screened for desired requirements [8]. Although most *non-Saccharomyces* yeasts are incapable of completing alcoholic fermentation due to low ethanol tolerance, they play a significant role in producing aroma compounds, such as esters, higher alcohols, acids, and monoterpenes [11]. The contributions to the wine aroma by *non-Saccharomyces* yeasts can occur through different mechanisms. The most important is probably the direct biosynthesis of volatile aroma compounds, and many of them have been identified. Some *non-Saccharomyces* yeasts produce certain glucosidase enzymes such as β-glucosidase or β-xylosidase that release volatile compounds from glycosidic precursors. Other metabolic products include terpenoids, esters, higher alcohols, acetaldehyde, organic acids, volatile fatty acids, carbonyl, and sulfur compounds. In addition, the application of *non-Saccharomyces* can also reduce the content of sulfur dioxide (SO_2) to lessen the hazards on human health and improve the content of total glycerol, which positively influences the wine taste by improving smoothness, sweetness, mouth-feel, and complexity [12]. In recent years, the inoculation of *non-Saccharomyces* yeasts as co-starters in wine production represents a promising winemaking method to ameliorate the wine aroma profiles. Besides the enhancement of wine aroma, they can also provide more essential functions, such as modulating the acidity, lowering the alcohol level, producing enzymes that optimize the winemaking process, such as clarification, filtration, and color extraction [13]. Numerous studies about co-fermentations of *Saccharomyces* yeasts with selected *non-Saccharomyces* yeasts in a controlled manner have proven to produce the wine with distinct aromatic profiles and higher complexity and quality of wine aromas, such as improving the content of glycerol, enhancing medium-chain fatty acid ethyl esters or producing higher contents of esters and β-damascenone [7,14].

Our previous study has screened and selected a wide diversity in an extensive collection of both *Saccharomyces* yeasts and *non-Saccharomyces* yeasts from multiple regions in China, giving suitable support and a solid foundation to this study. Here, we used Marselan as material, a young but promising variety in China. Marselan (*Vitis vinifera* L.) is a hybrid cultivar of two famous grape varieties, Cabernet Sauvignon and Grenache [15], combining the finesse and high quality of Cabernet Sauvignon, disease resistance, and high yield of Grenache, which has the capacity to produce high qualified wine with deep color, elegant aromas, and potential for aging [16]. However, little is understood about the aromatic profile of Marselan must and how Chinese autochthonous *Saccharomyces* yeasts inoculated with *non-Saccharomyces* yeasts perform on Marselan. In our previous work, three indigenous *Saccharomyces* yeast strains (FS36, YT28, and HL12) and three corresponding indigenous *non-Saccharomyces* (FS31, YT2, and HL9) were isolated from Fangshan (Beijing), Yantai (Shandong province), and Huailai (Hebei province), respectively. Therefore, the main objectives of the study were to firstly identify the characteristic aroma compounds of Marselan must by GC-MS; then to analyze the aroma profiles of Marselan wine fermented with autochthonous *Saccharomyces* yeasts and *non-Saccharomyces* yeasts by pure and mixed methods; and to evaluate the sensory attributes of different Marselan wine through panelists sensory evaluation, PCA and partial least squares regression (PLSR).

2. Materials and Methods

2.1. Marselan Must Samples

There are four Marselan must samples (China). Three were from the 2018 vintage, from Huailai (Hebei), Fangshan (Beijing), and Yantai (Shandong), respectively, shorted for 2018 HL, 2018 FS, and 2018 YT. Another sample was from 2017 vintage and Huailai, shorted for 2017 HL.

2.2. Yeast Strains

The detailed information about yeast strains was listed in Table 1, all selected and identified in our lab previously. Six autochthonous strains, including three *S. cerevisiae* strains (FS36, YT28, and HL12) and three matched *non-Saccharomyces* strains (FS31, YT2, and HL9). The strains were stored at −80 °C in yeast extract peptone dextrose medium (YPD, yeast extract 10 g/L, peptone 20 g/L, and glucose 20 g/L) with the addition of glycerol (20% *v*/*v* final concentration). Among them, HL12, YT28, and FS31 were deposited in China General Microbiological Culture Collection Center (CGMCC), and the numbers were No. 20632, No. 20633, and No. 20636, respectively.

Table 1. The information of yeast strains.

Name	Region	Species	Property
FS36	Fanghshan, Beijing	*Saccharomyces cerevisiae*	*Saccharomyces*
FS31	Fanghshan, Beijing	*Kazachstania exigua*	*Non-Saccharomyces*
YT28	Yantai, Shandong	*Saccharomyces cerevisiae*	*Saccharomyces*
YT2	Yantai, Shandong	*Candida glabrata*	*Non-Saccharomyces*
HL12	Huailai, Hebei	*Saccharomyces cerevisiae*	*Saccharomyces*
HL9	Huailai, Hebei	*Hanseniaspora opuntiae*	*Non-Saccharomyces*

2.3. Chemicals

Sulfuric acid, sulfuric acid, ammonia, sodium hydroxide, peptone, and hydrochloric acid were purchased from Beijing Chemical Reagent Factory (Beijing, China). Methanol and acetonitrile were purchased from Merck, Germany. Standards (glucose, fructose, glycerol, ethanol, citric acid, tartaric acid, pyruvic acid, malic acid, succinic acid, acetic acid, and ascorbic acid) were all purchased from Sigma-Aldrich, China Co. (St. Louis, MI, USA) with at least 97% purity.

2.4. HS-SPME-GC-MS Analysis

The volatile compounds of Marselan wine samples were extracted using the Head Space Solid Phase Microextraction (HS-SPME) method as described [17]. For each analysis, 2.5 g of sodium chloride (NaCl) and 10 μL of 2-octanol (internal standard) were added to 5 mL of must sample into a 20 mL headspace screw vial. Helium (He) was the carrier gas at a constant flow rate of 1 mL/min. The SPME fiber was inserted through the needle and exposed into the headspace of the vial to adsorb volatile compounds at 45 °C for 50 min and then immediately injected into the gas chromatography injection port at 250 °C for 2.5 min to desorb volatile compounds; afterward, GC-MS separation, analysis, and identification was performed.

The volatile compounds from the samples were analyzed by Gas Chromatography and Mass Spectrometry (GC-MS) according to a previous description with some modification [18]. GCMS-QP2010 Ultra (SHIMADZU, Kyoto, Japan) was used with a DB-Wax capillary column (30 mm length × 0.25 mm i.d. × 0.25 μm film thickness), and splitless injection mode was adopted. High purity helium, as a carrier gas, was used at a constant flow rate of 1 mL/min. The gas chromatographic oven was set at 50 °C for 5 min, increased until 230 °C at a rate of 5 °C/min, and finally maintained for 30 min. Other conditions included an interface temperature of 250 °C, an emission current of 100 μA, and mass spectra were obtained at 70 eV in the electron ionization + (EI+) mode. The mass spectral identification of volatile compounds in the samples was carried out by comparing to the National Institute of Standards and Technology (NIST) 2014 and Wiley 8.0 database. Qualitative analysis of mass spectral data was verified by comparing the retention indices and mass spectra of identified compounds. The qualitative aroma components were qualitatively determined by NIST2011 and the Demo library. Volatile compounds extraction (μg/L) = (peak area of volatile compounds/peak area of internal standard × mass of internal standard)/mass of sample.

2.5. Fermentation Methods

The 2017 HL Marselan grape must was used as material for all trails. The frozen grape juice was placed at 4 °C, centrifuged at high speed in 50 mL tubes. The supernatant was filtered with 0.65 then 0.45 μm filters to obtain the sterile grape juice. Yeasts were inoculated at 28 °C for 48 hours until the amount was 10^7 CFU/mL. A total of 300 mL sterile grape juice was measured. For single-strain fermentation groups, the pre-cultured *Saccharomyce* yeasts (FS36, YT28, and HL12) were inoculated. For mixed-strain fermentation groups, *non-Saccharomyce* yeasts (FS31, YT2, and HL9) were inoculated (with the ratio of 1%). Three parallels for each group were set. After the inoculation, all samples were placed in the incubator at 20 °C. On the 4th day, the pre-cultured *Saccharomyce* yeasts (FS36, YT28, and HL12) were added to the corresponding mixed groups (with the ratio of 1%). On the 0th, 2, 4, 6, 10, 14, and 19 days, the wine bottles were fully shaken, samples were taken to detect the yeast number, Brix, pH, residual sugar, acid, ethanol, glycerol, and volatile aroma components. When the yeast began to decline and the Brix value kept stable, the fermentation was terminated. The supernatant of wine was taken and stored at 4 °C. All the trials were demonstrated in Table 2: (1) single fermentation with FS36 (short for FS36); (2) mixed fermentation of FS36 and FS31 (short for FS mixed); (3) single fermentation with YT28 (short for YT28); (4) mixed fermentation of YT28 and YT2 (short for YT mixed); (5) single fermentation with HL12 (short for HL12); and (6) mixed fermentation of HL12 and HL9 (short for HL mixed).

Table 2. The method of fermentation trails.

Fermentation Trails	Yeasts	Method
FS36	FS36 alone	/
FS Mixed	FS31 + FS36	FS36 was inoculated on 4th day
YT28	YT28 alone	/
YT Mixed	YT2 + YT28	YT28 was inoculated on 4th day
HL12	HL12 alone	/
HL Mixed	HL9 + HL12	HL12 was inoculated on 4th day

2.6. Sensory Analysis

Wine aroma was evaluated in triplicates by a tasting panel consisting of four females and four males trained with a 54-aroma kit (Le Nez du Vin®, France) for three weeks. During the training, the performances were evaluated by aroma sense test every five days until their identification accuracy for each aroma reached above 95%. The analysis was conducted in a tasting room at 23 °C. Approximately 30 mL of wine (15 °C) was held in a black wine glass. Throughout the wine sensory analysis, the samples were presented in random order. The interval between the two samples was 5 min. The panelists evaluated wine aroma according to the following procedure: they smelled the aroma of wine sample for approximately 5–8 s, then shook the wine to smell the aroma for 5–10 s, defined aromas, and scored. The sensory descriptors were alcohol, floral, citrus, stone fruits, berries, dry fruits, herbs, and fermentative aromas. The samples were quantitatively measured on a 5-point interval scale to grade the intensity (1—very weak; 2—weak; 3—medium; 4—intense; 5—very intense).

2.7. Detection of Physicochemical Parameters

Glucose, fructose, ethanol, glycerol were all detected by high-performance liquid chromatography (HPLC) [19]. The chromatographic conditions: sugar analysis column Aminex HPX-87H (300 × 7.8 mm), column temperature 55 °C, differential detector (RID, Waters-2414), internal temperature 40 °C, 0.005 mol/L H_2SO_4, injection volume 10 μL, flow rate 0.5 mL/min isocratic elution, qualitative by retention time, and quantitative peak area. pH and Brix were detected by pH meter and Brix meter. Based on our previous method, the flow rate was modified to 0.8 mL/min, RP-HPLC was used to detect eight organic acids. The chromatographic conditions are: Tech Mate ST-C (4.6 × 250 mm, 5 μm), column temperature 25 °C, ultraviolet detector (PAD, Waters-2996), detection wavelength 210 nm, mobile phase: ammonium dihydrogen phosphate ($NH_4H_2PO_4$) 40 mmol/L (4.6 g/L), H_3PO_4 adjusted pH to 2.5, injection volume 10 μL, flow rate 0.8 mL/min, isocratic elution, qualitative retention time, and peak area quantification.

2.8. Statistical Analysis

Analysis of variance (ANOVA, least significant difference method at a significance level of $p \leq 0.05$) was used to evaluate differences between the samples by the SPSS 19.0 software (SPSS Inc., Chicago, IL, USA). Principal component analysis (PCA) was carried out by R studio to visualize the differences between wines fermented by different strains and inoculation methods. PLSR was applied to identify the contribution of the key aroma compounds to the flavor characteristics of wine samples, and it was conducted using Unscrambler X version (Camo, Trondheim, Norway). All the figures were carried out by Graphpad Prism 8 (GraphPad Software Inc., San Diego, CA, USA).

3. Results

3.1. Aroma Compounds of Marselan Must Detected by GC-MS

Previously, it has been identified that some key aromas such as β-damascenone, eugenol, and 2,3-butanedione in Marselan wine [20]. However, little research was found to investigate the characteristic aromas of Marselan must. There is no doubt that it is very essential to figure out the features of a certain variety before making a type of wine. Therefore, firstly we carried out the experiments to identify the characteristic varietal aromas in Marselan must, and those aromas could represent and distinguish the various properties. Table 3 showed the total volatile compounds that were identified and quantified by GC-MS with HS-SPME in four Marselan must samples. There were 30 aroma compounds detected in 2017 HL, 39 in 2018 FS, 42 in 2018 YT, and 41 in 2018 HL Marselan must, of which 20 aromas were detected in common. The threshold of C_{13}-norisoprenoids and terpenes in grape fruits is very low; however, they have great influence on grape aroma formation. The results showed that C_{13}-norisoprenoid compounds mainly included violine, β-damascenone, and calamenene; terpenes mainly included linalool, α-terpene, 4-terpineol, β-myrcene, and semenene. The aroma compounds detected in all groups included calamenene, β-damascenone, phenylethyl alcohol, ethyl octoate, ethyl hexanoate, ethyl benzoate, ethyl laurate, n-hexanal, benzaldehyde, furfural, ceremonene, 2,4-di-tert-butylphenol, etc. Furthermore, calamenene, β-damascenone, n-hexanal, furfural, and ceremonene were significantly different among the samples. Moreover, α-ionone was only detected in 2018 YT, and linalool was only found in 2018 HL. 2-Hexenal with 27% relative content only appeared in 2018 FS, accounting for a relatively high proportion. Hexanal, nonanal, and semenene could be detected in all samples. The relative content of 1-hexanol in 2018 YT and 2018 HL reached more than 60%; however, it was not detected in the rest samples; the relative content of n-hexanal in 2017 HL was the most abundant, accounting for 46.5%. All these results indicated that different must samples had verified aromatic profiles. To compare all the samples, the relative content of characteristic aroma compounds was showed in Figure S1. There were seven characteristic aromas as summarized: phenethyl alcohol, ethyl octoate, ethyl caproate, calamenene, β-damascenone, benzaldehyde, and 2,4-di-tert-butylphenol. In 2017 HL and 2018 HL, except for calamenene, all aroma compounds had significant differences. On the whole, except for β-damastonone that 2018 HL produced most, 2017 HL had more levels of aroma compounds than 2018 HL. Comparing the samples from 2018, significant differences were observed in all aroma compounds apart from β-damastonone. Grape aroma is principally influenced by sugar accumulation and transformation [21], which might explain the variation of the aroma profiles.

Table 3. The average peak area and relative peak area (RPA) of volatile aroma compounds of Marselan must.

	2017 HL		2018 FS		2018 YT		2018 HL	
Aroma Compound	**Peak Area * 10^6**	**RPA%**	**Peak Area * 10^6**	**RPA%**	**Peak Area * 10^6**	**RPA%**	**Peak Area * 10^6**	**RPA%**
C13-Norisoprenoids								
Violine	N.D.	0.00	5.42 ± 0.12^a	1.10	2.35 ± 0.09^b	0.24	2.17 ± 0.01^b	0.24
Calamenene	1.37 ± 0.33^d	0.42	3.45 ± 0.10^a	0.72	2.28 ± 0.39^b	0.23	1.77 ± 0.22^c	0.20
β-Damascone	18.35 ± 0.42^d	5.60	32.82 ± 1.01^a	6.83	30.69 ± 0.59^b	3.16	29.29 ± 0.72^c	3.25
Terpenes								
Linalol	N.D.	0.00	N.D.	0.00	N.D.	0.00	0.53 ± 0.01	0.06
4-terpineol	N.D.	0.00	N.D.	0.00	N.D.	0.00	0.57 ± 0.03	0.06
α-terpene	N.D.	0.00	N.D.	0.00	0.91 ± 0.01	0.09	N.D.	0.00
β-myrcene	N.D.	0.00	N.D.	0.00	N.D.	0.00	2.81 ± 0.13	0.31
Semenene	1.65 ± 0.11^c	0.50	2.35 ± 0.30^b	0.49	2.40 ± 0.14^b	0.25	3.34 ± 0.14^a	0.37
Alcohols								
Phenylethyl alcohol	4.46 ± 0.97^a	1.36	2.44 ± 0.26^{ab}	0.51	1.40 ± 0.10^b	0.14	1.19 ± 0.15^b	0.13
1-hexanol	N.D.	0.00	N.D.	0.00	659.27 ± 2.14^a	67.80	555.95 ± 3.67^b	61.70
1-hexene-3-ol	N.D.	0.00	N.D.	0.00	N.D.	0.00	6.59 ± 0.33	0.73
Esters								
Hexyl acetate	N.D.	0.00	N.D.	0.00	4.45 ± 0.12^a	0.37	2.48 ± 0.04^a	0.28
Ethyl octoate	1.50 ± 0.28^a	0.31	0.63 ± 0.01^a	0.13	0.40 ± 0.06^a	0.04	0.24 ± 0.00^a	0.03
Phenethyl acetate	1.42 ± 0.02	0.43	N.D.	0.00	N.D.	0.00	N.D.	0.00
Ethyl benzoate	0.52 ± 0.07^a	0.16	0.50 ± 0.07^a	0.10	0.34 ± 0.05^a	0.03	0.28 ± 0.01^a	0.03
Ethyl hexanoate	4.33 ± 0.21^a	0.99	0.82 ± 0.19^a	0.17	0.50 ± 0.01^a	0.05	0.53 ± 0.09^a	0.06
Ethyl phthalate	N.D.	0.00	0.14 ± 0.00^a	0.03	0.07 ± 0.00^b	0.01	N.D.	0.00
Ethyl myristate	1.88 ± 0.83^a	0.57	0.28 ± 0.05^a	0.06	0.09 ± 0.00^a	0.01	N.D.	0.00
Ethyl palmitate	1.85 ± 0.02^a	0.56	0.33 ± 0.11^b	0.07	0.14 ± 0.02^a	0.01	N.D.	0.00
Ethyl laurate	5.93 ± 0.31^a	1.81	1.56 ± 0.37^a	0.32	1.11 ± 0.01^a	0.11	0.36 ± 0.00^a	0.04
Others								
n-hexanal	152.28 ± 2.54^a	46.50	81.57 ± 0.06^b	17.00	53.60 ± 1.25^c	5.51	94.70 ± 0.22^b	10.50
2-hexenal	N.D.	0.00	129.77 ± 2.34	27.00	N.D.	0.00	N.D.	0.00
Benzaldehyde	6.01 ± 0.56^b	1.83	15.65 ± 0.22^a	3.26	14.02 ± 0.57^{ab}	1.44	2.73 ± 0.03^b	0.30
Nonanal	4.17 ± 0.03^a	0.97	1.77 ± 0.03^b	0.37	1.16 ± 0.16^c	0.12	1.00 ± 0.00^c	0.11
2-decanone	N.D.	0.00	0.93 ± 0.01^a	0.19	1.06 ± 0.05^a	0.11	1.02 ± 0.00^a	0.11
2,4-di-tert-butylphenol	30.08 ± 0.37^a	9.18	13.41 ± 0.07^b	2.79	10.34 ± 0.05^b	1.06	10.17 ± 1.20^b	1.13

N.D. means no detection. Different letters indicate the significant differences at $p < 0.05$.

3.2. Basic Parameters of Marselan Wine

The whole fermentation process was monitored, and the physiochemical parameters (alcohol, glycerol, glucose, fructose, and Brix) were detected (Table S1 and Figure 1). On the whole, the fermentation speed of single-strain groups (FS36, YT28, and HL12) was significantly faster than mixed-strain groups (FS mixed, YT mixed, and HL mixed), which was in agreement with numerous authors [8,22–24]. Among them, FS36 was the fastest, followed by HL12 and YT28, implying that FS36 had the best fermentation ability. The mixed-strain groups began to ferment rapidly after sequentially inoculated the *Saccharomyces* yeast strains on the 4th day. On the 19th day, the Brix value of FS36, YT28, and HL12 was 7.7, 7.5, and 7.3, respectively; FS mixed, YT mixed, and HL mixed was 8.0, 7.8, and 7.7, respectively. All groups reached basically the same level and remained stable, indicating that all these *Saccharomyces* yeasts were well capable of completing alcoholic fermentation, and *non-Saccharomyces* did not hinder the process.

During the fermentation process, the use of glucose is usually superior to fructose [22], which is consistent with our results. Single-strain groups consumed glucose faster, and the final content was less than 2 g/L on the 14th day. Mixed-strain groups used glucose from the 4th day and completed it on the 19th day. Finally, the residual glucose in all groups was about 1 g/L. It should be noticed that the difference in fructose consumption rate was minor than glucose. FS36 metabolized fructose fastest, and YT mixed was the slowest. HL mixed and YT mixed groups had the maximum residual fructose, which was 0.81 and 2.92 g/L, separately. Interestingly, no matter in glucose or fructose consumption, the speed rank was the same (FS36 > HL12 > YT28). Figure 1C exhibited that FS36 produced the most glycerol as well as with the fastest speed. Compared with mixed-strain groups, single-strain groups produced glycerol at a higher rate. In contrast, the mixed-strain groups (FS mixed and YT mixed) yielded much more glycerol in the end. Glycerol is a by-product of yeast metabolism during wine fermentation that had been previously discovered that mixed fermentation with *non-Saccharomyces* provides more glycerol, offering wine richer sweety and round tastes [7,8]. Single-strain groups produced alcohol faster; however, the final alcohol volume tended to be approximative, suggesting that the *non-Saccharomyces* applied in the study did not block the fermentation process because some research pointed that the addition of *non-Saccharomyces* could spoilage the alcoholic fermentation [23]. These results presented that all the selected *Saccharomyces* yeast strains have excellent fermentation property and all chemical parameters belonged to a normal range, thus providing a solid foundation for our further analysis.

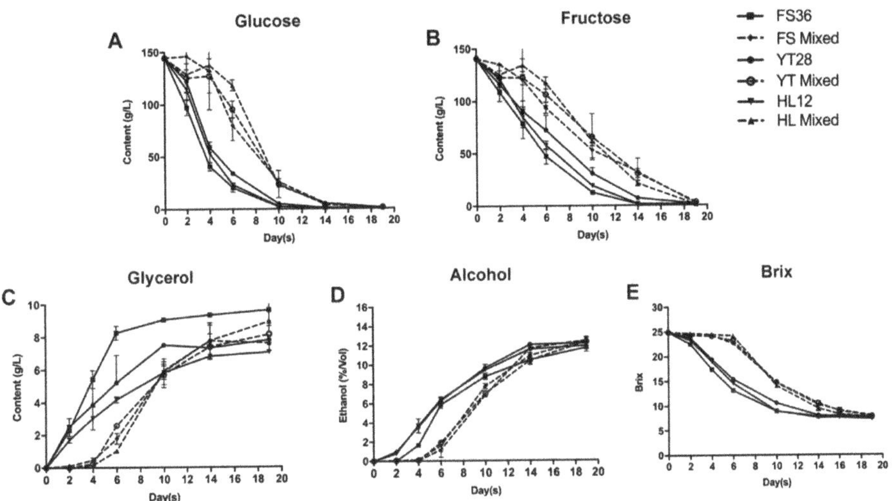

Figure 1. The basic parameters of six Marselan wines during fermentation.

3.3. Identification of Aroma Compounds in Marselan Wine by GC-MS

A total set of 53 aroma compounds were detected in FS36, 48 in YT28, 46 in HL12, 48 in FS mixed, 47 in the YT mixed, and 45 in the HL mixed, respectively. The main classes of volatile compounds were consistent with those analyzed by Lyu [24]. The results of quantitative GC-MS analysis for the aroma compounds were listed in Table 4 and summarized in Table S2.

Esters, including acetate esters and ethyl esters, are closely related to floral and fruity attributes formed by the esterification of fatty acids [25]. A total of 27 kinds of esters were detected, making them the most numerous groups in all aroma compounds. A total of 26 kinds of esters were detected in FS36, YT28, HL12 mixed, and FS mixed; a total of 27 and 25 kinds of esters were detected in YT mixed and HL mixed. Among them, ethyl decanoate was the single most abundant ester in all samples, with a significant difference between YT28 and YT mixed, giving fruity and pleasant floral flavors to the wine. Ethyl 9-decanoate ranked second, was less in HL groups (8.76% of single group and 7.73% of mixed group), but had above 16% relative content in the other groups. The content of ethyl octoate ranked third and was highest distributed in YT28, also accounting for 5% to 8% among all groups. Importantly, ethyl octoate offers a pleasant flower, fruit flavors such as apricot, pineapple, pear, and creamy mushroom with a threshold of only 0.01 mg/L, giving critical impacts on aroma profile [26]. Compared with grape must, the content of ethyl laurate was significantly increased as fermentation process. FS36 and FS mixed produced the most content of ethyl laurate, followed by YT and HL. Compared with the Marselan must samples, the number of esters in wine was remarkably increased, indicating that yeast biosynthesis is the main synthetic way to produce esters during fermentation.

Table 4. The average peak area of volatile aroma compounds of Marselan wine.

	FS36	YT28	HL12	FS Mixed	YT Mixed	HL Mixed
Aroma Compound	**Peak Area * 10^6**	**Peak Area * 10^6**	**Peak Area * 10^6**	**Peak Area * 10^6**	**Peak Area * 10^6**	**Peak Area * 10^6**
Norisoprenes						
Calamenene	N.D.	N.D.	N.D.	N.D.	1.17 ± 0.01 ^{b}	4.42 ± 0.03 ^{a}
β-Damascone	2.60 ± 0.20 ^{c}	2.20 ± 0.09 ^{d}	3.03 ± 0.15 ^{b}	3.18 ± 0.08 ^{ab}	2.77 ± 0.06 ^{c}	3.34 ± 0.18 ^{a}
Alcohols						
Phenethyl alcohol	284.27 ± 3.75 ^{de}	371.74 ± 1.97 ^{c}	287.04 ± 0.91 ^{de}	249.47 ± 0.89 ^{e}	462.46 ± 0.29 ^{a}	323.63 ± 1.02 ^{d}
Decyl alcohol	2.07 ± 0.07 ^{a}	N.D.	1.84 ± 0.07 ^{b}	N.D.	N.D.	N.D.
Trans β-farnesol	6.82 ± 0.70 ^{a}	3.68 ± 0.59 ^{b}	N.D.	1.33 ± 0.16 ^{c}	N.D.	1.06 ± 0.11 ^{d}
β-bisabolol	4.02 ± 0.81 ^{a}	1.28 ± 0.23 ^{b}	1.04 ± 0.21 ^{b}	2.02 ± 0.11 ^{ab}	1.32 ± 0.27 ^{b}	1.40 ± 0.21 ^{b}
Cis-α-bisabolol	0.70 ± 0.15 ^{a}	0.54 ± 0.10 ^{a}	0.50 ± 0.16 ^{a}	0.88 ± 0.07 ^{a}	0.11 ± 0.02 ^{a}	0.66 ± 0.32 ^{a}
Esters						
Ethyl hexanoate	33.62 ± 6.22 ^{b}	37.08 ± 2.62 ^{b}	67.43 ± 1.20 ^{a}	40.74 ± 0.76 ^{b}	43.93 ± 0.94 ^{b}	63.21 ± 1.39 ^{a}
Hexyl acetate	N.D.	N.D.	N.D.	2.45 ± 0.44 ^{a}	2.55 ± 0.87 ^{a}	3.40 ± 0.30 ^{a}
Ethyl octanoate	771.21 ± 5.13 ^{b}	792.82 ± 31.12 ^{b}	996.70 ± 8.57 ^{a}	732.79 ± 14.52 ^{b}	766.20 ± 1.45 ^{b}	777.31 ± 7.79 ^{b}
7-Ethyl caprylate	3.38 ± 0.78 ^{b}	3.11 ± 0.05 ^{b}	3.00 ± 0.85 ^{b}	2.52 ± 0.56 ^{b}	5.70 ± 0.82 ^{a}	4.56 ± 0.13 ^{ab}
3-Methylbutyl octanoate	3.24 ± 0.79 ^{b}	7.29 ± 1.42 ^{a}	6.27 ± 0.18 ^{ab}	4.95 ± 0.36 ^{ab}	6.20 ± 1.42 ^{ab}	3.87 ± 0.40 ^{b}
Ethyl benzoate	0.72 ± 0.13 ^{b}	1.31 ± 0.15 ^{ab}	3.42 ± 0.31 ^{a}	1.07 ± 0.23 ^{ab}	1.88 ± 0.38 ^{ab}	1.68 ± 0.44 ^{ab}
Diethyl succinate	2.07 ± 0.12 ^{bc}	1.83 ± 0.34 ^{bc}	3.06 ± 0.06 ^{ab}	1.27 ± 0.20 ^{c}	2.34 ± 0.44 ^{b}	3.56 ± 0.09 ^{a}
Phenethyl acetate	39.51 ± 6.42 ^{c}	51.78 ± 1.70 ^{bc}	40.29 ± 3.12 ^{c}	43.98 ± 0.75 ^{c}	72.26 ± 1.03 ^{a}	56.46 ± 0.56 ^{b}
2-Methyl propyl caprylate	2.03 ± 0.08 ^{a}	2.03 ± 0.12 ^{a}	2.05 ± 0.15 ^{a}	1.81 ± 0.40 ^{a}	2.04 ± 0.14 ^{a}	2.49 ± 0.06 ^{a}
Ethyl decanoate	3493.96 ± 43.58 ^{a}	3455.40 ± 23.77 ^{a}	3154.25 ± 24.12 ^{ab}	3051.76 ± 7.57 ^{ab}	2577.69 ± 19.87 ^{b}	2520.97 ± 15.39 ^{b}
Ethyl 9-decanoate	2670.03 ± 28.46 ^{a}	2093.40 ± 18.67 ^{b}	990.66 ± 7.37 ^{c}	1992.78 ± 17.91 ^{b}	2261.45 ± 3.09 ^{ab}	850.23 ± 8.75 ^{c}
Ethyl oleate	23.73 ± 1.82 ^{a}	6.22 ± 0.54 ^{bc}	4.82 ± 1.38 ^{bc}	11.11 ± 0.81 ^{b}	3.21 ± 1.26 ^{c}	3.39 ± 0.25 ^{c}
Ethyl propionate	N.D.	N.D.	N.D.	11.96 ± 1.20 ^{a}	3.28 ± 1.26 ^{a}	3.83 ± 1.40 ^{a}
Trans ethyl 4-decanoate	16.98 ± 3.64 ^{a}	16.65 ± 2.52 ^{a}	6.70 ± 1.07 ^{b}	11.88 ± 0.96 ^{ab}	4.10 ± 1.61 ^{b}	N.D.
Isoamyl caprylate	52.56 ± 6.71 ^{a}	48.57 ± 1.07 ^{a}	N.D.	42.43 ± 2.70 ^{a}	N.D.	N.D.
Propyl decanoate	4.91 ± 0.22 ^{a}	2.91 ± 0.48 ^{b}	2.08 ± 0.29 ^{c}	4.21 ± 2.74 ^{a}	2.35 ± 0.80 ^{bc}	1.67 ± 0.75 ^{c}
Ethyl undecanoate	2.48 ± 0.46 ^{ab}	2.29 ± 0.60 ^{b}	0.97 ± 0.02 ^{c}	2.90 ± 0.34 ^{a}	2.03 ± 0.09 ^{b}	0.93 ± 0.14 ^{c}
Ethyl laurate	65.84 ± 0.96 ^{ab}	51.71 ± 0.08 ^{b}	33.36 ± 0.84 ^{c}	71.22 ± 0.10 ^{a}	41.28 ± 0.15 ^{bc}	21.37 ± 0.23 ^{c}
Ethyl palmitate	149.85 ± 6.42 ^{a}	158.53 ± 1.18 ^{a}	68.69 ± 1.24 ^{b}	156.30 ± 1.62 ^{a}	129.55 ± 0.37 ^{a}	46.82 ± 1.17 ^{b}
Others						
Benzaldehyde	2.75 ± 0.21 ^{b}	4.99 ± 0.07 ^{ab}	4.20 ± 0.15 ^{b}	3.57 ± 0.04 ^{b}	3.47 ± 0.05 ^{b}	4.67 ± 0.06 ^{ab}
Phenylacetaldehyde	2.14 ± 0.05 ^{b}	5.39 ± 0.16 ^{a}	4.80 ± 0.04 ^{a}	5.55 ± 0.18 ^{a}	4.73 ± 0.08 ^{a}	3.80 ± 0.21 ^{ab}
16-octadecenal	1.63 ± 0.23	N.D.	N.D.	N.D.	N.D.	N.D.
Ethyl benzaldehyde	25.42 ± 0.91 ^{a}	11.71 ± 1.03 ^{b}	12.17 ± 0.67 ^{b}	10.21 ± 0.12 ^{b}	10.66 ± 0.30 ^{b}	8.49 ± 0.08 ^{b}
Curcumene	4.63 ± 0.10 ^{a}	5.10 ± 0.07 ^{a}	N.D.	5.50 ± 0.58 ^{a}	4.30 ± 0.09 ^{a}	6.15 ± 0.18 ^{a}
Cyclohexene	31.26 ± 1.21 ^{a}	N.D.	N.D.	16.76 ± 0.20 ^{b}	N.D.	N.D.
2,4-Di-tert-butylphenol	7.26 ± 0.05 ^{b}	8.11 ± 0.14 ^{b}	9.23 ± 0.13 ^{b}	6.35 ± 0.11 ^{b}	7.54 ± 0.01 ^{b}	5.82 ± 0.07 ^{b}

N.D. means no detection. Different letters indicate the significant differences at $p < 0.05$.

The higher alcohols were the main group of alcohol compounds in this study. Normally, the concentration of 300–400 mg/L is acceptable, and the optimal level gives a pleasant character [27]. It could be found that phenylethyl alcohol, contributing to "rose" and "sweet" notes for wine, distributed highest in all groups, and its content was improved greatly by yeast metabolism during fermentation. There was no significant difference between FS and FS mixed, but the content of phenylethyl alcohol in YT and HL mixed was higher than corresponding single groups, thus adding more floral flavors to mixed fermented wine. The amount of β-bisabolol was higher in FS36 and FS mixed groups; farnesol, bringing the wine with floral and greenwood flavor, was only detected in single FS36 and YT28.

Oxidation-related aldehydes with low sensory thresholds and apple-like odors are important to wine aroma, making wine smell fresher [28]. Four aldehydes were identified in this study. They were benzaldehyde, phenyl acetaldehyde, 16-octadecenal, and ethyl benzaldehyde. Compared with Marselan must, the types and content of aldehydes in wine dropped distinctly as the result of chemical reactions such as oxidation/reduction, enzymatic hydrolysis, or participating in yeast bio-metabolism to form new aroma compounds. Phenyl acetaldehyde, generating floral and honey odor in wine, was highest in YT28, lowest in FS36.

Norisoprenes and its derivatives usually have apple, raspberry, papaya, violet, and other floral and fruity features, with very low threshold, and have attracted much attention in wine aroma researches. β-Damascenone is a typical substance of C_{13}-norisoprene family, contributing lovely characteristics of apple, rose, and honey to wine [26]. The highest value was obtained in HL mixed; moreover, the concentration of each mixed-strain group was much higher than that of the corresponding single-strain groups, responsible for giving more sweety and floral flavors to the wine. Except for the major aroma compounds, a small amount of olefinic substances were also detected in Marselan must and wine. Among them, cyclohexene with the value of 31.26 and 16.76 was only detected in FS groups (single and mixed). Calamenene, as one of the most typical aromas in the Marselan grape, was detected only in YT mixed and HL mixed (latter was significantly higher than the former), indicating that HL mixed could preserve more varietal characteristics. All these results indicated that the aromatic profiles of single and mixed fermentation methods varied a lot, and the inoculation with *non-Saccharomyces* relatively retained more Marselan features.

3.4. Comparison of Aroma Compounds between Marselan Must and Wine

Compared with must, the relative content of phenethyl alcohol (50–100-fold change), ethyl caprylate (480–670-fold change), and ethyl caproate (10–20-fold change) was much higher in wine (Figure 2). These aromas were produced by yeasts during wine fermentation. However, some feature aromas of Marselan such as calamenene, only increased in HL mixed; β-damascenone reduced more than 80% in all groups; benzaldehyde diminished 54.26%, 16.92%, and 22.23% in FS36, YT28, and HL mixed, respectively (Tables 3 and 4); 2,4-di-tert-butylphenol dropped 68–81% in each group. In general, FS36 and FS mixed were worse on preserving characteristic aroma compounds, indicating that FS36 and FS mixed possibly were not capable of shaping the individual quality for Marselan wine. Esters were the main compound produced during fermentation, of which FS36 produced the most, yet HL mixed produced the least. FS36 also produced the most of new alcohols. Other fermentative aroma compounds included acids, aldehydes, etc., with a relatively minor effect on the wine, of which FS groups produced the most, followed by YT and HL groups. In conclusion, FS groups produced the most content of aromas in wine, followed by YT and HL groups. However, YT and HL groups were more capable of maintaining Marselan varietal characters such as calamenene and β-damascone.

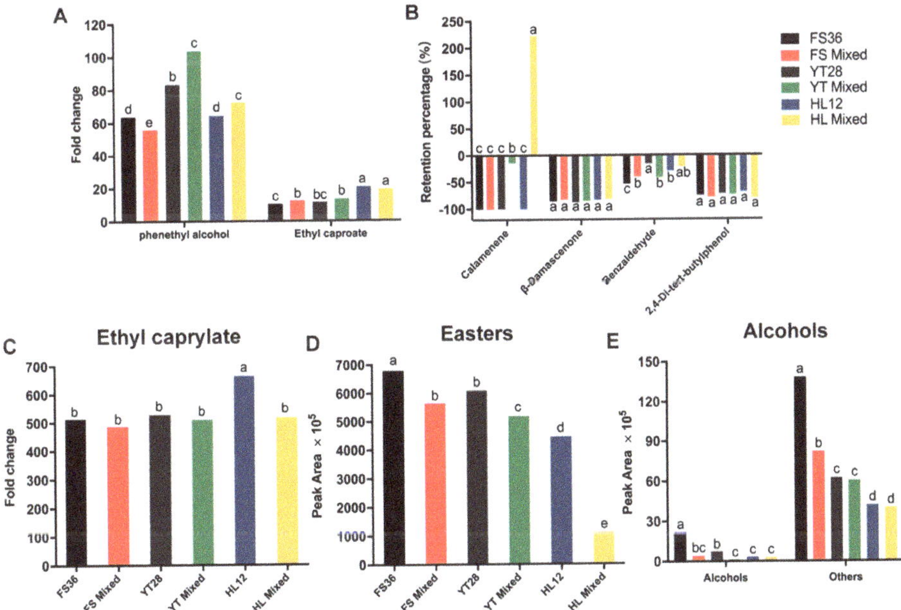

Figure 2. Changes of characteristic aromas in Marselan wine. Different letters indicate the significant differences at $p < 0.05$. (**A**,**C**) The fold change of phenethyl alcohol, ethyl caproate and ethyl caprylate in Marselan wines in comparison of must, respectively; (**B**) The retention percentage (%) of four varietal aromas in Marselan wines in comparison of must; (**D**) The peak area of easters of Marselan wines; (**E**) The peak area of alcohols of Marselan wines.

3.5. The PCA Analysis of Marselan Wine

The composition and abundance of aroma compounds in six groups were subjected to PCA analysis and explained 71.3% of the variability in the first two dimensions. Eight characteristic Marselan varieties and six newly produced aroma compounds were used to analyze. Figure 3 depicted that PCA1 accounted for the 43.0% and PCA2 for an additional 28.0% of the variability, which clearly separated single-strain groups from mixed-strain groups, proving that different fermentation methods had dissimilarity during the aromatic compound formation, the inoculation of *non-Saccharomyces* indeed changed the whole quality of wine aromas. Overall, characteristic variety aromas were preserved better in the mixed-strain groups, implying that mixed-strain fermentation was a better way to retain variety features. Reversely, single-strain groups were capable of producing new fermentative aroma compounds. Interestingly, HL mixed (with apparent benzaldehyde and β-damascenone) was located in a single quadrant, implying that the features of this group were totally diverse with others; obviously, fermentative aromas of this group were less than other groups. Ethyl octoate, ethyl 9-decanoate, and decyl alcohol coming from the fermentation process were mainly centralized on the quadrant of single-strain groups.

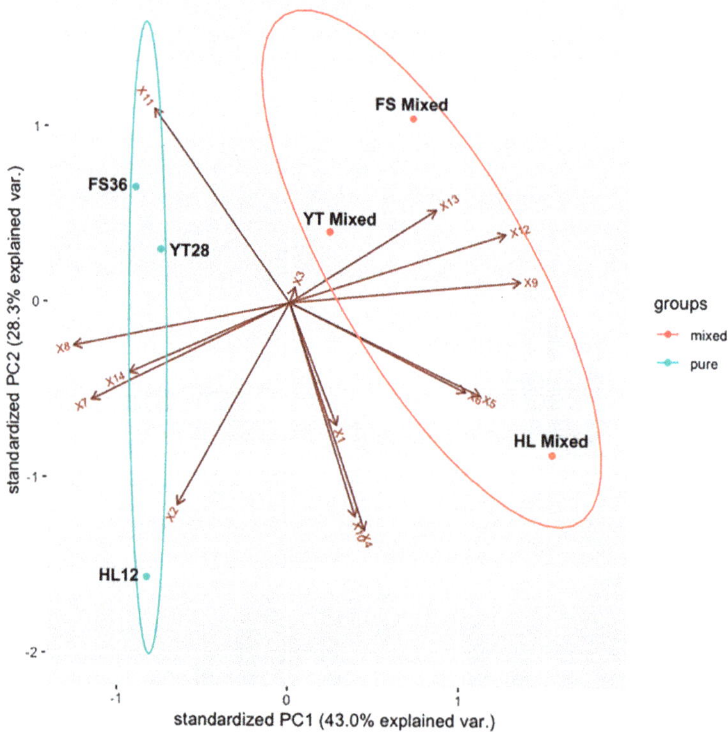

Figure 3. Principal Component Analysis (PCA) biplots of selected volatiles in six Marselan wines. (Notes X1: Benzaldehyde X2: Ethyl caprylate X3: Phenethyl alcohol X4: Ethyl caproate X5: Calamenene X6: β-Damascenone X7: 2,4-Di-tert-butylphenol X8: Phenethyl acetate X9: Hexyl acetate X10: Diethyl succinate X11: Ethyl caprate X12: Ethyl propionate X13: Trans β-farnesol X14: Decyl alcohol).

3.6. The Sensory Analysis of Marselan Wine

Eight aroma attributes, namely, ethanol, floral, citrus, stone fruits, berries, dry fruits, herbs, and fermentative aromas, contributed to the aroma profile of the wine. Figure 4 demonstrated the mean scores of the sensory attributes of wines. Fermentative, ethanol, and floral attributes, mostly produced during the fermentation process, were the dominating descriptors in single-strain groups. Apparently, HL mixed had the most intense attributes of citrus, herbs, berries, and dry fruits, implying that HL mixed group had a stronger and more complicated aroma profile than other groups. FS36 has the strongest fermentative flavor but a weaker perception of other attributes. On the whole, mixed-strain groups showed stronger notes of citrus, berries, and herbs, of which HL mixed had the highest scores in both terms. Thus, HL mixed showed the most promising potential to preserve Marselan varietal and also the richest sensation among all groups.

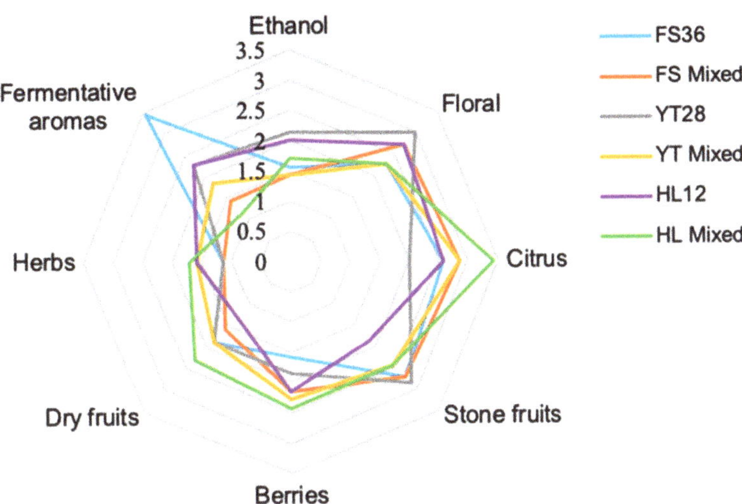

Figure 4. Radar plots of sensory evaluation of six Marselan wines.

In the PLSR model, 14 key aroma compounds were specified as the X-matrix, and the sensory attributes were specified as the Y-matrix. The correlation loading plot was shown in Figure 5. For the HL12, dried fruits, floral and alcoholic attributes were found to be associated with any of the key aroma compounds: specifically, with ethyl caprylate, β-damascenone, ethyl caproate, and diethyl succinate. For the HL mixed, strong correlations with aroma compounds were found for the citrus attribute. This was consistent with the sensory evaluations that HL mixed had the highest scores for citrus attribute, implying that citrus odor could be the characteristic sensory evaluation for HL mixed. The citrus attribute was correlated with several key aroma compounds, including benzaldehyde, trans-β-farnesol, ethyl propionate, hexyl acetate, and calamenene. For the FS36 and FS mixed, fermentation aromas and berries were the main attributes that correlated with the aroma compounds such as decyl alcohol, 2,4-di-tertbutyl phenol, and phenethyl acetate. For YT28 and YT mixed, both were related to phenethyl alcohol, but there was no related attribute.

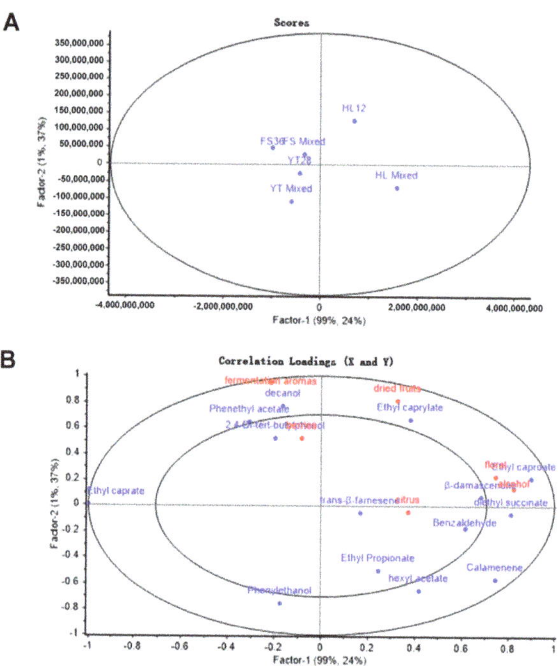

Figure 5. Partial least square regression correlation plot for six Marselan wines. The model was established by 14 key aroma compounds of GC-MS measurements as X, Y for the flavor attributes. (**A**) Scores of samples; (**B**) correlation loadings of X and Y.

4. Discussion

Aroma compounds are secondary metabolites that play an essential role in grape quality for enological purposes. Grape aromas were mostly coming from sugar. Many factors, including biotic and abiotic types, can influence the biosynthesis of aroma compounds in grape fruits, such as grape maturity, geographical locations, climates, and agronomic practice [29–31]. To the best of our knowledge, we were the first team to analyze the varietal aroma of Marselan grape must. Totally, there were 42 kinds of aromas identified, of which 2018 YT had the most. Three samples coming from the 2018 vintage had more types of aromas than the 2017 sample, probably due to its younger vintage. As wine ages, chemicals such as oxidation and reduction reactions will happen, and aging aromas will arise, but varietal aromas will decline. Compounds forming primary aromas belong to a limited number of chemical families, including methoxypyrazines, C_{13}-norisoprenoids, volatile sulfur compounds, and terpenes [32]. Twenty kinds of aromas were detected in common, and five of them (calamenene, β-damascenone, n-hexanal, furfural, and ceremonene) were significant among samples. β-ionone and β-damascenone are the main C_{13}-norisoprenoids compounds derived from carotenoids detected much in non-floral grapes. Terpenoids mostly existed in aromatic varieties such as Muscat, Gewürztraminer, and Rhine Riesling [33]. Particularly, five monoterpenoid alcohols, namely linalool, geraniol, nerol, citronellol, and α-terpineol, are the most abundant and the strongest contributors to wine aroma with floral sense and low odor thresholds [34]. Compared with other kinds of aromas, the concentrations of terpenes and C_{13}-norisoprenoids were much lower (Table 3), and our results were consistent with other studies [35–37].

It has been uncovered that phenols, terpenes, C_{13}-norisoprenoids, and non-terpenic alcohols are the most important aroma substances in grapes, and they can be detected as free volatiles or glycol-conjugated molecules, of which the non-volatile glycosylated group contributes the largest and presents in all varieties of Vitis vinifera. However, only volatile groups can be detected by GC-MS, so there still remains many aroma compounds that cannot be captured in the must. These important aroma precursors are linked to sugar molecules, mainly terpenol and C_{13}-norisoprenoid glycosides. Free styles could be released by hydrolysis during fermentation through the bioactivities of yeasts. There are two enzymes that existed in yeasts involving in aroma release, one is glycosidases, hydrolyzing the non-volatile glycosidic precursors, and the other is carbon-sulfur lyases, releasing volatile thiols from aroma-inactive cysteine-bound conjugates [38]. Even though all samples coming from the same region and vintage, the influences of agronomic practices cannot be excluded, such as irrigation, training systems, leaf removal, and bunch thinning [39]. Usually, agronomic practices do not always have uniform results, and each viticulturist has a distinct personal pattern; therefore, the metabolic and physiological changes resulting from agronomic practices were unknown. Furthermore, 20 characteristic aromas were found in common in all must samples, relatively representing the varietal aromas of the Marselan grape.

Marselan is one of the most popular grape varieties planted in China in recent years. The Marselan wine fermented by commercial *Saccharomyces* yeasts usually has a monotonous taste, lacking the varietal vigor and geographical characteristics. Conversely, indigenous wild yeasts are exhibiting more and more diversity and suitable brewing potential. However, limited studies have been carried out to improve its enological characteristics through the use of indigenous wine yeasts. Nowadays, the use of *non-Saccharomyces* yeasts co-inoculated with *Saccharomyces cerevisiae* is a popular strategy to improve the diversity and quality of wine aroma. The design of starters, including selected *non-Saccharomyces* yeasts with optimized biotechnological characteristics, has become one of the main challenges for researchers and oenologists. In this study, the effect of indigenous *non-Saccharomyces* inoculated with matched indigenous *S. cerevisiae* strains on aroma quality was investigated, with the aim to evaluate whether multiculture of specific *non-Saccharomyces* with local *S. cerevisiae* strains could produce high diversified aromatic quality. Naturally, *Saccharomyces* yeasts have a better ability to complete fermentation than *non-Saccharomyces* yeasts due to their adaptative genes to transform sugar and high tolerance of alcohol [40]. Consistent with above, all single-strain groups in our results fermented faster, which was in agreement with numerous studies [7,41–46]. FS36 was the fastest, implying that this strain has the best fermentation ability, followed by HL12 and YT28. There is a concern wherein the initial growth of *non-Saccharomyces* may be stuck or sluggish fermentations or reduce the production of detrimental compounds to the sensory properties of wine [47]. Like many successful *non-Saccharomyces* isolates, the selected *non-Saccharomyces* yeasts in our experiments did not suspend or affect the fermentation process, so these strains were well qualified to be used as starters [48–51]. All the groups showed suitable fermentation kinetics, implying that they have great potential to be used in industrial production.

Glycerol is one important product of yeast fermentation and is typically found at concentrations of 4–10 g/L in dry wine. In general, higher glycerol levels are considered to improve wine quality, contributing to smoothness, sweetness, and complexity for wines. From our results, except for FS36, mixed-strain groups produced more glycerol than single-strain groups. Previous studies discovered that wine fermented with *non-Saccharomyces* yeasts produced higher glycerol concentrations during alcoholic fermentation, which was in accordance with our results [8,52–54]. Generally, increased glycerol together with reduced alcohol in mixed fermentation has been observed in many research. The main reason is that some *non-Saccharomyces* cerevisiae is able to convert part of the carbon flux from the way of synthesizing alcohol to glycerol due to gene mutation or natural selection. However, from Table S1, the content of alcohol in three mixed-strain groups was not statistically significant from that of the single-strain groups, and there was even a slight increase trend, which was inconsistent with the above theory. In fact, many factors could affect the synthesis of alcohol or glycerol because numerous metabolic behaviors occur at the same time, although their carbon source all derives from sugar, the product of photosynthesis. For one thing, there is no absolute negative relation between the synthesis of glycerol and alcohol. For instance, the hybrid yeasts from *S. cerevisiae* × *S. uvarum* increased glycerol production compared to the *S. cerevisiae* parent; however, no reduction in ethanol concentration was observed [55]. Glycerol production in *Saccharomyces* cerevisiae can be induced by the addition of sulfite to the growth medium, which cannot be used for ethanol production due to the combination of sulfite and acetaldehyde. Thus, another study used sulfite to select yeasts with suitable adaption of high osmotic pressure, and this method successfully led to finding the yeasts that resulted in a 46% increase in glycerol yield. However, ethanol production was just a minor decrease [56]. For another, the production of glycerol largely depends on the redox balance within yeast cells, and both osmotic pressure and fermentation temperature can influence its content. In addition to alcohol, the synthesis rate of glycerol is widely adjusted by some metabolites such as organic acids, which can alter redox balance and/or influence yeast metabolism [56]. From Table S1, it could be found that the level of organic acids in the mixed group was lower (such as acetic acid, tartaric acid, and citric acid) than single ones, which partly explained why the alcohol content was not reduced because fractional carbon flux for organic acids has been transferred to the synthesis branches of alcohol, which further regulate the redox balance in the whole environment. Therefore, it is not credible to deduce the content of any one according to the other content. Furthermore, *non-Saccharomyces* yeasts have the ability to redirect the sugar consumption to produce alternative compounds, such as glycerol, with no apparent harm to wine or pyruvic acid produced via glycerol–pyruvic metabolisms. In conclusion, the inoculation with *non-Saccharomyces* yeasts as co-starters could be a promising technology to enhance the wine taste and final quality.

Previously, Lyu et al. identified several key aromas such as β-damascenone, eugenol, and 2,3-butanedione in Marselan wine, and found a suitable similarity of blackberry, green pepper, honey, raspberry, caramel, smoky, and cinnamon aroma attributes between the original Marselan wine and the reconstructed wine [20]. In this study, there were 53 aroma compounds detected in total, containing esters, higher alcohols, aldehydes, and norisoprenes. Esters are the most abundant compounds found in this study. Compared with the must, esters in wine were greatly improved due to the yeast metabolic activities. Ethyl decanoate was the single most abundant ester in all samples, associated with fruity and floral aromas [57]. Although esters provide wine with pleasant flavors, to preserve the varietal characteristics from the grape, the production of esters at proper concentrations is important to avoid masking the grape varietal aromas. The co-inoculation with some *non-Saccharomyces* yeast is an efficient way to produce wine with lower esters than *Saccharomyces* only. It has been found that the use of M. pulcherrima (a *non-Saccharomyces*) reduced the final total ester yield by approximately 33% [58]. Our results showed that the amount of esters in mixed-strain groups were lower than in single-strain groups, suggesting that the selected *non-Saccharomyces* reduced the easter aromas (Table S2). β-Damascenone is a typical substance of the C_{13}-norisoprene family, associated with apple, rose, and honey flavors [26]. The highest value was obtained in HL mixed; moreover, the concentration of β-damascenone in each mixed-strain group was much higher than that of the corresponding single-strain groups, implying that mixed fermentation was better to preserve the varietal aroma characteristics. Calamenene was another important aroma of Marselan must, which was also kept better in mixed groups. Actually, the promotion of flowery and fruity aromas by mixed fermentation with *non-Saccharomyces* have been confirmed by many studies. Zhang et al. used indigenous *Torulaspora delbrueckii* (TD12) co-fermented with *Saccharomyces* and found the mixed fermentation achieved higher aroma diversity, and generate higher intensity of fruity, flowery and sweet attributes of wine [59]. Another selected *Hanseniaspora uvarum* (*H. uvarum*) Yun268 (*non-Saccharomyces*) was found to improve the concentration of aromatic compounds via high levels of β-glucosidase activity and fatty acids [60]. The application of indigenous *non-Saccharomyces Hanseniaspora vineae* and *Metschnikowia pulcherrima* co-fermentation with *Saccharomyces cerevisiae* improved the aroma diversity in Vidal blanc ice wine [61]. Tristezza et al. used a local *S. cerevisiae* co-fermented with three *H.*

uvarum and found that *H. uvarum* contributed to increasing the wine organoleptic quality and simultaneously reducing the volatile acidity [61]. In addition, many studies came to similar conclusions [42,44–46,60,62]. These results explicated that mixed fermentation have more advantages than single fermentation.

Generally, there are two practices when using *non-Saccharomyces* yeasts in mixed fermentation. One is co-inoculation, the other is sequential inoculation. When fermentations are conducted with more than one yeast, complex interactions between organisms will happen. Both methods are feasible, although potential interactions between yeasts could determine which inoculation strategy is more appropriate. In this study, all groups were treated with sequential inoculation. Therefore, co-inoculation should be investigated further to figure out which method is better to improve aromas in Marselan wine. Interestingly, a study investigated both methods with *Torulaspora delbrueckii* and found sequential inoculation achieved higher aroma diversity than co-inoculation. The possible mechanism was the inhibitory effects on the growth of *non-Saccharomyces* in the initial fermentation [59]. Importantly, the composition, interaction, and diversity of the yeast micro-population significantly contribute to the sensory characteristics of wine, so more research needs to be performed to elucidate the relations between yeasts and final aromas. In this paper, we did not carry out the commercial fermentation trials because the main purpose of this study was to compare and explore the single or mixed fermentation started by indigenous *Saccharomyces* or *non-Saccharomyces* yeasts to figure out the effects on the aroma formation in Marselan wine. We hope to provide a theoretical basis for the industrial application of these promising yeasts. From the sensory analysis, it could be found that the features and uniqueness of the Marselan grape were most highlighted by HL mixed with the must from the same place. Previous studies have shown that the yeast populations have been found to be regionally distinct, a concept coined microbial terroir, which suggests that the certain aroma fingerprints formed between grapes and the environment are likely to be based on specific microbial populations, especially *non-Saccharomyces* cerevisiae. So indigenous yeasts from the same region as grape variety possibly are more conducive to preserve local variety characteristics than the yeasts from other areas.

Taken together, our results suggest that the wild indigenous yeasts from the same region as grape variety are possibly in favor of preserving varietal characteristics in wine, especially the mixed fermentation with non-*Saccharomycetes*, probably due to its better adaptability of local climate and geography, which could be applied as a clue to make personalized wine with outstanding characteristics.

5. Conclusions

Three autochthonous *Saccharomyces* yeasts had the excellent ability of fermentation, and the inoculation of matched *non-Saccharomyces* did not suspend or hinder the whole process. The sequential inoculation of *non-Saccharomyces* kept the varietal sensation better than single *Saccharomyces*, both in chemical compounds and sensory evaluation. HL mixed had the best performance to preserve Marselan grape features such as citrus, berries, dry fruits, and herbs characters, also with the least tedious fermentative flavors. Our study implies that the indigenous yeasts from the same region as the grape variety seem more conducive to preserve local variety characteristics.

Supplementary Materials: The following are available online at https://www.mdpi.com/article/10.3390/fermentation7030133/s1. Figure S1: The relative content of characteristic aroma compounds in Maeselan must. Table S1: The results of physicochenmical indexes of Marselan must and wine. Table S2: Summary of the volatile aroma components of Marselan wine.

Author Contributions: X.X.: Methodology, Formal analysis, Project administration, Writing—review and editing. A.X.: Methodology, Formal analysis, Project administration. Y.Y.: Investigation, Formal analysis, Funding acquisition, Writing—review and editing. W.H.: Conceptualization, Writing—review and editing, Funding acquisition. J.Z.: Conceptualization, Writing—review and editing, Supervision, Funding acquisition. All authors have read and agreed to the published version of the manuscript.

Funding: This work was supported by the National "Thirteenth Five-Year" Plan for Science and Technology Support (2016YFD0400500).

Institutional Review Board Statement: Not applicable.

Informed Consent Statement: Not applicable.

Data Availability Statement: Not applicable.

Acknowledgments: We genuinely thank programmer Shuai Peng for the PCA coding with R Studio software.

Conflicts of Interest: The authors declare no conflict of interest.

References

1. Styger, G.; Prior, B.; Bauer, F.F. Wine flavor and aroma. *J. Ind. Microbiol. Biotechnol.* **2011**, *38*, 1145–1159. [CrossRef]
2. Van Wyk, N.; Grossmann, M.; Wendland, J.; Von Wallbrunn, C.; Pretorius, I.S. The Whiff of Wine Yeast Innovation: Strategies for Enhancing Aroma Production by Yeast during Wine Fermentation. *J. Agric. Food Chem.* **2019**, *67*, 13496–13505. [CrossRef]
3. Borren, E.; Tian, B. The Important Contribution of Non-Saccharomyces Yeasts to the Aroma Complexity of Wine: A Review. *Foods* **2021**, *10*, 13. [CrossRef]
4. Alperstein, L.; Gardner, J.M.; Sundstrom, J.F.; Sumby, K.M.; Jiranek, V. Yeast bioprospecting versus synthetic biology—Which is better for innovative beverage fermentation? *Appl. Microbiol. Biotechnol.* **2020**, *104*, 1–15. [CrossRef]
5. Meier-Dörnberg, T.; Kory, O.I.; Jacob, F.; Michel, M.; Hutzler, M. Saccharomyces cerevisiae variety diastaticus friend or foe? Spoilage potential and brewing ability of different Saccharomyces cerevisiae variety diastaticus yeast isolates by genetic, phenotypic and physiological characterization. *FEMS Yeast Res.* **2018**, *18*, foy023. [CrossRef] [PubMed]
6. Lin, M.M.-H.; Boss, P.K.; Walker, M.E.; Sumby, K.M.; Grbin, P.R.; Jiranek, V. Evaluation of indigenous Non-Saccharomyces yeasts isolated from a South Australian vineyard for their potential as wine starter cultures. *Int. J. Food Microbiol.* **2020**, *312*, 108373. [CrossRef] [PubMed]
7. Binati, R.L.; Lemos, J.W.J.; Luzzini, G.; Slaghenaufi, D.; Ugliano, M.; Torriani, S. Contribution of Non-Saccharomyces yeasts to wine volatile and sensory diversity: A study on Lachancea thermotolerans, Metschnikowia spp. and Starmerella bacillaris strains isolated in Italy. *Int. J. Food Microbiol.* **2020**, *318*, 108470. [CrossRef]
8. Escribano-Viana, R.; González-Arenzana, L.; Portu, J.; Garijo, P.; López-Alfaro, I.; López, R.; Santamaría, P.; Gutiérrez, A.R. Wine aroma evolution throughout alcoholic fermentation sequentially inoculated with Non-Saccharomyces/Saccharomyces yeasts. *Food Res. Int.* **2018**, *112*, 17–24. [CrossRef] [PubMed]
9. Liu, P.-T.; Lu, L.; Duan, C.-Q.; Yan, G.-L. The contribution of indigenous Non-Saccharomyces wine yeast to improved aromatic quality of Cabernet Sauvignon wines by spontaneous fermentation. *LWT* **2016**, *71*, 356–363. [CrossRef]
10. Morata, A.; Escott, C.; Bañuelos, M.A.; Loira, I.; Del Fresno, J.M.; González, C.; Suárez-lepe, J.A. Contribution of Non-Saccharomyces yeasts to wine freshness. A review. *Biomolecules* **2020**, *10*, 34. [CrossRef]
11. Jolly, N.P.; Varela, C.; Pretorius, I.S. Not your ordinary yeast: Non-Saccharomyces yeasts in wine production uncovered. *FEMS Yeast Res.* **2014**, *14*, 215–237. [CrossRef] [PubMed]
12. Comitini, F.; Gobbi, M.; Domizio, P.; Romani, C.; Lencioni, L.; Mannazzu, I.; Ciani, M. Selected Non-Saccharomyces wine yeasts in controlled multistarter fermentations with Saccharomyces cerevisiae. *Food Microbiol.* **2011**, *28*, 873–882. [CrossRef] [PubMed]
13. Petruzzi, L.; Capozzi, V.; Berbegal, C.; Corbo, M.R.; Bevilacqua, A.; Spano, G.; Sinigaglia, M. Microbial Resources and Enological Significance: Opportunities and Benefits. *Front. Microbiol.* **2017**, *8*, 995. [CrossRef]
14. Chen, K.; Escott, C.; Loira, I.; del Fresno, J.M.; Morata, A.; Tesfaye, W.; Calderon, F.; Suárez-Lepe, J.A.; Han, S.; Benito, S. Use of Non-Saccharomyces yeasts and oenological tannin in red winemaking: Influence on colour, aroma and sensorial properties of young wines. *Food Microbiol.* **2018**, *69*, 51–63. [CrossRef] [PubMed]
15. Carrau, F.M.; Rivas-Gonzalo, J.C.; Alcalde-eon, C.; Boido, E.; Dellacassa, E. Pigment Profiles in Monovarietal Wines Produced in Uruguay. *Am. J. Enol. Vitic.* **2006**, *57*, 449–459.
16. Robinson, J.; Harding, J.; Vouillamoz, J. *Wine Grapes: A Complete Guide to 1368 Vine Varieties, Including Their Origins and Flavours*; Penguin: London, UK, 2013.
17. Sagratini, G.; Maggi, F.; Caprioli, G.; Cristalli, G.; Ricciutelli, M.; Torregiani, E.; Vittori, S. Comparative study of aroma profile and phenolic content of Montepulciano monovarietal red wines from the Marches and Abruzzo regions of Italy using HS-SPME–GC–MS and HPLC–MS. *Food Chem.* **2012**, *132*, 1592–1599. [CrossRef] [PubMed]
18. Suzzi, G.; Arfelli, G.; Schirone, M.; Corsetti, A.; Perpetuini, G.; Tofalo, R. Effect of grape indigenous Saccharomyces cerevisiae strains on Montepulciano d'Abruzzo red wine quality. *Food Res. Int.* **2012**, *46*, 22–29. [CrossRef]
19. Sun, Y.; Li, E.; Qi, X.; Liu, Y. Changes of diversity and population of yeasts during the fermentations by pure and mixed inoculation of Saccharomyces cerevisiae strains. *Ann. Microbiol.* **2015**, *65*, 911–919. [CrossRef]
20. Lyu, J.; Ma, Y.; Xu, Y.; Nie, Y.; Tang, K. Characterization of the Key Aroma Compounds in Marselan Wine by Gas Chromatography-Olfactometry, Quantitative Measurements, Aroma Recombination, and Omission Tests. *Molecules* **2019**, *24*, 2978. [CrossRef]

21. Conde, C.; Agasse, A.; Glissant, D.; Tavares, R.; Gerós, H.; Delrot, S. Pathways of Glucose Regulation of Monosaccharide Transport in Grape Cells. *Plant Physiol.* **2006**, *141*, 1563–1577. [CrossRef]
22. Tronchoni, J.; Gamero, A.; Arroyo-López, F.N.; Barrio, E.; Querol, A. Differences in the glucose and fructose consumption profiles in diverse Saccharomyces wine species and their hybrids during grape juice fermentation. *Int. J. Food Microbiol.* **2009**, *134*, 237–243. [CrossRef] [PubMed]
23. Loureiro, V.; Malfeito-Ferreira, M. Spoilage yeasts in the wine industry. *Int. J. Food Microbiol.* **2003**, *86*, 23–50. [CrossRef]
24. Kelebek, H.; Kesen, S.; Sonmezdag, A.S.; Cetiner, B.; Kola, O.; Selli, S. Characterization of the key aroma compounds in tomato pastes as affected by hot and cold break process. *J. Food Meas. Charact.* **2018**, *12*, 2461–2474. [CrossRef]
25. Comuzzo, P.; Tat, L.; Tonizzo, A.; Battistutta, F. Yeast derivatives (extracts and autolysates) in winemaking: Release of volatile compounds and effects on wine aroma volatility. *Food Chem.* **2006**, *99*, 217–230. [CrossRef]
26. González-Barreiro, C.; Rial-Otero, R.; Cancho-Grande, B.; Simal-Gándara, J. Wine Aroma Compounds in Grapes: A Critical Review. *Crit. Rev. Food Sci. Nutr.* **2015**, *55*, 202–218. [CrossRef] [PubMed]
27. Rapp, A.; Versini, G. Influence of nitrogen compounds in grapes on aroma compounds of wines. *Dev. Food Sci.* **1995**, *37*, 1659–1694. [CrossRef]
28. Culleré, L.; Cacho, J.; Ferreira, V. An assessment of the role played by some oxidation-related aldehydes in wine aroma. *J. Agric. Food Chem.* **2007**, *55*, 876–881. [CrossRef]
29. Spayd, S.E.; Tarara, J.M.; Mee, D.L.; Ferguson, J.C. Separation of sunlight and temperature effects on the composition of Vitis vinifera cv. Merlot berries. *Am. J. Enol. Vitic.* **2002**, *53*, 171–182. [CrossRef]
30. Hernández-Orte, P.; Cersosimo, M.; Loscos, N.; Cacho, J.; Garcia-Moruno, E.; Ferreira, V. The development of varietal aroma from non-floral grapes by yeasts of different genera. *Food Chem.* **2008**, *107*, 1064–1077. [CrossRef]
31. Ewart, A.; Brien, C.; Soderlund, R.; Smart, R. The effects of light pruning, irrigation and improved soil management on wine quality of the Vitis vinifera cv. Riesling. *Vitis Geilweilerhof.* **1985**, *24*, 209–217.
32. Ebeler, S.E.; Thorngate, J.H. Wine chemistry and flavor: Looking into the crystal glass. *J. Agric. Food Chem.* **2009**, *57*, 8098–8108. [CrossRef]
33. King, A.; Dickinson, J.R. Biotransformation of monoterpene alcohols by Saccharomyces cerevisiae, Torulaspora delbrueckii and Kluyveromyces lactis. *Yeast* **2000**, *16*, 499–506. [CrossRef]
34. Mateo, J.J.; Jiménez, M. Monoterpenes in grape juice and wines. *J. Chromatogr. A* **2000**, *881*, 557–567. [CrossRef]
35. Pogorzelski, E.; Wilkowska, A. Flavour enhancement through the enzymatic hydrolysis of glycosidic aroma precursors in juices and wine beverages: A review. *Flavour Fragr. J.* **2010**, *22*, 251–254. [CrossRef]
36. Liang, Z.; Fang, Z.; Pai, A.; Luo, J.; Gan, R.; Gao, Y.; Lu, J.; Zhang, P. Glycosidically bound aroma precursors in fruits: A comprehensive review. *Crit. Rev. Food Sci. Nutr.* **2020**, 1–29. [CrossRef]
37. Yang, Y.; Jin, G.-J.; Wang, X.-J.; Kong, C.-L.; Liu, J.; Tao, Y.-S. Chemical profiles and aroma contribution of terpene compounds in Meili (*Vitis vinifera* L.) grape and wine. *Food Chem.* **2019**, *284*, 155–161. [CrossRef]
38. Holt, S.; Miks, M.H.; De Carvalho, B.T.; Foulquié-Moreno, M.R.; Thevelein, J.M. The molecular biology of fruity and floral aromas in beer and other alcoholic beverages. *FEMS Microbiol. Rev.* **2019**, *43*, 193–222. [CrossRef] [PubMed]
39. Alem, H.; Rigou, P.; Schneider, R.; Ojeda, H.; Torregrosa, L. Impact of agronomic practices on grape aroma composition: A review. *J. Sci. Food Agric.* **2019**, *99*, 975–985. [CrossRef] [PubMed]
40. Padilla, B.; Gil, J.V.; Manzanares, P. Past and Future of Non-Saccharomyces Yeasts: From Spoilage Microorganisms to Biotechnological Tools for Improving Wine Aroma Complexity. *Front. Microbiol.* **2016**, *7*, 411. [CrossRef]
41. Padilla, B.; Zulian, L.; Ferreres, À.; Pastor, R.; Esteve-zarzoso, B.; Beltran, G.; Mas, A. Sequential Inoculation of Native Non-Saccharomyces and Saccharomyces cerevisiae Strains for Wine Making. *Front. Microbiol.* **2017**, *8*, 1293. [CrossRef]
42. Renault, P.; Coulon, J.; de Revel, G.; Barbe, J.C.; Bely, M. Increase of fruity aroma during mixed T. delbrueckii/S. cerevisiae wine fermentation is linked to specific esters enhancement. *Int. J. Food Microbiol.* **2015**, *207*, 40–48. [CrossRef]
43. Tofalo, R.; Patrignani, F.; Lanciotti, R.; Perpetuini, G.; Schirone, M.; Di Gianvito, P.; Pizzoni, D.; Arfelli, G.; Suzzi, G. Aroma profile of montepulciano d'abruzzo wine fermented by single and co-culture starters of autochthonous Saccharomyces and Non-Saccharomyces yeasts. *Front. Microbiol.* **2016**, *7*, 610. [CrossRef]
44. Medina, K.; Boido, E.; Fariña, L.; Gioia, O.; Gomez, M.E.; Barquet, M.; Gaggero, C.; Dellacassa, E.; Carrau, F. Increased flavour diversity of Chardonnay wines by spontaneous fermentation and co-fermentation with Hanseniaspora vineae. *Food Chem.* **2013**, *141*, 2513–2521. [CrossRef]
45. Terrell, E.; Cliff, M.A.; Van Vuuren, H.J.J. Functional characterization of individual- and mixed-Burgundian Saccharomyces cerevisiae isolates for fermentation of pinot noir. *Molecules* **2015**, *20*, 5112–5136. [CrossRef]
46. Hu, K.; Jin, G.-J.; Mei, W.-C.; Li, T.; Tao, Y.-S. Increase of medium-chain fatty acid ethyl ester content in mixed H. uvarum/S. cerevisiae fermentation leads to wine fruity aroma enhancement. *Food Chem.* **2018**, *15*, 495–501. [CrossRef]
47. Du Toit, M.; Pretorius, I.S. Microbial Spoilage and Preservation of Wine: Using Weapons from Nature's Own Arsenal. *S. Afr. J. Enol. Vitic.* **2000**, *21*, 74–76. [CrossRef]
48. Minnaar, P.P.; Du Plessis, H.W.; Jolly, N.P.; Van Der Rijst, M.; Du Toit, M. Non-Saccharomyces yeast and lactic acid bacteria in Co-inoculated fermentations with two Saccharomyces cerevisiae yeast strains: A strategy to improve the phenolic content of Syrah wine. *Food Chem.* **2019**, *11*, 100070. [CrossRef] [PubMed]

49. Vilela, A. Lachancea thermotolerans, the Non-Saccharomyces yeast that reduces the volatile acidity of wines. *Fermentation* **2018**, *4*, 56. [CrossRef]
50. Ivit, N.N.; Kemp, B. The impact of Non-Saccharomyces yeast on traditional method sparkling Wine. *Fermentation* **2018**, *4*, 73. [CrossRef]
51. Dutraive, O.; Benito, S.; Fritsch, S.; Beisert, B.; Patz, C.-D.; Rauhut, D. Effect of sequential inoculation with Non-Saccharomyces and Saccharomyces yeasts on Riesling wine chemical composition. *Fermentation* **2019**, *5*, 79. [CrossRef]
52. Rantsiou, K.; Dolci, P.; Giacosa, S.; Torchio, F.; Tofalo, R.; Torriani, S.; Suzzi, G.; Rolle, L.; Cocolin, L. Candida zemplinina can reduce acetic acid produced by Saccharomyces cerevisiae in sweet wine fermentations. *Appl. Environ. Microbiol.* **2012**, *78*, 1987–1994. [CrossRef]
53. Domizio, P.; Liu, Y.; Bisson, L.F.; Barile, D. Cell wall polysaccharides released during the alcoholic fermentation by SchizoSaccharomyces pombe and S. japonicus: Quantification and characterization. *Food Microbiol.* **2017**, *61*, 136–149. [CrossRef]
54. Di Gianvito, P.; Perpetuini, G.; Tittarelli, F.; Schirone, M.; Arfelli, G.; Piva, A.; Patrignani, F.; Lanciotti, R.; Olivastri, L.; Suzzi, G. Impact of Saccharomyces cerevisiae strains on traditional sparkling wines production. *Food Res. Int.* **2018**, *109*, 552–560. [CrossRef]
55. Coloretti, F.; Zambonelli, C.; Tini, V. Characterization of flocculent Saccharomyces interspecific hybrids for the production of sparkling wines. *Food Microbiol.* **2006**, *23*, 672–676. [CrossRef]
56. Kutyna, D.R.; Varela, C.; Stanley, G.A.; Borneman, A.R.; Henschke, P.A.; Chambers, P.J. Adaptive evolution of Saccharomyces cerevisiae to generate strains with enhanced glycerol production. *Appl. Microbiol. Biotechnol.* **2012**, *93*, 1175–1184. [CrossRef]
57. Lambrechts, M.G.; Pretorius, I.S. Yeast and its Importance to Wine Aroma-A Review. *S. Afr. J. Enol. Vitic.* **2019**, *21*, 97–129. [CrossRef]
58. Benito, S.; Hofmann, T.; Laier, M.; Lochbühler, B.; Schüttler, A.; Ebert, K.; Fritsch, S.; Röcker, J.; Rauhut, D. Effect on quality and composition of Riesling wines fermented by sequential inoculation with Non-Saccharomyces and Saccharomyces cerevisiae. *Eur. Food Res. Technol.* **2015**, *241*, 707–717. [CrossRef]
59. Zhang, B.Q.; Luan, Y.; Duan, C.Q.; Yan, G.L. Use of Torulaspora delbrueckii Co-fermentation with two Saccharomyces cerevisiae Strains with different aromatic characteristic to improve the diversity of red wine aroma profile. *Front. Microbiol.* **2018**, *5*, 606. [CrossRef] [PubMed]
60. Hu, K.; Jin, G.-J.; Xu, Y.-H.; Tao, Y.-S. Wine aroma response to different participation of selected Hanseniaspora uvarum in mixed fermentation with Saccharomyces cerevisiae. *Food Res. Int.* **2018**, *108*, 119–127. [CrossRef]
61. Zhang, B.-Q.; Shen, J.-Y.; Duan, C.-Q.; Yan, G.-L. Use of indigenous Hanseniaspora vineae and Metschnikowia pulcherrima co-fermentation with Saccharomyces cerevisiae to improve the aroma diversity of Vidal Blanc icewine. *Front. Microbiol.* **2018**, *9*, 2303. [CrossRef]
62. Gobbi, M.; Comitini, F.; Domizio, P.; Romani, C.; Lencioni, L.; Mannazzu, I.; Ciani, M. Lachancea thermotolerans and Saccharomyces cerevisiae in simultaneous and sequential co-fermentation: A strategy to enhance acidity and improve the overall quality of wine. *Food Microbiol.* **2013**, *33*, 271–281. [CrossRef] [PubMed]

 *fermentation*

Review

Varietal Aromas of Sauvignon Blanc: Impact of Oxidation and Antioxidants Used in Winemaking

Pei-Chin Tsai, Leandro Dias Araujo and Bin Tian *

Faculty of Agriculture and Life Sciences, Lincoln University, Lincoln 7647, New Zealand
* Correspondence: bin.tian@lincoln.ac.nz

Abstract: Key varietal characteristics of Sauvignon Blanc, including the descriptors of 'green' and 'tropical fruit', are mostly attributed to methoxypyrazines and volatile thiols, while monoterpenes, higher alcohols, esters, fatty acids, and other volatile compounds also add complexity and fruity notes to the wines. During the winemaking and ageing period, oxidation decreases the concentrations of these compounds and diminishes the flavours derived from this aromatic grape variety. Therefore, antioxidants, such as sulfur dioxide, are commonly utilized in Sauvignon Blanc wine production for better preservation of those beneficial primary aromas. This review focuses on key varietal aromas in Sauvignon Blanc wine and how they are influenced by oxidation, and SO_2 alternatives, including ascorbic acid, glutathione, and glutathione-enriched inactivated dry yeasts, that can be used in winemaking as antioxidants.

Keywords: antioxidants; glutathione; glutathione-enriched inactivated dry yeasts; methoxypyrazines; oxidation; Sauvignon Blanc; thiols

Citation: Tsai, P.-C.; Araujo, L.D.; Tian, B. Varietal Aromas of Sauvignon Blanc: Impact of Oxidation and Antioxidants Used in Winemaking. *Fermentation* **2022**, *8*, 686. https://doi.org/10.3390/fermentation8120686

Academic Editor: Antonio Morata

Received: 31 October 2022
Accepted: 22 November 2022
Published: 28 November 2022

Publisher's Note: MDPI stays neutral with regard to jurisdictional claims in published maps and institutional affiliations.

Copyright: © 2022 by the authors. Licensee MDPI, Basel, Switzerland. This article is an open access article distributed under the terms and conditions of the Creative Commons Attribution (CC BY) license (https://creativecommons.org/licenses/by/4.0/).

1. Introduction

Sauvignon Blanc is one of the most popular white wines around the world. Originating from the Loire Valley of France, Sauvignon Blanc was considered a wild weed before winemakers started to turn grapes into wine hence the name Sauvignon Blanc came from the meaning of 'wild whites' in French. Some of the best grape-growing regions for Sauvignon Blanc in the world are namely Loire Valley (France), Bordeaux (France), Marlborough (New Zealand), California (United States), and Casablanca (Chile). The flavours and styles of Sauvignon Blanc may vary by region due to different terroir and winemaking practices, but typically, Sauvignon Blanc wines are made in a dry, still, and light-bodied style with high acidity and aromatic characteristics dominated by grape derived fruity flavours.

Sauvignon Blanc vines in New Zealand were first commercially planted in Marlborough in the mid-1970s and the wines have gained worldwide recognition since then. Until now, Sauvignon Blanc is the most widely planted grape variety in New Zealand, with over 26,000 hectares of vineyard land devoted to growing the grape. The majority of which is planted in Marlborough (23k ha), followed by Hawke's Bay (1k ha) and Nelson (0.6k ha) [1]. Sauvignon Blanc accounts for 72% of New Zealand's wine production and makes up 86% of total wine exportation from New Zealand. The advantages of New Zealand's growing environment, including cool climate, low rainfall, long sunshine hours, large diurnal temperatures, and mixed soil types, aid in reaching the freshness and crispiness of Sauvignon Blanc grapes. Reductive winemaking, which aims at minimizing oxygen exposure during vinification, is commonly adopted for Sauvignon Blanc to reduce the loss of primary aromas and limit the development of oxidative characters.

2. Important Aroma Compounds in Sauvignon Blanc

Sauvignon Blanc is an aromatic variety typically having pronounced aromas of 'green' and 'tropical' characters. Green characters, including leafy, herbaceous, and grassy notes, are generally related to a specific group of volatile compounds, methoxypyrazines, while tropical characters with aromas of gooseberry, grapefruit, and passion fruit are attributed to volatile thiols. Despite the varietal aromas of Sauvignon Blanc being dominated by methoxypyrazines and volatile thiols (Table 1), other aroma compounds (Table 2) derived from sugar and amino acid metabolisms during alcoholic fermentation, such as higher alcohols, fatty acids, esters, and to a lesser content, acetaldehyde, play a supportive role [2]. These volatile metabolites enhance wine complexity and impart fruity aromas that are not varietal specific to Sauvignon Blanc [3,4].

Table 1. Key varietal compounds in Sauvignon Blanc wines.

Compound		Sensory Description	Concentrations in Wine (ng/L)	Sensory Threshold (ng/L)
Methoxypyrazines	2-methoxy-3-isobutylpyrazine (IBMP)	Asparagus, capsicum	0.4–56.3 [5,6]	2 [7] [1]
	2-methoxy-3-isopropylpyrazine (IPMP)	Earth, leaf	<0.03–13.7 [5,6,8]	2 [7] [1]
	2-methoxy-3-sec-butylpyrazine (SBMP)	Earth, leaf	<0.03–11.2 [5,9]	2 [7] [1]
Volatile thiols	4-mercapto-4-methylpentan-2-one (4MMP)	Box tree, passion fruit, black currant bud	4–40 [8,10]	0.8 [11] [2]
	3-mercaptohexan-1-ol (3MH)	Passion fruit, grapefruit, gooseberry, guava	200–18,000 [8,10]	60 [11] [2]
	3-mercaptohexyl acetate (3MHA)	Passion fruit, box tree	0–2500 [8,10]	4.2 [11] [2]
	Benzenemethanethiol (BMT)	Mineral, flint, smoke, burnt wood	10–15 [12]	0.3 [12] [2]

[1] determined in water; [2] determined in model wine.

2.1. Methoxypyrazines

Methoxypyrazines are commonly associated with the 'green' character in Sauvignon Blanc must and wine [13]. Methoxypyrazines are nitrogen-containing heterocyclic substances (Figure 1) biosynthesised as the secondary metabolites of amino acids, which are considered grape-derived aroma compounds [14]. Three methoxypyrazines in Sauvignon Blanc have very low perceptive thresholds in wine with general concentrations well above these threshold values [15].

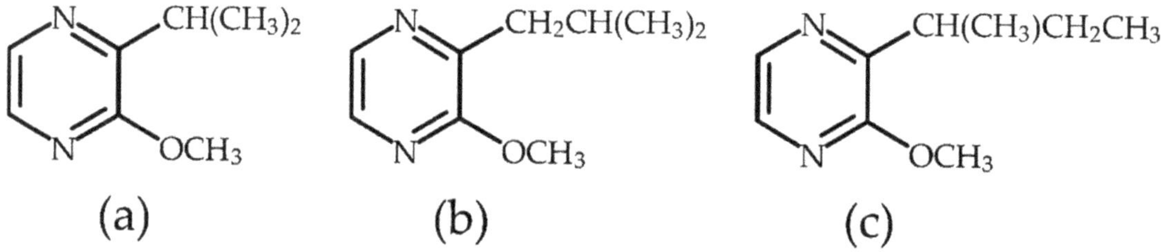

Figure 1. Methoxypyrazines in Sauvignon Blanc grapes and wine. (**a**) 2-methoxy-3-isopropylpyrazine, IPMP; (**b**) 2-methoxy-3-isobutylpyrazine, IBMP; (**c**) 2-methoxy-3-sec-butylpyrazine, SBMP.

The 2-methoxy-3-isobutylpyrazine (IBMP) with asparagus or capsicum-like aromas has the highest concentration recorded at 307 ng/L in grapes and 56.3 ng/L in wines, and thus it is considered the main contributor to the green character of Sauvignon Blanc wine [13]. However, the correlation between IBMP and the 'capsicum' attribute is found to be surprisingly weak [8], the explanation behind this may lean on the masking effect of other volatile compounds, such as varietal thiols, C6-alcohols, and dimethyl sulfide [8,16]. In addition, the study indicated when IBMP, (Z)-3-hexenol, and 1-hexanol were added together, the aroma presented in wine was changed from a veggie and earthy character to a pepper odour nuance, which showed a synergistic interaction among these compounds [17].

Both 2-methoxy-3-isopropylpyrazine (IPMP) and 2-methoxy-3-sec-butylpyrazine (SBMP) mainly contribute to earthy and leafy notes, however, they make subtle contributions to the green character due to their low concentrations present in grapes and wine. The aroma imparted by pyrazines was also found to depend on the presence of other compounds. In terms of the concentrations of pyrazines, up to 48.7 ng/L and 13.7 ng/L of IPMP were found in grapes and wines, respectively, and to a lesser content, the highest concentration of SBMP was recorded at 11.2 ng/L in both grapes and wines [8].

Studies conducted on red wine grapes showed that IBMP begins to accumulate in berries 10 days after anthesis, and declines rapidly during the maturation afterwards [18,19]. Final concentration of IBMP is determined before véraison [20], while early basal removal from the fruiting zones was found to reduce IBMP accumulation effectively [21]. The concentration of methoxypyrazines also depends on the climate of growing regions as the accumulation and degradation of these compounds are sensitive to temperature and sunlight exposure [20,22,23].

2.2. *Volatile Thiols*

Volatile sulfur compounds in wine can be divided into two classes based on the aromas perceived by humans, either negative or positive. On the one hand, some sulfur-containing compounds are responsible for off-flavours, such as rotten eggs ascribed to H2S produced by wine yeast [24,25], and cooked vegetables caused by thioacetic acid esters and mercaptans under low redox potential in wine [25]. On the other hand, certain sulfur- containing compounds can contribute to positive odours such as the notes of gooseberry, passion fruit, grapefruit, and guava [3]. These compounds considered to be the impact odorants in New Zealand's Sauvignon Blanc are referred to as volatile thiols or polyfunctional mercaptans [26]. Volatile thiols contain one or more sulfhydryl groups with additional functional groups such as ketones, alcohols, and esters in their molecules [3], and they can be perceived by the human nose at very low concentrations [27].

Three important volatile thiols, 4-mercapto-4-methylpentan-2-one (4MMP), 3-mercaptohexan-1-ol (3MH), and, 3-mercaptohexyl acetate (3MHA) (Figure 2), are key aroma compounds of New Zealand Sauvignon Blanc, i.e., imparting aromas of box tree and black currant bud contributed by 4MMP [28], and aromas of passionfruit and grapefruit contributed by 3MH and 3MHA [11]. The perceptive thresholds of 4MMP, 3MH, and 3MHA in model wine are 0.8 ng/L, 60 ng/L, and 4.2 ng/L [11], and their concentrations in Sauvignon Blanc wines from France and New Zealand are reported in the range of 4–40 ng/L, 200–18,000 ng/L, and 0–2500 ng/L, respectively [29]. If present at high concentrations, these compounds can emit sweaty aromas of cat urine character [30]. The biosynthesis of 3MHA is the result of the esterification of 3MH with acetic acid during fermentation, which is controlled by yeast ester-forming alcohol acetyltransferase encoded by the ATF1 gene [31]. The amount of 3MHA depends on the conversion rate differed in yeast strains [32], which is normally up to 10% of 3MH [33]. Previous studies have proposed three biogenesis pathways to explain the formation of 3MH and 4MMP in wine.

The first pathway involves cysteinylated precursors, S-3-(hexan-1-ol)-cysteine (Cys3MH) and S-3-(4-mercapto-4-methylpentan-2-one)-cysteine (Cys4MMP), which originate from grapes and are cleaved of the carbon-sulfur linkage by wine yeast through its beta-lyase activity during fermentation [33]. The second biogenesis pathway involves glutathionylated precursors, S-3-(hexan-1-ol)-glutathione (Glu-3MH) and S-3-(4-mercapto- 4-methylpentan-2-one)-glutathione (Glu-4MMP), which also originate from grapes and act as the precursors of both volatile thiols, 3MH and 4MMP, respectively [34]. Apart from this, Glu-4MMP also acts as the precursor of Cys3MH deciphering part of the Cys3MH biogenesis pathway [35]. The third biogenesis pathway involves the direct reaction of H_2S and unsaturated C_6 compounds, such as (E)-2-hexenal and (E)-2-hexenol [36]. H_2S can be produced by yeast metabolism of different sulfur sources, including elemental sulfur. The reduction of elemental sulfur by grape enzymes or other reducing agents, i.e., without the involvement of microorganisms, has also been suggested [37]. C6 compounds are formed through the enzymatic breakdown of polyunsaturated fatty acids catalysed by lipoxygenase, hydroperoxide lyase and isomerase [38]. While the importance of this pathway remains to be determined, it has been demonstrated that suppling H_2S to grape juice directly (Harsch paper again) or indirectly via elemental sulfur additions [37] invariably leads to an increase in 3MH and 3MHA concentrations in wine.

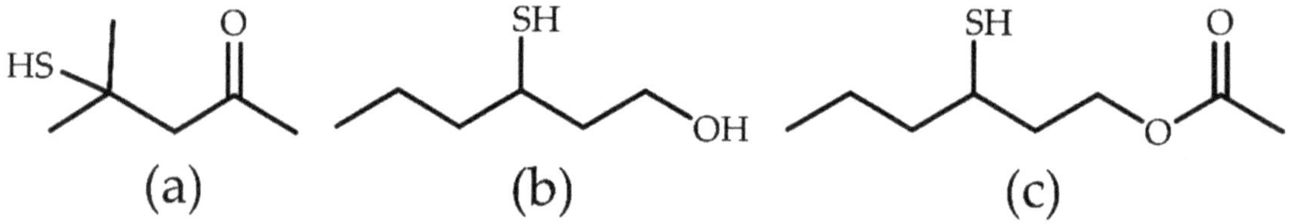

Figure 2. Volatile thiols found in Sauvignon Blanc wines. (**a**) 4-mercapto-4-methylpentan-2-one, 4MMP; (**b**) 3-mercaptohexan-1-ol, 3MH; (**c**) 3-mercaptohexyl acetate, 3MHA.

Apart from three varietal thiols that contributed mainly to tropical fruit aromas, there is another volatile thiol, benzenemethanethiol (BMT) that is worth mentioning. BMT was found to be responsible for some minerality characteristics, including flint, wet stones, and smoke in white wines [12,39], especially for a gun flint aroma presenting in Sauvignon Blanc wine [12]. However, the mineral character perceived in wine is considered a complex issue as it can be associated with various aspects of wine composition, including other aroma compounds [40], acidity [41], and reductive character [42]. The contribution of BMT to minerality in wine needs further investigation.

Table 2. Other volatile compounds in Sauvignon Blanc wines.

Compounds		**Sensory Description**	**Concentrations in Wine**	**Sensory Threshold**
Terpenes	Linalool	floral, citrus	7.2–24.3 μg/L [43,44]	100 μg/L [1] [45]
	Geraniol	freshly cut grass	1.0–2.8 μg/L [46]	130 μg/L [1] [45]
Higher alcohols	2-phenylethanol (2-PE)	Rose, honey, spice	43.4–436 μg/L [3] [43,47]	200 mg/L [2] [48]
	Isoamyl alcohol	Solvent, whiskey, malt, burnt	80–300 mg/L [10,46]	60 mg/L [2] [48]
Esters	Isoamyl acetate	Banana, pear	2080–2880 μg/L [3] [43]	50 μg/L [2] [48]
	Ethyl hexanoate	Apple, banana, violets	999–2892 μg/L [46]	45 μg/L [2] [46]
	2-phenylethyl acetate (2-PEA)	Rose, fruity, honey	0.21 mg/L [49]	1.8 mg/L [2] [48]
Volatile fatty acids	Acetic acid	Vinegar	150–900 mg/L [3] [50]	1130 mg/L [3] [51]
Others	Acetaldehyde	Bruised apple, grass, nut, sherry	7–240 mg/L [3] [52]	100 mg/L [3] [53]
	Methionol	Asparagus, potato, tomato	529–728 μg/L [46]	500 μg/L [2] [54]
	2-furanmethanethiol	Roast coffee	0.42–0.44 ng/L [55]	0.4 ng/L [2] [55]
	Sotolon	Curry	n.d.–36 μg/L [56]	8 μg/L [3] [57]

[1] determined in Muscat Alexandre, Muscat Blanc wines; [2] determined in model wine; [3] determined in dry white wine.

2.3. *Terpenes*

Monoterpenes and their oxygen-containing derivatives, monoterpene alcohols, are C_{10} compounds commonly found in grapes. Within more than 50 monoterpenes identified in grapes, linalool, citronellol, geraniol, nerol, and α-terpineol are the most abundant monoterpenes, especially in aromatic grape varieties, such as Muscat grapes, Riesling, and Gewürztraminer [58,59]. They contribute to floral (rose-like in particular), fruity, citrus, and perfume odours [60]. Even though the concentrations of terpenes are generally below the sensory threshold in Sauvignon Blanc, terpenes can still have a synergistic impact on the overall aromas present in wines [61].

Monoterpenes can be present in grapes in free form, polyhydroxylated form, or glycosidically bounded form [59]. Free forms of monoterpenes are odour-active compounds that include additional monoterpene ethyl esters and acetate esters present in wine. Polyhydroxylated forms of monoterpenes are free odourless polyols that make no direct contribution to wine aroma but can break down to release pleasant volatiles [59], i.e., diendiol can form hotrienol and nerol oxide (Williams, Strauss, & Wilson, 1980); (E)-3,7-dimethyl octa-2,5-dien-1,7-diol can form cis-rose oxide [62]. The majority of monoterpenes in grapes are glycosidically conjugated and are considered potential aroma compounds as they can be hydrolyzed into aroma-activated forms by acid or enzymatic hydrolysis during fermentation [63,64].

Monoterpene biosynthesis firstly involves the production of isopentenyl diphosphate (IPP) and its allylic isomer, dimethylallyl pyrophosphate (DMAPP), through the 2-C-methyl-D-erythritol 4-phosphate (MEP) pathway catalysed by terpene synthases (TPSs) and monoterpenol β-D-glucosyltransferases (GTs) in grape [65]. Then, IPP and DMAPP are condensed to form geranyl pyrophosphate (GPP) under the action of geranyl diphosphate synthase [66]. Finally, GPP is catalyzed by terpene synthases to form monoterpenes [67].

2.4. Higher Alcohols

Among all the yeast-derived volatile compounds, higher alcohols and esters are the two most abundant groups in wines [68]. Higher alcohols, also known as fusel alcohols, are alcohols that have more than two carbon atoms, and are mostly derived from amino acid metabolism during fermentation via the Ehrlich pathway, which includes transamination, decarboxylation, and reduction. After the initial transamination, amino acids are transferred into α-keto acids, which are then decarboxylated into fusel aldehydes and reduced into fusel alcohol [69]. Higher alcohols can be classified into aliphatic and aromatic alcohols based on amino acids that they are assimilated from [70]. Aliphatic alcohols are derived from linear- or branched-chain amino acids and have a strong solvent-like odour. Aromatic alcohols contain at least one benzene ring, which can enhance wine complexity and contribute positively to wine aromas when present at around 300 mg/L in white wines [71–73]. However, excessive higher alcohols (over 400 mg/L), including both aromatic and aliphatic alcohols, could contribute to a pungent smell and taste, and thus negatively affect the quality and character of wines [32,74]. The most important aromatic alcohol found in wines is 2-phenylethanol (2-PE), which is derived from phenylalanine and is associated with a rose-like aroma [75,76]. The 2-PE is naturally present in grapes at an undetectable concentration [49], however, certain yeast strains can enhance the production of 2-PE by increasing the production of its precursor, phenylalanine [77]. Apart from yeast strain selection, there are other factors during alcoholic fermentation that can affect the production of higher alcohols, such as fermentation temperature, juice clarity, oxygen levels, and amino acid composition [71,73]. Grape variety also has an impact on the concentration of higher alcohols, both general trends, and individuals. Overall, white wine varieties contain lower concentrations of higher alcohols than red wine varieties; individually, for example, more isoamyl alcohol is detected in Sauvignon Blanc than in Chardonnay [78].

2.5. Esters

Esters are found in wine through the reactions between alcohols and fatty acids. Yeast-derived esters can be categorised into acetate esters and ethyl esters [30]. In white wines, acetate esters largely contribute to the floral and tropical-fruity character of the wines [79], and ethyl esters contribute to wine complexity with the hints of tree fruits aromas at low concentrations but forming off-flavour like wax and honey at high concentrations [71]. High concentrations of esters in Sauvignon Blanc can be produced by selected yeast strains, which could have a masking effect over the green notes pertinent to methoxypyrazines [80]. The formation of acetate esters is mainly influenced by the yeast strains and the concentration of higher alcohols, while the formation of ethyl esters largely depends on the fermentation rate, which can be affected by the fermentation temperature and oxygen exposure [81]. Among these significant esters, the most critical esters identified in white wines that contribute to wine aromas are acetate esters of higher alcohols, such as isoamyl acetate, ethyl hexanoate and 2-phenylethyl acetate (2-PEA) [82]. Isoamyl acetate relates to banana and pear nuances, while ethyl hexanoate is associated with apple, banana, and violets notes. 2-PEA formed from the esterification of 2-PE contributes to the aromas of rose, fruity, and honey in wines [32]. Similar concentration ranges of most esters are found in Sauvignon Blanc to those in other white wine varieties except for a few compounds, such as ethyl hexanoate and ethyl isovalerate with higher concentrations, and ethyl isobutyrate with lower concentrations are quantified in Sauvignon Blanc [46].

2.6. Volatile Fatty Acids

Volatile fatty acids in wines are referred to a group of short- or medium-chain organic acids, including those responsible for volatile acidity (VA). The concentrations of volatile acids normally range from 500 to 1000 mg/L, and about 90% of them are acetic acid [50,83]. Acetic acid possesses a vinegar-like aroma when the concentration is near its perception threshold (0.7–1.1 g/L) [73]. Excessive production of acetic acid is usually associated with microbial spoilage of Acetobacter and Gluconobacter [84].

Medium-chain fatty acids in wine, such as hexanoic, octanoic, and decanoic fatty acids, are found to play an essential role in the fruity notes of wine [3]. These fatty acids are believed to be produced by yeast as intermediates in the biosynthesis of long-chain fatty acids [73]. The biosynthesis starts with the formation of acetyl coenzyme A (acetyl-CoA) from the oxidative decarboxylation of pyruvic acid, followed by the action of the fatty acid synthase complex [85].

The factors affecting the concentrations of volatile fatty acids include yeast strains, fermentation temperature, and juice nutrients. As for higher alcohols, the concentration of volatile fatty acids in wine is also depending on grape varieties, and differences in the concentrations of several volatile acids have been reported between Sauvignon Blanc and Chardonnay wines. Significantly higher concentrations of acetic acid, decanoic acid, octanoic acid, and hexanoic acid are found in Sauvignon Blanc than in Chardonnay wines [86].

2.7. Other Volatile Compounds

Volatile aldehydes contribute to different flavours, including notes of citrus, apple, grassy, pungent and nutty depending on the chemical structures [24,87], and they are of importance to wine aroma and flavours as they have low sensory thresholds. Volatile aldehydes are the intermediates in the Ehrlich pathway involving the formation of higher alcohols from amino acids and sugar [69], therefore, conditions favour the production of higher alcohols and also favour the formation of aldehydes [88]. Main aldehydes found in wine include acetaldehyde, 3-(methylthio)-propionaldehyde (methional), and phenylacetaldehyde, which are mainly generated from the oxidation of corresponding alcohols [89,90].

Acetaldehyde formed predominantly from the oxidation of ethanol via the coupled auto-oxidation of certain phenolic compounds [91], accounting for more than 90% of total aldehydes contents in wines [70]. The concentration of acetaldehyde in dry white wines ranges from 7 to 240 mg/L [52]. Methional and phenylacetaldehyde have also been found to occur via the Strecker reaction in which dicarbonyl compounds react with amino acids, methionine, and phenylalanine, respectively [89,90]. They have perception thresholds of 0.5 μg/L and 1 μg/L [89,92] with the concentrations in young Sauvignon Blanc wine at bottling below 0.5 and 5 μg/L, respectively [93].

Sauvignon Blanc wines are normally produced in a crisp and acidic way with fruit-driven aromas, however, oak-related aroma compounds can still be found in Sauvignon Blanc wines matured in barrels. For example, 2-furanmethanethiol contributing to roast coffee aroma is found in barrel-aged Sauvignon Blanc as the biogenesis of furanthiol derivatives requires furfural extracted from oak [94]. Sotolon (4,5-dimethyl-3-hydroxy-2(5)H-furanone), a volatile furanone with an intense curry odour that causes a detrimental effect on dry white wines [57,95], is also found in barrel-aged wines. Its concentration even exceeds the perceptive thresholds (8 μg/L) in Sauvignon Blanc wines aged in barrels without yeast lees, as the formation of sotolon in wines involves oxygen which can be easily reached without the protection from yeast lees during the ageing period [57].

3. Oxidation of Aroma Compounds in Sauvignon Blanc

Grape must and wine are inevitably exposed to oxygen during winemaking, and the amount of oxygen dissolved in wine can be up to 8 mg/L at cellar temperatures and air pressure [96]. Either insufficient or excessive amounts of oxygen in wine can lead to negative effects on wine quality. Too little oxygen exposure is associated with the formation of reductive characters in wine, while too much oxygen exposure causes the oxidation of both grape- and fermentation-derived aroma compounds, the development of undesirable oxidative odours, and the acceleration of colour browning [97,98]. Oxidation is detrimental for white wine, especially for Sauvignon Blanc wine, which is typically pale in colour and with pronounced fresh and fruity aromas.

3.1. Oxidation Mechanism for Volatile Thiols

The cysteinylated and glutathionylated precursors of volatile thiols cannot be directly oxidised during fermentation, as the C–S bond in the precursors is quite stable towards oxidation [27]. However, after alcoholic fermentation, volatile thiols have been released and their concentrations in wine can diminish largely by various mechanisms, including oxidation and nucleophilic addition, as they are highly reactive and chemically unstable [3]. First, volatile thiols can be easily oxidised into the corresponding disulfides in the presence of oxygen and trace amounts of metal ions [99,100]. Second, these compounds are nucleophilic and can undergo substitution reactions with polymeric phenolic compounds, with detailed results proposed by Nicolantonaki et al. that (−)-epicatechin was more reactive with thiols than (+)-catechin and the presence of Fe (III) favoured the oxidation reactions [101]. Third, these compounds can also take part in Michael addition reaction with phenolic oxidation products, such as *o*-quinones [102].

In the third mechanism, phenolic compounds react directly with RSO (reactive species of oxygen) which are activated oxygen species formed by the reduction of O_2 in the presence of metals, such as Fe^{2+} [101]. Two types of RSO are found related to phenolic oxidation, free superoxide ($O_{2\bullet}^{-}$) and peroxide (O_2^{2-}) radicals, while peroxide (O_2^{2-}) radicals are present in hydrogen peroxide form at wine pH [103,104]. Then, the oxidation of phenolic compounds undergoes a series of chemical transformations and forms semiquinone radicals and quinones, which are both catechol derivatives [101]. The formation of these catechol derivatives initiates in two pathways, oxidation by the intermediates from the reduction of oxygen to hydrogen peroxide [103,105] and direct oxidation by Fe^{3+} under acid conditions [101]. The acid conditions provided by wine preferentially go through the pathway of direct oxidation by Fe^{3+} to reduce the formal reduction potential of the Fe^{3+}/Fe^{2+} couple caused by RSO formation and further facilitate the redox cycling [103]. Finally, the phenol oxidation products, which are electrophilic, unstable, and highly reactive, can react with other compounds with low redox potentials, such as phenolic molecules, SO_2, and thiol-containing compounds, including glutathione and amino acids [106].

Quinones can react directly with nucleophilic thiols by a Michael addition reaction [104,107], and the thiols are degraded to an odourless form causing a loss of fruit character in the wine [108]. Specifically, the *o*-quinones derived from caftaric acid or (+)-catechin reacting with 3MH are most likely to be responsible for the loss of 3MH in wine under oxidative conditions [109].

3.2. Oxidation during Winemaking

Winemaking practices from harvesting, grape crushing, skin contact, and pressing, to alcoholic fermentation can input various amounts of oxygen into the musts. Winemaking decisions such as different harvesting methods, skin contact time, and pressing pressure applied should be made into consideration when producing Sauvignon Blanc wines, as the precursor content and the extraction rate in the juice leading to different concentrations of volatile compounds in the corresponding wines will be affected [110].

During the winemaking process, a high oxygen consumption rate, related to enzymatic activity, occurs with grape crushing [111]. A study demonstrated that adding SO_2 and ascorbic acids at crushing produces more pleasant wines with richer and more delicate aromas [111]. Skin contact and pressing under nitrogen increase the concentrations of glutathione (a natural antioxidant present in grapes) in Sauvignon Blanc wine [112,113]. Although longer skin contact time increases Cys3MH in the juice and larger pressure applied during pressure aids the release of volatile compounds [114], the oxidative potential of the juice also increases, which affects the concentrations of phenolic compounds dramatically, leading to decreased concentrations of 3MH and 3MHA in the resulting Sauvignon Blanc wines [115].

Methoxypyrazines are less prone to oxidation during winemaking. The concentration of IPMP was found to decrease at the stage of juice settling [15] and increase during the first day of maceration as methoxypyrazines are mainly found in the skins, seeds, and stems of the grapes and these compounds are extracted from grape musts to juice [116]. After racking, methoxypyrazines in Sauvignon Blanc wines level off regardless of the O_2/SO_2 additions at the juice stage [117,118].

The effects of the fermentation vessels on the chemical, physical, and sensory attributes of Sauvignon Blanc wines were compared, showing that different materials of the vessels affect the carbon dioxide and the oxygen dissolved in juice. As a result, juice fermented in stainless steel tanks with less oxygen dissolved produced higher concentrations of volatile fatty acids and esters, while juice in polyethylene tanks and clay jar received lower levels of these volatile compounds [113].

3.3. Oxidation during Ageing

Despite oak barrels have an antioxidant capacity that influences the oxidation potential of the wine and a positive correlation was demonstrated between the oxidative stability of the wine and barrel ageing [119], barrel maturation is not a common winemaking practice to produce Sauvignon Blanc wines, except for a specific style known as Fumé Blanc, developed in the United States. However, some Sauvignon Blanc wines will go through bottle ageing due to transportation or storage required from time to time, and oxidation during this ageing period can affect the aroma compounds significantly.

Closures with consistently low oxygen transfer rates, such as screw caps and micro-agglomerated corks, were found to be effective in detaining the oxidation of Sauvignon Blanc wine during long-term ageing [120], the oxidative status of the bottle-ageing wine between closures can be visually indicated by the degree of browning [121].

The concentration of methoxypyrazines during ageing was studied on Riesling and Cabernet Franc wine regarding the closure effects. The result showed that the concentration of IBMP decreased by 30% after 12 months of bottle-ageing, with the highest concentration retained in wines bottled with corks, followed by screw caps, synthetic corks, and Tetrapak cartons decrease the most after 18 months of bottle-ageing [122,123]. However, most studies suggested that bottle ageing had no measurable effect on the concentrations of methoxypyrazines in Sauvignon Blanc wine [124], and they remain stable even under progressive oxidative storage [118].

Volatile thiols are particularly prone to oxidation during bottle-ageing, and they can easily oxidise to corresponding disulfides [100]. During the bottle-ageing period of Sauvignon Blanc wines, 3MHA is considered to be the least stable one that decreases steadily throughout the first year of bottle-ageing [100] while 3MH first level off then increases after 3-month ageing, and the hydrolysis of 3MHA to 3MH is pointed to be the reason [125]. The concentrations of 3MH decrease after 6-month bottle ageing and continue declining as the ageing periods extended [93].

Closure and packaging options affecting the amounts of oxygen transferred through the bottles are also important to the concentrations of volatile thiols preserved in wines. 3MH was found to be better preserved in wines ageing in glass bottles compared to other configurations [93], while wines sealed with screw caps and bottle ampules displayed higher contents of volatile thiols compared to wines bottled using closure with higher permeability to oxygen [52].

The evolutions of other volatile compounds during ageing are mostly illustrated in other white wines as they are not specific to Sauvignon Blanc and are less prominent compared to methoxypyrazines and volatile thiols. Monoterpenes, such as linalool and α-terpineol, were found to decrease during storage [126,127], while increases of monoterpene alcohol derivatives, such as linalool oxides, nerol oxide, and hotrienol, were observed during ageing [128].

Wine aromas change drastically during the ageing period, in which the concentrations of yeast-derived compounds are significantly affected by oxygen exposure. Higher alcohols are supposed to decrease as they can be oxidised to form aldehydes during ageing [129], however, the concentrations of higher alcohols were reported to be stable [130,131] while hexanol was found to increase probably from the oxidation of linoleic and linolenic acids [132]. Ester concentrations decrease during bottle ageing due to chemical hydrolysis [126,127] or ester interactions with *o*-quinones or oxidation caused by a direct attack by hydroxyl radicals [97,133]. On the other hand, the concentrations of fatty acids are supposed to increase due to the hydrolysis of the corresponding ethyl ester, however, some compounds were reported to increase while others decrease or remain stable during ageing [122,134].

Oxidative aromas of white wines described as honey, woody, and cooked vegetables are mainly attributed to aldehydes, lactones, and acetals [95,97]. Under extreme storage conditions, such as higher dissolved oxygen concentrations and higher storage temperatures [95], the concentrations of methional and phenylacetaldehyde in dry white wines were found to increase significantly. Packaging materials with different gas exchange rates can influence the wine even under a short period of ageing. Higher concentrations of methional, phenylacetaldehyde, and sotolon were found in wines ageing in Bag in Box® and PET (polyethylene terephthalate) monolayer bottles compared to wines ageing in PET multi-layer bottles and glass bottles as oxygen is allowed to permeate and the wines are easily oxidised [93].

4. Antioxidants Used in Wine Production

Sulfur dioxide has been used as the main antioxidant in wine production at various winemaking stages, even though it occurs naturally in the wines as a by-product of yeast metabolism during fermentation [135]. Wines without any SO_2 additions will still have 10–20 mg/L of total SO_2 at the end of the alcoholic fermentation originating from yeast metabolism of amino acids [136]. The majority of SO_2 is mainly added in the form of potassium bisulfite, with typical concentration of free SO_2 ranging from 20 to 40 mg/L to protect juice or wine from oxidation, as well as, to keep the molecular SO_2 concentration below its sensory threshold of 2 mg/L.

Volatile thiols could be well preserved in red wines and model wines by the use of SO_2 under oxidative winemaking conditions in the presence of polyphenols [101,102,137]. The maximum amounts of 3MH and 3MHA were found in the final Sauvignon Blanc wines when SO_2 was added at around 120 mg/L at harvest [138]. However, the excessive use of SO_2 can have negative effects on human health, such as hives, swelling, headaches, stomach pain, and diarrhea [139]. In addition, the total SO_2 in wine is regulated and the use of SO_2 should be declared on the wine label in most winemaking countries. Therefore, there is a trend that the use of SO_2 is minimized in wine production and the wine industry has always been seeking for alternatives to reduce the usage of SO_2.

Common SO_2 alternatives used in wine production that have shown the efficacy of protecting juice and wine against oxidation are ascorbic acid and glutathione, to a lesser extent, glutathione-enriched inactivated dry yeast. Most of these alternatives have combined effects when used together with SO_2 while some of them need to be used complementary with SO_2. Various combinations of these antioxidants have been studied to find the most effective and efficient way to protect the wines while keeping the SO_2 additions to a minimum level.

4.1. Ascorbic Acid

Ascorbic acid has been utilized as an antioxidant with its capability to protect wines from oxidation by preferentially reacting with oxygen before the auto-oxidation of phenolic compounds happens and by reducing *o*-quinones back to the original phenolic compounds [103,140]. The oxidation of ascorbic acid generates hydrogen peroxide and dehydroascorbic acid, which is unstable and can degrade to a wide range of products [141]. Therefore, ascorbic acid is recommended to be used in conjunction with sulfur dioxide in wines as the presence of sulfur dioxide is essential to sufficiently remove hydrogen peroxide [142,143] and bind the dehydroascorbic acid and its degradation products [103].

Ascorbic acid protects white wines against oxidation and minimizes the browning of the wine under regular oxygen concentrations [144]. The combination of ascorbic acid and SO_2 better preserves fruity aromas as well as reduces oxidative aromas of the wines than using SO_2 alone [145]. Sauvignon Blanc wines supplemented with ascorbic acids and SO_2 were shown to contain higher levels of varietal thiols [146], and present higher intensities of fruity, grass, and green pepper aromas compared to wines added with SO_2 alone [147]. In addition, the presence of ascorbic acid decreases the requirement of SO_2 for a given amount of oxygen consumed in wines and thus extends the shelf-time of the wines [148,149]. However, in wines exposed to excessive amounts of oxygen either from poor bottling practices, inadequate closures, or long-term ageing, the role of ascorbic acid might convert from anti-oxidant to pro-oxidant [150,151]. This pro-oxidant activity relies on the rapid oxygen consumption of ascorbic acid together with SO_2, which will result in browning and a shortened lifetime compared to ascorbic acid-free wines, and can even contribute to spoilage of the wine over a longer time [150]. Conversely, Sauvignon Blanc wines added with ascorbic acids but sealed with closures with low oxygen transmission rate were found to develop reductive characters after bottle ageing [52]. Therefore, TPO (total package oxygen) concentrations indicating the oxidation potential should be considered when bottling [150].

4.2. Glutathione

Glutathione (GSH) is a tripeptide consisting of L-glutamate, L-cysteine, and glycine. The unique reducing and nucleophilic properties of glutathione that allow protein thiolation and modification of protein structure and function are ascribed to its free sulfhydryl moiety of the cysteine residue [152]. When acting as an antioxidant, glutathione can be oxidised enzymatically to glutathione disulfide (GSSG) [153]. Under unstressed conditions, GSSG can then be reduced back to glutathione by glutathione reductase, resulting in over 90% of glutathione existing in a reduced form [154,155].

Glutathione is naturally present in grapes and accumulates during grape maturation [156]. The concentration varies due to different grape varieties and is affected by both viticultural and oenological practices [107,114,157]. It can also be supplemented into musts or wine during the vinification process. A study revealed that glutathione added to Sauvignon Blanc juice shortly after crushing can provide antioxidant protection to the juice, reduce the use of SO_2, and produce more volatile thiols in wines [158]. However, the amount of glutathione supplementation is regulated by the Organisation Internationale de la Vigne et du Vin (OIV) with a limited dose of no more than 20 mg/L in the must [159]. The antioxidant properties of glutathione rely on the capability of its high affinity for oxygen which allows it preferentially oxidise its thiol group into a disulfuric group and form grape reaction products (2-S-glutathionyl caftaric acid), which terminates the oxidation process and thus protect other molecules from the attack of reactive oxygen species [160].

The effect of glutathione on the stability of wine flavour and wine colour has been demonstrated. It limits the formation of browning pigments by trapping *o*-quinones in a colourless form [161] and by limiting the production of xanthylium cation pigment precursors and *o*-quinone-derived phenolic compounds [162]. Additionally, glutathione protects varietal thiols from oxidation during bottle ageing [106], while its cysteinyl residue can be used as a source of sulfur to increase the concentration of polyfunctional mercaptans by reacting with trans-2-hexenal to form Glut-3MHal [163]. In addition, glutathione decreases the degradation of polyfunctional mercaptans during storage. Glutathione inhibits the decrease of several aromatic esters and terpene alcohols, such as isoamyl acetate, ethyl hexanoate, linalool, and α-terpineol [164], as well as limits the accumulation of acetaldehyde [165] during storage. Furthermore, it suppresses the formation of stolon and 2-aminoacetophenone (2-AAP), which release unpleasant odours in wines and contribute to atypical wine ageing defects [166]. However, glutathione is found to favour the accumulation of hydrogen sulfide and methyl mercaptan, especially in the presence of copper under low oxygen conditions [167].

The use of glutathione in conjunction with SO_2 has a combined effect on protecting volatile thiols against oxidation [26], also, the addition of SO_2 slows down the enzymatic reduction of trans-2-hexenal and inhibits the enzymatic oxidative loss of glutathione [163]. The combination of glutathione and ascorbic acids strengthens the antioxidant capabilities and protects the phenolic compounds from oxidation [26], in addition, glutathione delays the loss of ascorbic acids and inhibits the reaction of ascorbic acids degradation products and (+)-catechin [162].

4.3. Glutathione-Enriched Inactivated Dry Yeast

Despite glutathione has good antioxidant property, the application of glutathione to winemaking is still limited as the amount of glutathione supplementation is regulated [168]. The recommended approach of adding glutathione to juice or wine is through the addition of inactivated dry yeast (IDY) with guaranteed glutathione levels [169]. The preparations of glutathione-enriched inactive dry yeast (GSH-IDY) are manufactured from the thermal inactivation of *Saccharomyces cerevisiae*, which is cultivated under specific conditions (highly concentrated sugar medium) to stimulate the intracellular accumulation of glutathione. Commercial GSH-IDY were claimed to boost glutathione content either by liberating glutathione into the wine or by allowing the yeast to assimilate glutathione precursors during alcoholic fermentation for increased glutathione production [170]. Sauvignon Blanc wines supplemented with GSH-IDY preparations are shown to increase the concentration of certain volatile compounds, including thiols, higher alcohols, fatty acids, esters, and monoterpenes, and lead to higher intensities of aromas associated with riper tropical fruit than GSH-added wines [171]. The reasons that influenced the aroma profile of the wine are ascribed to the release of compounds other than glutathione by yeast products [26,171]. Previous studies that supplemented GSH-IDYs into the wines had some promising findings (Table 3), most of them affirming the antioxidant capability provided by glutathione released into the wine. More research is needed to better understand the full potential of this prospective antioxidant in white wine production, e.g., optimization of glutathione accumulation process, dosage rate, and timing of addition.

Table 3. Summary of previous studies on glutathione-enriched inactivated dry yeast (GSH-IDY).

Wine Matrix	Addition Timing	Key Findings	Reference
Sauvignon Blanc	Before fermentation	Increased thiols, higher alcohols, fatty acids, esters, and monoterpenes, leading to higher intensities of riper tropical fruity notes in wines.	[171]
Model wine	N.A.	Decreased the loss of typical wine terpenes.	[172]
Model wine	N.A.	Both yeast strain and glutathione accumulation process in preparation of GSH-IDY played an important role in the modulation of glutathione released into wine.	[173]
Model wine	N.A.	Yeast derivatives enriched with glutathione were more efficient at quenching radical species than those without glutathione enrichment.	[174]
Grenache Rosé	Before fermentation	More intense in fruity aromas (strawberry, banana) and less intense in yeast notes after 9 months ageing.	[175]
Sauvignon Blanc	After fermentation	Increased the release of polysaccharides into wines, and positive effects on the wine colour and on the prevention of wine oxidation.	[176]

5. Conclusions and Future Study

With low perception thresholds and distinctive aroma characters, methoxypyrazines and volatile thiols constitute the key varietal characteristics of Sauvignon Blanc wines. In addition to that, terpenes and other fermentative-derived volatile compounds are also important in adding complexity and imparting fruity notes to the wine. Being a fruit-driven wine dominated by primary aromas, one of the most important issues for Sauvignon Blanc wine production is oxidation during winemaking and ageing.

Methoxypyrazines are shown to remain stable after racking while oxidation condition including the additions of O_2/SO_2 plays a minor role. However, inconsistent reviews present in the concentration changes during ageing. The discrepancy between the results is speculated to be the variables in the ranges of ageing periods, the grape varieties, or the ageing conditions, and thus further investigation is needed to confirm the influencing factors behind it. Compared to methoxypyrazines, volatile thiols are more prone to oxidation after alcoholic fermentation. They can either be oxidised into corresponding disulfides or be degraded to an odourless form. The oxidation mechanisms for volatile thiols are well studied, but more research is needed to build a comprehensive perspective of the concentration changes for 4MMP during ageing, as most studies only focused on the concentrations of 3MH and 3MHA. Additionally, further investigation is needed to understand the whole picture of biogenesis pathways and to fill the gap between the theoretical amounts of thiols backtracking from the precursors and the actual concentrations present in wine. During ageing, some monoterpenes are oxidized into corresponding terpene alcohol derivatives; esters concentrations decrease due to hydrolysis or oxidation; higher alcohols remain stable in concentrations as they can be oxidized to aldehydes but also be released from the oxidation of acids. The concentrations of certain volatiles such as aldehydes, lactones, and acetals increase, while some other volatile compounds evolve without a general trend during ageing.

In general, SO_2 is still the most commonly used antioxidant in wine production. Other SO_2 antioxidants, such as ascorbic acid and glutathione, have been commonly accepted to exhibit their antioxidant capabilities when added at harvest. Given that SO_2 has been confirmed to have negative effects on sulfite-sensitive individuals and can cause a detrimental effect on wine quality and thus should be limited in addition, finding antioxidants other than SO_2 to protect wines from oxidation during storage and transportation is especially crucial to the aromatic expression of New Zealand Sauvignon Blanc.

Inactivated dry yeasts enriched with glutathione with promising potential have been researched, and products have been developed and sold to commercial wineries. The suggested timing of supplement by the manufacturers is before alcoholic fermentation as the yeast derivatives are found to improve alcoholic fermentation and protect wines from oxidation at the early stage. However, previous experiments regarding GSH-IDYs were insufficient and the supplemented timing of GSH-IDYs was studied to a limited extent. Further investigation is needed to compare the difference between adding the glutathione-enriched yeasts before and after fermentation and to study the antioxidant capability of GSH-IDYs protecting wines during ageing.

Author Contributions: Conceptualization, P.-C.T., L.D.A. and B.T.; writing—original draft preparation, P.-C.T.; writing—review and editing, L.D.A. and B.T.; supervision, L.D.A. and B.T.; project administration, B.T. All authors have read and agreed to the published version of the manuscript.

Funding: This research received no external funding.

Institutional Review Board Statement: Not applicable.

Informed Consent Statement: Not applicable.

Data Availability Statement: Not applicable.

Conflicts of Interest: The authors declare no conflict of interest.

References

1. *New Zealand Winegrowers Annual Reports*; New Zealand Winegrowers: Auckland, New Zealand, 2022.
2. Sponholz, W.; Kliewer, M.; Rapp, A.; Versini, G. Influence of nitrogen compounds in grapes on aroma compounds of wine. In Proceedings of the International Symposium on Nitrogen in Grapes and Wine, Seattle, WA, USA, 18–19 June 1991; pp. 156–164.
3. Coetzee, C.; du Toit, W.J. A comprehensive review on Sauvignon blanc aroma with a focus on certain positive volatile thiols. *Food Res. Int.* **2012**, *45*, 287–298. [CrossRef]
4. Louw, L.; Roux, K.; Tredoux, A.; Tomic, O.; Naes, T.; Nieuwoudt, H.H.; Van Rensburg, P. Characterization of selected South African young cultivar wines using FTMIR spectroscopy, gas chromatography, and multivariate data analysis. *J. Agric. Food Chem.* **2009**, *57*, 2623–2632. [CrossRef] [PubMed]
5. Alberts, P.; Stander, M.A.; Paul, S.O.; de Villiers, A. Survey of 3-Alkyl-2-methoxypyrazine content of South African Sauvignon blanc wines using a novel LC−APCI-MS/MS method. *J. Agric. Food Chem.* **2009**, *57*, 9347–9355. [CrossRef]
6. Lacey, M.J.; Allen, M.S.; Harris, R.L.; Brown, W.V. Methoxypyrazines in Sauvignon blanc grapes and wines. *Am. J. Enol. Vitic.* **1991**, *42*, 103–108.
7. Seifert, R.M.; Buttery, R.G.; Guadagni, D.G.; Black, D.R.; Harris, J. Synthesis of some 2-methoxy-3-alkylpyrazines with strong bell pepper-like odors. *J. Agric. Food Chem.* **1970**, *18*, 246–249. [CrossRef]
8. Lund, C.M.; Thompson, M.K.; Benkwitz, F.; Wohler, M.W.; Triggs, C.M.; Gardner, R.; Heymann, H.; Nicolau, L. New Zealand Sauvignon blanc distinct flavor characteristics: Sensory, chemical, and consumer aspects. *Am. J. Enol. Vitic.* **2009**, *60*, 1–12. [CrossRef]
9. Sala, C.; Busto, O.; Guasch, J.; Zamora, F. Contents of 3-alkyl-2-methoxypyrazines in musts and wines from *Vitis vinifera* variety Cabernet Sauvignon: Influence of irrigation and plantation density. *J. Sci. Food Agric.* **2005**, *85*, 1131–1136. [CrossRef]
10. Ribéreau-Gayon, P.; Glories, Y.; Maujean, A.; Dubourdieu, D. *Handbook of Enology: The Chemistry of Wine and Stabilization and Treatments*; John Wiley & Sons, Ltd.: Hoboken, NJ, USA, 2006; pp. 205–221.
11. Tominaga, T.; Furrer, A.; Henry, R.; Dubourdieu, D. Identification of new volatile thiols in the aroma of *Vitis vinifera* L. var. Sauvignon blanc wines. *Flavour Fragr. J.* **1998**, *13*, 159–162. [CrossRef]
12. Tominaga, T.; Guimbertau, G.; Dubourdieu, D. Contribution of benzenemethanethiol to smoky aroma of certain *Vitis vinifera* L. wines. *J. Agric. Food Chem.* **2003**, *51*, 1373–1376. [CrossRef]
13. Marais, J. Sauvignon blanc cultivar aroma—A review. *S. Afr. J. Enol. Vitic.* **1994**, *15*, 41–45. [CrossRef]
14. Maga, J. Sensory and stability properties of added methoxypyrazines to model and authentic wines. *Dev. Food Sci.* **1990**, *28*, 61–70.
15. Kotseridis, Y.; Spink, M.; Brindle, I.D.; Blake, A.; Sears, M.; Chen, X.; Soleas, G.; Inglis, D.; Pickering, G. Quantitative analysis of 3-alkyl-2-methoxypyrazines in juice and wine using stable isotope labelled internal standard assay. *J. Chromatogr. A* **2008**, *1190*, 294–301. [CrossRef]
16. Van Wyngaard, E. Volatiles Playing an Important Role in South African Sauvignon Blanc Wines. Master's Thesis, Stellenbosch University, Stellenbosch, South Africa, 2013.
17. Escudero, A.; Campo, E.; Fariña, L.; Cacho, J.; Ferreira, V. Analytical characterization of the aroma of five premium red wines. Insights into the role of odor families and the concept of fruitiness of wines. *J. Agric. Food Chem.* **2007**, *55*, 4501–4510. [CrossRef]

18. De Boubée, D.R.; Cumsille, A.M.; Pons, M.; Dubourdieu, D. Location of 2-methoxy-3-isobutylpyrazine in Cabernet Sauvignon grape bunches and its extractability during vinification. *Am. J. Enol. Vitic.* **2002**, *53*, 1–5. [CrossRef]
19. Hashizume, K.; Samuta, T. Grape maturity and light exposure affect berry methoxypyrazine concentration. *Am. J. Enol. Vitic.* **1999**, *50*, 194–198.
20. Ryona, I.; Pan, B.S.; Intrigliolo, D.S.; Lakso, A.N.; Sacks, G.L. Effects of cluster light exposure on 3-isobutyl-2-methoxypyrazine accumulation and degradation patterns in red wine grapes (*Vitis vinifera* L. cv. Cabernet Franc). *J. Agric. Food Chem.* **2008**, *56*, 10838–10846. [CrossRef]
21. Scheiner, J.J.; Sacks, G.L.; Pan, B.; Ennahli, S.; Tarlton, L.; Wise, A.; Lerch, S.D.; Heuvel, J.E.V. Impact of severity and timing of basal leaf removal on 3-isobutyl-2-methoxypyrazine concentrations in red winegrapes. *Am. J. Enol. Vitic.* **2010**, *61*, 358–364. [CrossRef]
22. Koch, A.; Ebeler, S.E.; Williams, L.E.; Matthews, M.A. Fruit ripening in *Vitis vinifera*: Light intensity before and not during ripening determines the concentration of 2-methoxy-3-isobutylpyrazine in Cabernet Sauvignon berries. *Physiol. Plant.* **2012**, *145*, 275–285. [CrossRef]
23. Tschiersch, C.; Nikfardjam, M.P.; Schmidt, O.; Schwack, W. Degree of hydrolysis of some vegetable proteins used as fining agents and its influence on polyphenol removal from red wine. *Eur. Food Res. Technol.* **2010**, *231*, 65–74. [CrossRef]
24. Henschke, P. Yeast-metabolism of nitrogen compounds. *Wine Microbiol. Biotechnol.* **1993**, *4*, 77–163.
25. Rauhut, D. Yeasts-production of sulfur compounds. *Wine Microbiol. Biotechnol.* **1993**, *6*, 183–223.
26. Lyu, X.; Del Prado, D.; Araujo, L.; Quek, S.Y.; Kilmartin, P. Effect of glutathione addition at harvest on Sauvignon Blanc wines. *Aust. J. Grape Wine Res.* **2021**, *27*, 431–441. [CrossRef]
27. Roland, A.; Schneider, R.; Razungles, A.; Cavelier, F. Varietal thiols in wine: Discovery, analysis and applications. *Chem. Rev.* **2011**, *111*, 7355–7376. [CrossRef] [PubMed]
28. Darriet, P.; Tominaga, T.; Lavigne, V.; Boidron, J.N.; Dubourdieu, D. Identification of a powerful aromatic component of *Vitis vinifera* L. var. Sauvignon wines: 4-mercapto-4-methylpentan-2-one. *Flavour Fragr. J.* **1995**, *10*, 385–392. [CrossRef]
29. Ribéreau-Gayon, P.; Dubourdieu, D.; Donèche, B.; Lonvaud, A. *Handbook of Enology, Volume 1: The Microbiology of Wine and Vinifications*; John Wiley & Sons: Hoboken, NJ, USA, 2006; Volume 1.
30. Swiegers, J.; Francis, I.; Herderich, M.; Pretorius, I. Meeting consumer expectations through management in vineyard and winery. *Wine Ind. J.* **2006**, *21*, 34–43.
31. Swiegers, J.H.; Capone, D.L.; Pardon, K.H.; Elsey, G.M.; Sefton, M.A.; Francis, I.L.; Pretorius, I.S. Engineering volatile thiol release in *Saccharomyces cerevisiae* for improved wine aroma. *Yeast* **2007**, *24*, 561–574. [CrossRef]
32. Swiegers, J.; Bartowsky, E.; Henschke, P.; Pretorius, I. Yeast and bacterial modulation of wine aroma and flavour. *Aust. J. Grape Wine Res.* **2005**, *11*, 139–173. [CrossRef]
33. Dubourdieu, D.; Tominaga, T.; Masneuf, I.; des Gachons, C.P.; Murat, M.L. The role of yeasts in grape flavor development during fermentation: The example of Sauvignon blanc. *Am. J. Enol. Vitic.* **2006**, *57*, 81–88. [CrossRef]
34. Roland, A.l.; Schneider, R.; Razungles, A.; Le Guerneve, C.; Cavelier, F. Straightforward synthesis of deuterated precursors to demonstrate the biogenesis of aromatic thiols in wine. *J. Agric. Food Chem.* **2010**, *58*, 10684–10689. [CrossRef]
35. Thibon, C.; Cluzet, S.; Mérillon, J.M.; Darriet, P.; Dubourdieu, D. 3-Sulfanylhexanol precursor biogenesis in grapevine cells: The stimulating effect of Botrytis cinerea. *J. Agric. Food Chem.* **2011**, *59*, 1344–1351. [CrossRef]
36. Harsch, M.J.; Benkwitz, F.; Frost, A.; Colonna-Ceccaldi, B.; Gardner, R.C.; Salmon, J.-M. New Precursor of 3-Mercaptohexan-1-ol in Grape Juice: Thiol-Forming Potential and Kinetics during Early Stages of Must Fermentation. *J. Agric. Food Chem.* **2013**, *61*, 3703–3713. [CrossRef]
37. Araujo, L.D.; Vannevel, S.; Buica, A.; Callerot, S.; Fedrizzi, B.; Kilmartin, P.A.; du Toit, W.J. Indications of the prominent role of elemental sulfur in the formation of the varietal thiol 3-mercaptohexanol in Sauvignon blanc wine. *Food Res. Int.* **2017**, *98*, 79–86. [CrossRef]
38. Matsui, K. Green leaf volatiles: Hydroperoxide lyase pathway of oxylipin metabolism. *Curr. Opin. Plant Biol.* **2006**, *9*, 274–280. [CrossRef]
39. Capone, D.; Barker, A.; Williamson, P.; Francis, I. The role of potent thiols in Chardonnay wine aroma. *Aust. J. Grape Wine Res.* **2018**, *24*, 38–50. [CrossRef]
40. Tominaga, T.; Baltenweck-Guyot, R.; Des Gachons, C.P.; Dubourdieu, D. Contribution of volatile thiols to the aromas of white wines made from several *Vitis vinifera* grape varieties. *Am. J. Enol. Vitic.* **2000**, *51*, 178–181.
41. Ross, J. Minerality: Rigorous or romantic. *Pract. Winery Vineyard J.* **2012**, *33*, 50–57.
42. Goode, J. Minerality in wine. *Sommel. J.* **2012**, *24*, 63–67.
43. Luan, F.; Mosandl, A.; Gubesch, M.; Wüst, M. Enantioselective analysis of monoterpenes in different grape varieties during berry ripening using stir bar sorptive extraction-and solid phase extraction-enantioselective-multidimensional gas chromatography-mass spectrometry. *J. Chromatogr. A* **2006**, *1112*, 369–374. [CrossRef]
44. Mateo, J.; Jiménez, M. Monoterpenes in grape juice and wines. *J. Chromatogr. A* **2000**, *881*, 557–567. [CrossRef]
45. Marais, J. Terpenes in the aroma of grapes and wines: A review. *S. Afr. J. Enol. Vitic.* **1983**, *4*, 49–58. [CrossRef]
46. Ribéreau-Gayon, P.; Glories, Y.; Maujean, A.; Dubourdieu, D. Phenolic compounds. *Handb. Enol.* **2006**, *2*, 141–203.
47. Koslitz, S.; Renaud, L.; Kohler, M.; Wüst, M. Stereoselective formation of the varietal aroma compound rose oxide during alcoholic fermentation. *J. Agric. Food Chem.* **2008**, *56*, 1371–1375. [CrossRef] [PubMed]

48. Schwab, W.; Davidovich-Rikanati, R.; Lewinsohn, E. Biosynthesis of plant-derived flavor compounds. *Plant J.* **2008**, *54*, 712–732. [CrossRef]
49. Wen, Y.-Q.; Zhong, G.-Y.; Gao, Y.; Lan, Y.-B.; Duan, C.-Q.; Pan, Q.-H. Using the combined analysis of transcripts and metabolites to propose key genes for differential terpene accumulation across two regions. *BMC Plant Biol.* **2015**, *15*, 240. [CrossRef] [PubMed]
50. Luan, F.; Wüst, M. Differential incorporation of 1-deoxy-D-xylulose into (3S)-linalool and geraniol in grape berry exocarp and mesocarp. *Phytochemistry* **2002**, *60*, 451–459. [CrossRef] [PubMed]
51. Oldfield, E.; Lin, F.Y. Terpene biosynthesis: Modularity rules. *Angew. Chem. Int. Ed.* **2012**, *51*, 1124–1137. [CrossRef]
52. Matarese, F.; Cuzzola, A.; Scalabrelli, G.; D'Onofrio, C. Expression of terpene synthase genes associated with the formation of volatiles in different organs of *Vitis vinifera*. *Phytochemistry* **2014**, *105*, 12–24. [CrossRef]
53. Kozina, B.; Karoglan, M.; Herjavec, S.; Jeromel, A.; Orlic, S. Influence of basal leaf removal on the chemical composition of Sauvignon Blanc and Riesling wines. *J. Food Agric. Environ.* **2008**, *6*, 28.
54. Yue, X.; Ma, X.; Tang, Y.; Wang, Y.; Wu, B.; Jiao, X.; Zhang, Z.; Ju, Y. Effect of cluster zone leaf removal on monoterpene profiles of Sauvignon Blanc grapes and wines. *Food Res. Int.* **2020**, *131*, 109028. [CrossRef]
55. Ribéreau-Gayon, P.; Boidron, J.; Terrier, A. Aroma of Muscat grape varieties. *J. Agric. Food Chem.* **1975**, *23*, 1042–1047. [CrossRef]
56. Benkwitz, F.; Tominaga, T.; Kilmartin, P.A.; Lund, C.; Wohlers, M.; Nicolau, L. Identifying the Chemical Composition Related to the Distinct Aroma Characteristics of New Zealand Sauvignon blanc Wines. *Am. J. Enol. Vitic.* **2012**, *63*, 62–72. [CrossRef]
57. Ferreira, V.; Rapp, A.; Cacho, J.F.; Hastrich, H.; Yavas, I. Fast and quantitative determination of wine flavor compounds using microextraction with Freon 113. *J. Agric. Food Chem.* **1993**, *41*, 1413–1420. [CrossRef]
58. Peinado, R.A.; Moreno, J.; Bueno, J.E.; Moreno, J.A.; Mauricio, J.C. Comparative study of aromatic compounds in two young white wines subjected to pre-fermentative cryomaceration. *Food Chem.* **2004**, *84*, 585–590. [CrossRef]
59. Rodrıguez-Bencomo, J.; Conde, J.; Rodrıguez-Delgado, M.; Garcıa-Montelongo, F.; Pérez-Trujillo, J. Determination of esters in dry and sweet white wines by headspace solid-phase microextraction and gas chromatography. *J. Chromatogr. A* **2002**, *963*, 213–223. [CrossRef]
60. Radler, F. Yeasts-metabolism of organic acids. *Wine Microbiol. Biotechnol.* **1993**, *5*, 165–181.
61. Corison, C.; Ough, C.; Berg, H.; Nelson, K. Must acetic acid and ethyl acetate as mold and rot indicators in grapes. *Am. J. Enol. Vitic.* **1979**, *30*, 130–134.
62. Lopes, P.; Silva, M.A.; Pons, A.; Tominaga, T.; Lavigne, V.; Saucier, C.; Darriet, P.; Teissedre, P.-L.; Dubourdieu, D. Impact of oxygen dissolved at bottling and transmitted through closures on the composition and sensory properties of a Sauvignon blanc wine during bottle storage. *J. Agric. Food Chem.* **2009**, *57*, 10261–10270. [CrossRef]
63. Zoecklein, B.; Fugelsang, K.C.; Gump, B.H.; Nury, F.S. *Wine Analysis and Production*; Springer Science & Business Media: Berlin/Heidelberg, Germany, 2013.
64. Guth, H. Quantitation and sensory studies of character impact odorants of different white wine varieties. *J. Agric. Food Chem.* **1997**, *45*, 3027–3032. [CrossRef]
65. Tominaga, T.; Blanchard, L.; Darriet, P.; Dubourdieu, D. A Powerful Aromatic Volatile Thiol, 2-Furanmethanethiol, Exhibiting Roast Coffee Aroma in Wines Made from Several *Vitis vinifera* Grape Varieties. *J. Agric. Food Chem.* **2000**, *48*, 1799–1802. [CrossRef]
66. Gabrielli, M.; Buica, A.; Fracassetti, D.; Stander, M.; Tirelli, A.; du Toit, W.J. Determination of sotolon content in South African white wines by two novel HPLC–UV and UPLC–MS methods. *Food Chem.* **2015**, *169*, 180–186. [CrossRef]
67. Lavigne, V.; Pons, A.; Darriet, P.; Dubourdieu, D. Changes in the sotolon content of dry white wines during barrel and bottle aging. *J. Agric. Food Chem.* **2008**, *56*, 2688–2693. [CrossRef]
68. Vilanova, M.; Genisheva, Z.; Graña, M.; Oliveira, J. Determination of odorants in varietal wines from international grape cultivars (*Vitis vinifera*) grown in NW Spain. *S. Afr. J. Enol. Vitic.* **2013**, *34*, 212–222. [CrossRef]
69. Hazelwood, L.A.; Daran, J.-M.; Van Maris, A.J.; Pronk, J.T.; Dickinson, J.R. The Ehrlich pathway for fusel alcohol production: A century of research on *Saccharomyces cerevisiae* metabolism. *Appl. Environ. Microbiol.* **2008**, *74*, 2259–2266. [CrossRef]
70. Nykänen, L.; Nykänen, I. Production of esters by different yeast strains in sugar fermentations. *J. Inst. Brew.* **1977**, *83*, 30–31. [CrossRef]
71. Coetzee, C. Oxygen and Sulphur Dioxide Additions to Sauvignon Blanc: Effect on Must and Wine Composition. Master's Thesis, University of Stellenbosch, Stellenbosch, South Africa, 2011.
72. Ferreira, V. Volatile aroma compounds and wine sensory attributes. In *Managing Wine Quality*; Elsevier: Amsterdam, The Netherlands, 2010; pp. 3–28.
73. Lambrechts, M.; Pretorius, I. Yeast and its importance to wine aroma—A review. *S. Afr. J. Enol. Vitic.* **2000**, *21*, 97–129. [CrossRef]
74. Pretorius, I.S.; Høj, P.B. Grape and wine biotechnology: Challenges, opportunities and potential benefits. *Aust. J. Grape Wine Res.* **2005**, *11*, 83–108. [CrossRef]
75. Fang, Y.; Qian, M. Aroma compounds in Oregon Pinot Noir wine determined by aroma extract dilution analysis (AEDA). *Flavour Fragr. J.* **2005**, *20*, 22–29. [CrossRef]
76. Lilly, M.; Bauer, F.F.; Styger, G.; Lambrechts, M.G.; Pretorius, I.S. The effect of increased branched-chain amino acid transaminase activity in yeast on the production of higher alcohols and on the flavour profiles of wine and distillates. *FEMS Yeast Res.* **2006**, *6*, 726–743. [CrossRef]

77. Bordiga, M.; Lorenzo, C.; Pardo, F.; Salinas, M.; Travaglia, F.; Arlorio, M.; Coïsson, J.D.; Garde-Cerdán, T. Factors influencing the formation of histaminol, hydroxytyrosol, tyrosol, and tryptophol in wine: Temperature, alcoholic degree, and amino acids concentration. *Food Chem.* **2016**, *197*, 1038–1045. [CrossRef]
78. Weldegergis, B.T.; de Villiers, A.; Crouch, A.M. Chemometric investigation of the volatile content of young South African wines. *Food Chem.* **2011**, *128*, 1100–1109. [CrossRef]
79. Ferreira, V.; Fernandez, P.; Pena, C.; Escudero, A.; Cacho, J.F. Investigation on the Role Played by Fermentation Esters in the Aroma of Young Spanish Wines by Multivariate-Analysis. *J. Sci. Food Agric.* **1995**, *67*, 381–392. [CrossRef]
80. Marais, J. Effect of grape temperature and yeast strain on Sauvignon blanc wine aroma composition and quality. *S. Afr. J. Enol. Vitic.* **2001**, *22*, 47–50. [CrossRef]
81. Guth, H. Identification of character impact odorants of different white wine varieties. *J. Agric. Food Chem.* **1997**, *45*, 3022–3026. [CrossRef]
82. Thurston, P.; Taylor, R.; Ahvenainen, J. Effects of linoleic acid supplements on the synthesis by yeast of lipids and acetate esters. *J. Inst. Brew.* **1981**, *87*, 92–95. [CrossRef]
83. Fleet, G. Yeasts-growth during fermentation. *Wine Microbiol. Biotechnol.* **1993**, *2*, 27–54.
84. Boulton, R.B.; Singleton, V.L.; Bisson, L.F.; Kunkee, R.E. *Principles and Practices of Winemaking*; Springer Science & Business Media: Berlin/Heidelberg, Germany, 2013.
85. Lynen, F. The role of biotin-dependent carboxylations in biosynthetic reactions. *Biochem. J.* **1967**, *102*, 381. [CrossRef]
86. Louw, L.; Tredoux, A.G.J.; Van Rensburg, P.; Kidd, M.; Naes, T.; Nieuwoudt, H.H. Fermentation-derived Aroma Compounds in Varietal Young Wines from South Africa. *South Afr. J. Enol. Vitic.* **2010**, *31*, 213–225. [CrossRef]
87. Miyake, T.; Shibamoto, T. Quantitative analysis of acetaldehyde in foods and beverages. *J. Agric. Food Chem.* **1993**, *41*, 1968–1970. [CrossRef]
88. Berry, D.; Slaughter, J. Alcoholic beverage fermentations. In *Fermented Beverage Production*; Springer: Berlin/Heidelberg, Germany, 2003; pp. 25–39.
89. Escudero, A.; Cacho, J.; Ferreira, V. Isolation and identification of odorants generated in wine during its oxidation: A gas chromatography–olfactometric study. *Eur. Food Res. Technol.* **2000**, *211*, 105–110. [CrossRef]
90. Pripis-Nicolau, L.; De Revel, G.; Bertrand, A.; Maujean, A. Formation of flavor components by the reaction of amino acid and carbonyl compounds in mild conditions. *J. Agric. Food Chem.* **2000**, *48*, 3761–3766. [CrossRef]
91. Wildenradt, H.; Singleton, V. The production of aldehydes as a result of oxidation of polyphenolic compounds and its relation to wine aging. *Am. J. Enol. Vitic.* **1974**, *25*, 119–126.
92. Culleré, L.; Cacho, J.; Ferreira, V. An assessment of the role played by some oxidation-related aldehydes in wine aroma. *J. Agric. Food Chem.* **2007**, *55*, 876–881. [CrossRef]
93. Ghidossi, R.; Poupot, C.; Thibon, C.; Pons, A.; Darriet, P.; Riquier, L.; De Revel, G.; Peuchot, M.M. The influence of packaging on wine conservation. *Food Control* **2012**, *23*, 302–311. [CrossRef]
94. Blanchard, L.; Tominaga, T.; Dubourdieu, D. Formation of furfurylthiol exhibiting a strong coffee aroma during oak barrel fermentation from furfural released by toasted staves. *J. Agric. Food Chem.* **2001**, *49*, 4833–4835. [CrossRef] [PubMed]
95. Silva Ferreira, A.C.; Hogg, T.; Guedes de Pinho, P. Identification of key odorants related to the typical aroma of oxidation-spoiled white wines. *J. Agric. Food Chem.* **2003**, *51*, 1377–1381. [CrossRef] [PubMed]
96. Singleton, V.; Salgues, M.; Zaya, J.; Trousdale, E. Caftaric acid disappearance and conversion to products of enzymic oxidation in grape must and wine. *Am. J. Enol. Vitic.* **1985**, *36*, 50–56.
97. Escudero, A.; Asensio, E.; Cacho, J.; Ferreira, V. Sensory and chemical changes of young white wines stored under oxygen. An assessment of the role played by aldehydes and some other important odorants. *Food Chem.* **2002**, *77*, 325–331. [CrossRef]
98. Oliveira, C.M.; Silva Ferreira, A.C.; Guedes de Pinho, P.; Hogg, T.A. Development of a potentiometric method to measure the resistance to oxidation of white wines and the antioxidant power of their constituents. *J. Agric. Food Chem.* **2002**, *50*, 2121–2124. [CrossRef]
99. Jocelyn, P.C. *Biochemistry of the SH Group*; Academic Press: London, UK, 1972; Volume 10.
100. Kotseridis, Y.; Ray, J.-L.; Augier, C.; Baumes, R. Quantitative determination of sulfur containing wine odorants at sub-ppb levels. 1. Synthesis of the deuterated analogues. *J. Agric. Food Chem.* **2000**, *48*, 5819–5823. [CrossRef]
101. Nikolantonaki, M.; Chichuc, I.; Teissedre, P.-L.; Darriet, P. Reactivity of volatile thiols with polyphenols in a wine-model medium: Impact of oxygen, iron, and sulfur dioxide. *Anal. Chim. Acta* **2010**, *660*, 102–109. [CrossRef]
102. Blanchard, L.; Darriet, P.; Dubourdieu, D. Reactivity of 3-mercaptohexanol in red wine: Impact of oxygen, phenolic fractions, and sulfur dioxide. *Am. J. Enol. Vitic.* **2004**, *55*, 115–120. [CrossRef]
103. Danilewicz, J.C. Review of reaction mechanisms of oxygen and proposed intermediate reduction products in wine: Central role of iron and copper. *Am. J. Enol. Vitic.* **2003**, *54*, 73–85. [CrossRef]
104. Singleton, V.L. Oxygen with phenols and related reactions in musts, wines, and model systems: Observations and practical implications. *Am. J. Enol. Vitic.* **1987**, *38*, 69–77.
105. Waterhouse, A.L.; Laurie, V.F. Oxidation of wine phenolics: A critical evaluation and hypotheses. *Am. J. Enol. Vitic.* **2006**, *57*, 306–313. [CrossRef]
106. Nikolantonaki, M.; Waterhouse, A.L. A method to quantify quinone reaction rates with wine relevant nucleophiles: A key to the understanding of oxidative loss of varietal thiols. *J. Agric. Food Chem.* **2012**, *60*, 8484–8491. [CrossRef]

107. Cheynier, V.F.; Trousdale, E.K.; Singleton, V.L.; Salgues, M.J.; Wylde, R. Characterization of 2-S-glutathionyl caftaric acid and its hydrolysis in relation to grape wines. *J. Agric. Food Chem.* **1986**, *34*, 217–221. [CrossRef]
108. Kreitman, G.Y.; Laurie, V.F.; Elias, R.J. Investigation of ethyl radical quenching by phenolics and thiols in model wine. *J. Agric. Food Chem.* **2013**, *61*, 685–692. [CrossRef]
109. Nikolantonaki, M.; Jourdes, M.; Shinoda, K.; Teissedre, P.-L.; Quideau, S.; Darriet, P. Identification of adducts between an odoriferous volatile thiol and oxidized grape phenolic compounds: Kinetic study of adduct formation under chemical and enzymatic oxidation conditions. *J. Agric. Food Chem.* **2012**, *60*, 2647–2656. [CrossRef]
110. Coetzee, C.; Du Toit, W. Sauvignon blanc wine: Contribution of ageing and oxygen on aromatic and non-aromatic compounds and sensory composition-A review. *S. Afr. J. Enol. Vitic.* **2015**, *36*, 347–365. [CrossRef]
111. Antonelli, A.; Arfelli, G.; Masino, F.; Sartini, E. Comparison of traditional and reductive winemaking: Influence on some fixed components and sensorial characteristics. *Eur. Food Res. Technol.* **2010**, *231*, 85–91. [CrossRef]
112. Pons, A.; Lavigne, V.; Darriet, P.; Dubourdieu, D. Glutathione preservation during winemaking with *Vitis vinifera* white varieties: Example of Sauvignon blanc grapes. *Am. J. Enol. Vitic.* **2015**, *66*, 187–194. [CrossRef]
113. I Cortiella, M.G.; Úbeda, C.; Covarrubias, J.I.; Peña-Neira, Á. Chemical, physical, and sensory attributes of Sauvignon blanc wine fermented in different kinds of vessels. *Innov. Food Sci. Emerg. Technol.* **2020**, *66*, 102521. [CrossRef]
114. Maggu, M.; Winz, R.; Kilmartin, P.A.; Trought, M.C.; Nicolau, L. Effect of skin contact and pressure on the composition of Sauvignon Blanc must. *J. Agric. Food Chem.* **2007**, *55*, 10281–10288. [CrossRef] [PubMed]
115. Patel, P.; Herbst-Johnstone, M.; Lee, S.A.; Gardner, R.C.; Weaver, R.; Nicolau, L.; Kilmartin, P.A. Influence of juice pressing conditions on polyphenols, antioxidants, and varietal aroma of Sauvignon blanc microferments. *J. Agric. Food Chem.* **2010**, *58*, 7280–7288. [CrossRef]
116. Sala, C.; Busto, O.; Guasch, J.; Zamora, F. Influence of vine training and sunlight exposure on the 3-alkyl-2-methoxypyrazines content in musts and wines from the *Vitis vinifera* variety Cabernet Sauvignon. *J. Agric. Food Chem.* **2004**, *52*, 3492–3497. [CrossRef] [PubMed]
117. Coetzee, C.; Lisjak, K.; Nicolau, L.; Kilmartin, P.; du Toit, W.J. Oxygen and sulfur dioxide additions to Sauvignon blanc must: Effect on must and wine composition. *Flavour Fragr. J.* **2013**, *28*, 155–167. [CrossRef]
118. Coetzee, C.; Van Wyngaard, E.; Suklje, K.; Silva Ferreira, A.C.; Du Toit, W.J. Chemical and sensory study on the evolution of aromatic and nonaromatic compounds during the progressive oxidative storage of a Sauvignon blanc wine. *J. Agric. Food Chem.* **2016**, *64*, 7979–7993. [CrossRef]
119. Nikolantonaki, M.; Daoud, S.; Noret, L.; Coelho, C.; Badet-Murat, M.-L.; Schmitt-Kopplin, P.; Gougeon, R.g.D. Impact of oak wood barrel tannin potential and toasting on white wine antioxidant stability. *J. Agric. Food Chem.* **2019**, *67*, 8402–8410. [CrossRef]
120. Pons, A.; Lavigne, V.; Thibon, C.; Redon, P.; Loisel, C.; Dubourdieu, D.; Darriet, P. Impact of closure OTR on the volatile compound composition and oxidation aroma intensity of Sauvignon Blanc wines during and after 10 years of bottle storage. *J. Agric. Food Chem.* **2021**, *69*, 9883–9894. [CrossRef]
121. Cantu, A.; Guernsey, J.; Anderson, M.; Blozis, S.; Bleibaum, R.; Cyrot, D.; Waterhouse, A.L. Wine Closure Performance of Three Common Closure Types: Chemical and Sensory Impact on a Sauvignon Blanc Wine. *Molecules* **2022**, *27*, 5881. [CrossRef]
122. Blake, A.; Kotseridis, Y.; Brindle, I.D.; Inglis, D.; Sears, M.; Pickering, G.J. Effect of closure and packaging type on 3-alkyl-2-methoxypyrazines and other impact odorants of Riesling and Cabernet Franc wines. *J. Agric. Food Chem.* **2009**, *57*, 4680–4690. [CrossRef]
123. Blake, A.; Kotseridis, Y.; Brindle, I.; Inglis, D.; Pickering, G. Effect of light and temperature on 3-alkyl-2-methoxypyrazine concentration and other impact odourants of Riesling and Cabernet Franc wine during bottle ageing. *Food Chem.* **2010**, *119*, 935–944. [CrossRef]
124. Alberts, P.; Kidd, M.; Stander, M.; Nieuwoudt, H.; Tredoux, A.; De Villiers, A. Quantitative Survey of 3-alkyl-2-methoxypyrazines and First Confirmation of 3-ethyl-2-methoxypyrazine in South African Sauvignon blanc Wines. *S. Afr. J. Enol. Vitic.* **2013**, *34*, 54–67. [CrossRef]
125. Herbst-Johnstone, M.; Nicolau, L.; Kilmartin, P.A. Stability of varietal thiols in commercial Sauvignon blanc wines. *Am. J. Enol. Vitic.* **2011**, *62*, 495–502. [CrossRef]
126. Ferreira, V.; Escudero, A.; Fernandez, P.; Cacho, J.F. Changes in the profile of volatile compounds in wines stored under oxygen and their relationship with the browning process. *Z. Lebensm. Unters.-Forsch. A Food Res. Technol.* **1997**, *205*, 392–396. [CrossRef]
127. Lambropoulos, I.; Roussis, I.G. Inhibition of the decrease of volatile esters and terpenes during storage of a white wine and a model wine medium by caffeic acid and gallic acid. *Food Res. Int.* **2007**, *40*, 176–181. [CrossRef]
128. Bordiga, M.; Rinaldi, M.; Locatelli, M.; Piana, G.; Travaglia, F.; Coïsson, J.D.; Arlorio, M. Characterization of Muscat wines aroma evolution using comprehensive gas chromatography followed by a post-analytic approach to 2D contour plots comparison. *Food Chem.* **2013**, *140*, 57–67. [CrossRef]
129. Marais, J.; Pool, H. Effect of Storage Time and Temperature on the Volatile Composition and Quality of Dry White Table Wines. Oenological and Viticultural Research Institute, Stellenbosch, South Africa. *Vitis* **1980**, *19*, 151–164.
130. Marais, J. The effect of pH on Esters and Quality of Colombar Wine during Maturation. Oenological and Viticultural Research Institute, Stellenbosch, South Africa. *Vitis* **1978**, *17*, 396–403.
131. Roussis, I.G.; Lambropoulos, I.; Papadopoulou, D. Inhibition of the decline of volatile esters and terpenols during oxidative storage of Muscat-white and Xinomavro-red wine by caffeic acid and N-acetyl-cysteine. *Food Chem.* **2005**, *93*, 485–492. [CrossRef]

132. Oliveira, J.M.; Faria, M.; Sá, F.; Barros, F.; Araújo, I.M. C6-alcohols as varietal markers for assessment of wine origin. *Anal. Chim. Acta* **2006**, *563*, 300–309. [CrossRef]
133. Patrianakou, M.; Roussis, I. Decrease of wine volatile aroma esters by oxidation. *S. Afr. J. Enol. Vitic.* **2013**, *34*, 241–245. [CrossRef]
134. Câmara, J.; Alves, M.A.; Marques, J.C. Changes in volatile composition of Madeira wines during their oxidative ageing. *Anal. Chim. Acta* **2006**, *563*, 188–197. [CrossRef]
135. Rankine, B.; Pocock, K. Influence of yeast strain on binding of sulphur dioxide in wines, and on its formation during fermentation. *J. Sci. Food Agric.* **1969**, *20*, 104–109. [CrossRef] [PubMed]
136. Fleet, G.H. *Wine Microbiology and Biotechnology*; CRC Press: Boca Raton, FL, USA, 1993.
137. Laurie, V.F.; Zúñiga, M.C.; Carrasco-Sánchez, V.; Santos, L.S.; Cañete, Á.; Olea-Azar, C.; Ugliano, M.; Agosin, E. Reactivity of 3-sulfanyl-1-hexanol and catechol-containing phenolics in vitro. *Food Chem.* **2012**, *131*, 1510–1516. [CrossRef]
138. Makhotkina, O.; Herbst-Johnstone, M.; Logan, G.; du Toit, W.; Kilmartin, P.A. Influence of sulfur dioxide additions at harvest on polyphenols, C6-compounds, and varietal thiols in Sauvignon blanc. *Am. J. Enol. Vitic.* **2013**, *64*, 203–213. [CrossRef]
139. Vally, H.; Misso, N.L. Adverse reactions to the sulphite additives. *Gastroenterol. Hepatol. Bed Bench* **2012**, *5*, 16–23.
140. Bisson, L. *Yeast and Biochemistry of Ethanol Formation in Principles and Practices of Winemaking*; Boulton, R.B., Singleton, V.L., Bisson, L.F., Kunkee, R.E., Eds.; Chapman & Hall: New York, NY, USA, 1996; pp. 448–473.
141. Bradshaw, M.P.; Barril, C.; Clark, A.C.; Prenzler, P.D.; Scollary, G.R. Ascorbic acid: A review of its chemistry and reactivity in relation to a wine environment. *Crit. Rev. Food Sci. Nutr.* **2011**, *51*, 479–498. [CrossRef]
142. McArdle, J.V.; Hoffmann, M.R. Kinetics and mechanism of the oxidation of aquated sulfur dioxide by hydrogen peroxide at low pH. *J. Phys. Chem.* **1983**, *87*, 5425–5429. [CrossRef]
143. Peng, Z.; Duncan, B.; Pocock, K.; Sefton, M. The effect of ascorbic acid on oxidative browning of white wines and model wines. *Aust. J. Grape Wine Res.* **1998**, *4*, 127–135. [CrossRef]
144. Skouroumounis, G.K.; Kwiatkowski, M.; Francis, I.; Oakey, H.; Capone, D.; Peng, Z.; Duncan, B.; Sefton, M.; Waters, E. The influence of ascorbic acid on the composition, colour and flavour properties of a Riesling and a wooded Chardonnay wine during five years' storage. *Aust. J. Grape Wine Res.* **2005**, *11*, 355–368. [CrossRef]
145. Morozova, K.; Schmidt, O.; Schwack, W. Effect of headspace volume, ascorbic acid and sulphur dioxide on oxidative status and sensory profile of Riesling wine. *Eur. Food Res. Technol.* **2015**, *240*, 205–221. [CrossRef]
146. Makhotkina, O.; Araujo, L.D.; Olejar, K.; Herbst-Johnstone, M.; Fedrizzi, B.; Kilmartin, P.A. Aroma impact of ascorbic acid and glutathione additions to Sauvignon blanc at harvest to supplement sulfur dioxide. *Am. J. Enol. Vitic.* **2014**, *65*, 388–393. [CrossRef]
147. Swart, E.; Marais, J.; Britz, T. Effect of ascorbic acid and yeast strain on Sauvignon blanc wine quality. *S. Afr. J. Enol. Vitic.* **2001**, *22*, 41–46. [CrossRef]
148. Barril, C.; Clark, A.C.; Scollary, G.R. Chemistry of ascorbic acid and sulfur dioxide as an antioxidant system relevant to white wine. *Anal. Chim. Acta* **2012**, *732*, 186–193. [CrossRef]
149. Danilewicz, J.C.; Seccombe, J.T.; Whelan, J. Mechanism of interaction of polyphenols, oxygen, and sulfur dioxide in model wine and wine. *Am. J. Enol. Vitic.* **2008**, *59*, 128–136. [CrossRef]
150. Barril, C.; Rutledge, D.N.; Scollary, G.R.; Clark, A.C. Ascorbic acid and white wine production: A review of beneficial versus detrimental impacts. *Aust. J. Grape Wine Res.* **2016**, *22*, 169–181. [CrossRef]
151. Bradshaw, M.P.; Cheynier, V.; Scollary, G.R.; Prenzler, P.D. Defining the ascorbic acid crossover from anti-oxidant to pro-oxidant in a model wine matrix containing (+)-catechin. *J. Agric. Food Chem.* **2003**, *51*, 4126–4132. [CrossRef]
152. Shenton, D.; Perrone, G.; Quinn, K.A.; Dawes, I.W.; Grant, C.M. Regulation of protein S-thiolation by glutaredoxin 5 in the yeast *Saccharomyces cerevisiae*. *J. Biol. Chem.* **2002**, *277*, 16853–16859. [CrossRef]
153. Fahey, R.C. Novel thiols of prokaryotes. *Annu. Rev. Microbiol.* **2001**, *55*, 333–356. [CrossRef]
154. Li, Y.; Wei, G.; Chen, J. Glutathione: A review on biotechnological production. *Appl. Microbiol. Biotechnol.* **2004**, *66*, 233–242. [CrossRef] [PubMed]
155. Noctor, G.; Veljovic-Jovanovic, S.; Foyer, C.H. Peroxide processing in photosynthesis: Antioxidant coupling and redox signalling. *Philos. Trans. R. Soc. Lond. Ser. B Biol. Sci.* **2000**, *355*, 1465–1475. [CrossRef] [PubMed]
156. Adams, D.O.; Liyanage, C. Glutathione increases in grape berries at the onset of ripening. *Am. J. Enol. Vitic.* **1993**, *44*, 333–338.
157. Lacroux, F.; Trégoat, O.; van Leeuwen, C.; Pons, A.; Tominaga, T.; Lavigne-Cruege, V.; Dubourdieu, D. Effect of foliar nitrogen and sulphur application on aromatic expression of *Vitis vinifera* L. cv. Sauvignon blanc. *OENO One* **2008**, *42*, 125–132. [CrossRef]
158. Lyu, X.; Araujo, L.D.; Quek, S.-Y.; Kilmartin, P.A. Effects of antioxidant and elemental sulfur additions at crushing on aroma profiles of Pinot Gris, Chardonnay and Sauvignon Blanc wines. *Food Chem.* **2021**, *346*, 128914. [CrossRef]
159. OIV. Resolutions OIV-OENO 445-2015. In Proceedings of the 13th OIV General Assembly, Mainz, Germany, 15 July 2015. Available online: https://www.oiv.int/public/medias/1686/oiv-oeno-445-2015-en.pdf (accessed on 23 September 2022).
160. Antoce, O. Oenology. In *Chemistry and Sensory Analysis*; Universitaria Printing House: Craiova, Romania, 2007.
161. Singleton, V.L.; Cilliers, J.J. *Phenolic Browning: A Perspective from Grape and Wine Research*; ACS Publications: Washington, DC, USA, 1995.
162. Sonni, F.; Clark, A.C.; Prenzler, P.D.; Riponi, C.; Scollary, G.R. Antioxidant action of glutathione and the ascorbic acid/glutathione pair in a model white wine. *J. Agric. Food Chem.* **2011**, *59*, 3940–3949. [CrossRef]
163. Clark, A.C.; Deed, R.C. The chemical reaction of glutathione and trans-2-hexenal in grape juice media to form wine aroma precursors: The impact of pH, temperature, and sulfur dioxide. *J. Agric. Food Chem.* **2018**, *66*, 1214–1221. [CrossRef]

164. Papadopoulou, D.; Roussis, I.G. Inhibition of the decrease of volatile esters and terpenes during storage of a white wine and a model wine medium by glutathione and N-acetylcysteine. *Int. J. Food Sci. Technol.* **2008**, *43*, 1053–1057. [CrossRef]
165. Webber, V.; Dutra, S.V.; Spinelli, F.R.; Carnieli, G.J.; Cardozo, A.; Vanderlinde, R. Effect of glutathione during bottle storage of sparkling wine. *Food Chem.* **2017**, *216*, 254–259. [CrossRef]
166. Lavigne-Cruège, V.; Dubourdieu, D. Role of glutathione on development of aroma defects in dry white wines. In Proceedings of the 13th International Enology Symposium, Montpellier, France, 9–12 June 2002; pp. 331–347.
167. Ugliano, M.; Kwiatkowski, M.; Vidal, S.; Capone, D.; Siebert, T.; Dieval, J.-B.; Aagaard, O.; Waters, E.J. Evolution of 3-mercaptohexanol, hydrogen sulfide, and methyl mercaptan during bottle storage of Sauvignon blanc wines. Effect of glutathione, copper, oxygen exposure, and closure-derived oxygen. *J. Agric. Food Chem.* **2011**, *59*, 2564–2572. [CrossRef]
168. Andújar-Ortiz, I.; Chaya, C.; Martín-Álvarez, P.J.; Moreno-Arribas, M.; Pozo-Bayón, M. Impact of using new commercial glutathione enriched inactive dry yeast oenological preparations on the aroma and sensory properties of wines. *Int. J. Food Prop.* **2014**, *17*, 987–1001. [CrossRef]
169. OIV. Resolutions OIV-OENO 532-2017. In Proceedings of the 15th OIV General Assembly, Sofia, Bulgaria, 6 June 2017. Available online: https://www.oiv.int/public/medias/5365/oiv-oeno-532-2017-en.pdf (accessed on 23 September 2022).
170. Kritzinger, E.C.; Bauer, F.F.; Du Toit, W.J. Role of glutathione in winemaking: A review. *J. Agric. Food Chem.* **2013**, *61*, 269–277. [CrossRef] [PubMed]
171. Gabrielli, M.; Aleixandre-Tudo, J.; Kilmartin, P.; Sieczkowski, N.; Du Toit, W. Additions of glutathione or specific glutathione-rich dry inactivated yeast preparation (DYP) to sauvignon blanc must: Effect on wine chemical and sensory composition. *S. Afr. J. Enol. Vitic.* **2017**, *38*, 18–28. [CrossRef]
172. Rodríguez-Bencomo, J.J.; Andújar-Ortiz, I.; Moreno-Arribas, M.V.; Simo, C.; Gonzalez, J.; Chana, A.; Davalos, J.; Pozo-Bayón, M.A.N. Impact of glutathione-enriched inactive dry yeast preparations on the stability of terpenes during model wine aging. *J. Agric. Food Chem.* **2014**, *62*, 1373–1383. [CrossRef]
173. Bahut, F.; Liu, Y.; Romanet, R.; Coelho, C.; Sieczkowski, N.; Alexandre, H.; Schmitt-Kopplin, P.; Nikolantonaki, M.; Gougeon, R.D. Metabolic diversity conveyed by the process leading to glutathione accumulation in inactivated dry yeast: A synthetic media study. *Food Res. Int.* **2019**, *123*, 762–770. [CrossRef]
174. Bahut, F.; Romanet, R.; Sieczkowski, N.; Schmitt-Kopplin, P.; Nikolantonaki, M.; Gougeon, R.D. Antioxidant activity from inactivated yeast: Expanding knowledge beyond the glutathione-related oxidative stability of wine. *Food Chem.* **2020**, *325*, 126941. [CrossRef]
175. Andujar-Ortiz, I.; Pozo-Bayón, M.Á.; Moreno-Arribas, M.; Martín-Álvarez, P.J.; Rodríguez-Bencomo, J.J. Reversed-phase high-performance liquid chromatography–fluorescence detection for the analysis of glutathione and its precursor γ-glutamyl cysteine in wines and model wines supplemented with oenological inactive dry yeast preparations. *Food Anal. Methods* **2012**, *5*, 154–161. [CrossRef]
176. Barrio Galán, R.d.; Úbeda Aguilera, C.; Cortiella, G.I.; Sieczkowski, N.; Peña Neira, Á. Different application dosages of a specific inactivated dry yeast (SIDY): Effect on the polysaccharides, phenolic and volatile contents and color of Sauvignon blanc wines. *OENO One Int. J. Vine Wine Sci.* **2018**, *52*, 333–346. [CrossRef]

Review

The Impact of Vineyard Mechanization on Grape and Wine Phenolics, Aroma Compounds, and Sensory Properties

Qun Sun *, Craig Ebersole, Deborah Parker Wong and Karley Curtis

Department of Viticulture and Enology, California State University, 2360 E. Barstow Avenue, Fresno, CA 93740, USA; ebersole@mail.fresnostate.edu (C.E.); deborahparkerwong@mail.fresnostate.edu (D.P.W.); knicole@mail.fresnostate.edu (K.C.)
* Correspondence: qsun@csufresno.edu

Abstract: Grapes are one of the most valuable fruit crops in the United States and can be processed into a variety of products. The grape and wine industry contributes to and impacts the U.S. agricultural economy. However, rising labor costs and global competition pose challenges for the grape and wine industry. Vineyard mechanization is a promising strategy to increase efficiency and address the labor shortage and cost issues. Recent studies have focused on the impact of vineyard mechanization on general grape and wine quality. Wine phenolics, aroma compounds, and sensory characteristics are the key indicators of wine quality and consumer preference. This article aims to review the impact of vineyard mechanization, specifically mechanical harvesting, mechanical leaf removal, mechanical shoot thinning, cluster thinning, and mechanical pruning on grape and wine phenolics, and aroma compounds and sensory profile. Studies have shown that vineyard mechanization significantly affects phenolic and aroma compounds, especially grape-derived aroma compounds such as volatile thiols, terpenes, C13-norpentadiene, and methoxypyrazine. Mechanically processed grapes can produce wines of the same or better quality than wines made from hand-operated grapes. Vineyard mechanization could be a promising strategy for grape growers to reduce operating costs and maintain or improve grape and wine quality. Future research directions in the area of vineyard mechanization were discussed. It provides a comprehensive view and information on the topic to both grape growers and winemakers in the application of vineyard mechanization.

Keywords: vineyard mechanization; phenolics; aroma compounds; sensory properties

Citation: Sun, Q.; Ebersole, C.; Wong, D.P.; Curtis, K. The Impact of Vineyard Mechanization on Grape and Wine Phenolics, Aroma Compounds, and Sensory Properties. *Fermentation* **2022**, *8*, 318. https://doi.org/10.3390/fermentation8070318

Academic Editors: Bin Tian and Jicheng Zhan

Received: 8 June 2022
Accepted: 5 July 2022
Published: 6 July 2022

Publisher's Note: MDPI stays neutral with regard to jurisdictional claims in published maps and institutional affiliations.

Copyright: © 2022 by the authors. Licensee MDPI, Basel, Switzerland. This article is an open access article distributed under the terms and conditions of the Creative Commons Attribution (CC BY) license (https://creativecommons.org/licenses/by/4.0/).

1. Introduction

Grapes are one of the most valuable crops in the American agricultural sector. The grape and wine industry is the main economic engine of the U.S. agricultural economy. According to a market analysis report, the U.S. wine market was worth $63.69 billion in 2021 and is predicted to grow at a compounded annual growth rate (CAGR) of 6.8% through 2030 [1]. California leads the country in grape production, and 2021 grape acreage of this region totaled 881,000 acres [2]. The crop is mainly used for wine, raisins, table grapes, concentrated grape juice, and distillate. It is well known that grape production is highly labor-intensive. Labor costs account for about 60% of the annual cost of wine grape production [3]. Grape harvesting, pruning, canopy management, grapevines tying, and suckering are the most labor-intensive practices. However, over the last several years, labor-related issues are becoming increasingly challenging due to labor shortage [4,5]. Labor safety issues such as respiratory problems, high temperatures, and seasonal rainfall, and the recent effects of COVID-19 negatively affect the production cost and stable labor supply for future seasons. Short-term strategies to deal with labor shortages include raising wages, reducing workload or using immigrant labor. In the long run, adopting vineyard mechanization to manage labor demand and production costs and improving working conditions for farm workers are sustainable solutions. In addition, the increased competition from global markets with inexpensive labor is also one of the big drivers behind the faster adoption of mechanization in the vineyard to ensure competitiveness.

Vineyard mechanization has been around for about 70 years. In the early 1950s, the University of California, Davis, began to study the mechanization of vineyards [6]. In 1957, a few researchers at Cornell University's New York State Agricultural Experiment Station in Geneva, NY developed the Geneva double curtain (GDC) trellis, which facilitated the mechanical harvesting of Concord grapes [7]. Today, mechanization is used in vineyards for different purposes, from simple tillage to harvesting, pruning, defoliation, shoot positioning, and shoot and cluster thinning throughout the vine growing season. Well-trained operators can perform tasks more efficiently [8].

An increasing number of grape growers are interested in adopting mechanization in their vineyards, especially in the regions focusing on grape yield, such as the Central Valley of California. It is critical to understand better the relationship between mechanization and berry/wine quality. Vineyard mechanization is a rapidly evolving area of research and development. A number of studies have focused on the effects of vineyard mechanization of harvesting, pruning, leaf removal, and shoot thinning on berry and wine phenolics, aroma compounds, and sensory changes. Wine phenolics (polyphenols, phenols) are a diverse group of compounds that share a phenol ring in their primary chemical structure. Important groups of wine phenolics include anthocyanins, flavan-3-ols, and their derivatives, which contribute to wine color, wine texture, mouthfeel, and aging potential. The majority of phenols in wine are grape-derived, and a small proportion may be contributed by oak barrels or oak adjuncts. Aroma compounds are important secondary metabolites. Although they only accounts for a small part of grape and wine, these compounds play an essential role in shaping wine's identity. They can be grape-derived, such as terpenes, C13-norisoprene, methoxypyrazine, or microbial-derived esters. From the consumers' perspective, taste is generally considered the most important factor influencing purchase preference. This article reviews the impact of vineyard mechanization on grape and wine phenolics, aroma compounds and sensory profiles. It will help researchers, grape growers, and winemakers better understand the benefit and potential application of vineyard mechanization.

2. Mechanical Harvesting

The wine industry has been using hand-picking as a differentiator when promoting the quality and style of wine. Despite the lack of evidence to support a relationship between this practice and wine quality, it has made consumers believe that mechanically harvested grapes produce inferior wines [9].

Mechanical harvesting of wine grapes was developed in the 1950s [6,7] and widely studied in the United States in the 1970s [10]. The practice was widely adopted in Italy as early as 1980 [11]. The rudimentary technology of early mechanical harvesters led to increased mechanical stress on the fruit and could not sufficiently eliminate materials other than grapes (MOG) during the harvest. The extent of mechanical stress grapes undergo during harvesting depends on several factors, including the degree of ripeness and the overall condition and health of the berries at harvest. Technologies in mechanical harvesting have evolved over the past four decades, particularly with the addition of sorting and destemming techniques to remove MOG. Typically, the purpose of hand-sorting at the winery is to eliminate undesirable fruit picked by mechanical harvesters, but it is tedious and requires tremendous resources as inspection of individual berries is necessary to sort the already destemmed fruit. Optical sorters, however, are well-suited to sort destemmed grapes rapidly. Sorting is based on parameters, such as berry size, color, and shape, and whether there is foreign material. Nowadays, optical berry sorters have become more common in commercial wine production. There are a few studies investigating their effects on the chemical and sensory properties of wine. In a study on synergistic effects of harvest method (hand and machine) and optical sorting or none, differences in the cv. Pinot Noir grape and wine composition that were attributable to the harvest method were reduced or eliminated with the use of optical sorting. The machine-harvested grapes had higher levels of β-damascenone, linalool, β-myrcene, and -terpinene, potentially caused by glycosidic hydrolysis triggered by berry damage during harvest or from induced synthesis as a wounding response [12].

There is growing evidence that wines made from mechanically harvested grapes are of comparable quality to wines made from hand-picked grapes. One study used cv. Chardonnay wines were made with machine- and hand-harvested grapes and found that there was no significant difference between the two for both young wines and wines aged 18 months made from grapes harvested by the two methods [13]. Another study evaluated red and white grape varieties (cv. Petit Syrah, French Colombard, and Chenin Blanc) and found no difference in wines made from grapes harvested with either method [14]. In some studies, wines made from machine-picked grapes were even superior to wines made from hand-picked grapes [15]. Kaltbach et al. (2022) investigated the effects of manual and mechanical harvesting on the composition of Merlot musts and wines produced in the Campanha Gaúcha region of Brazil through comprehensive testing of physicochemical parameters. Merlot wines made from mechanically harvested grapes had slightly higher levels of pH, ethanol, and magnesium. Mechanical harvesting significantly reduced caffeic and coumaric acids in wines. All other parameters did not show significant differences. Manual and mechanical harvesting of grapes can produce identical wines [16].

The fruity aroma of Sauvignon blanc wine is dependent upon concentrations of varietal thiols, various esters, higher alcohols, methoxypyrazines, and terpenes. The intense passion fruit aroma has been associated with the varietal thiols, which includes 3-Mercaptohexanol (3MH), 3-mercaptohexyl acetate (3MHA), 4-mercapto-4-methylpentan-2-one (4MMP), 4-mercapto-4-methylpentan-2-ol (4MMPOH), 3-mercapto-3-methylbutan-1-ol (3MMB) [17]. 3MH and 3MHA are key compounds of the aroma typicity of young Sauvignon blanc wines and generated during the fermentation process by yeast from precursors initially present in the grape. Cys-3MH and glutathionyl-3MH (Glut-3MH) are two longer-lived precursors, which were the first 3MH precursors to be identified [16]. Bonnaffoux et al. (2018) identified and quantified two short-lived precursors such as 3S-cysteinylglycinylhexan-1-ol (CysGly-3SH) and 3S-glutamylcysteinylhexan-1-ol (-GluCys-3SH) in juice and wine samples [18].

Jouanneau (2011) has found that the 3-MH concentrations in commercial Marlborough cv. Sauvignon blanc made from machine-harvest grapes are 5–10 times higher than the values obtained in experimental Sauvignon blanc wines made from hand-picked fruits [19]. Allen et al. (2011) and Herbst-Johnstone et al. (2013) showed that the harvesting of cv. Sauvignon blanc by machine harvester increased certain varietal aroma compounds [20,21]. The ability to increase the aroma precursor compounds during the harvesting process provides an opportunity to alter maceration techniques to enhance the passion fruit aroma in the finished wines. Olejar et al. (2015) sourced hand and mechanically harvested grapes of cv. Sauvignon blanc from Marlborough, New Zealand [22]. There was an increase in varietal thiol content for wines made from juices that had been machine harvested compared to the hand-picked samples. Herbst-Johnstone et al. (2013) sourced cv. Sauvignon blanc grapes from five locations in Marlborough and at five stages during the harvesting process. Grapes were picked by hand and mechanical harvester. Commercial free run was pressed at one bar pressure. The study found that varietal thiols were present at relatively higher concentrations in wines made from machine-harvested fruit compared with wines made from hand-picked grapes [21]. A possible explanation is that berry damage that occurs during mechanical harvesting can release glycosidases, which may lead to higher concentrations of varietal thiols in mechanically harvested crops.

3. Mechanical Leaf Removal

Canopy management practices are carried out annually in vineyards to establish and maintain healthy canopies. Leaf removal is an integral part of canopy management practices in wine grapes (*Vitis vinifera*) to remove leaves around grape clusters, which commonly occurs in cool climate vineyards to accelerate air movement in the cluster zone, prevent disease, and promotes the biosynthesis of several important grape constituents, improve berry maturation, color and flavor [23]. Defoliation is typically performed between fruit set and veraison, but early leaf removal is a good cultural practice for yield management in grapevines and usually occurs around bloom. Due to the link between yield and the availability of carbohydrates at pre-bloom, early leaf removal suppresses yield [24], the effects of which are known to improve wine quality parameters. Leaf removal can be performed manually or mechanically. Vineyard mechanization can decrease labor costs and reduce exposure to the impact of labor shortages, but its influence on wine organoleptic properties is still not fully understood. With regard to leaf removal, research has focused on changes to red wine phenolics [25–28], white wine aroma compounds [29–31], hydroxycinnamates [32,33], and wine sensory quality [34–36], but few studies connect the effects of mechanization to wine sensory characteristics.

The timing of intervention plays a key role in the complex mechanisms involved in grape and wine aroma since it will affect the extent of berry sunlight exposure. Diago et al. (2010) investigated the effect of defoliation timing (before flowering and fruit set) on cv. Tempranillo grapes and wines. The effect of manual and mechanical defoliation on the aroma and sensory attributes of wines was compared in a vineyard in La Rioja, Spain, over two consecutive seasons, 2007 and 2008. The study confirms the effectiveness of both manual and mechanical early leaf removal. Pre-blooming leaf removal resulted in smaller berries and improved fruit health by reducing the occurrence of botrytis occurrence. Significant differences in wine aroma attributes were observed between the control and mechanical defoliation treatments [37].

Guidoni et al. (2008) studied the effects of mechanical and manual defoliation on the cv. Barbera in northwest Italy. In moderate climates, such as in northwestern Italy, leaf removal is employed to increase spray penetration and allow more light into the canopy to achieve physiological ripeness. There were significant differences in climatic conditions during the three years the experiment was conducted, which explains most of the variation in results. In terms of phenolic compounds, total phenols, proanthocyanidins, and anthocyanins were measured, but no significant differences between mechanical and hand defoliated grapes were found. Proanthocyanidins contribute to astringency in wine, a major factor for mouthfeel. From a sensory perspective, significant differences were not found between the treatment and control groups. This study suggests that the application of mechanical defoliation may vary. Climate, vineyard weather conditions, and vintage are key factors to consider when using mechanical defoliation. Under unfavorable ripening conditions, mechanical defoliation is more effective in improving grape health and quality. Excessive direct sunlight exposure needs to be prevented since it will cause berries to sunburn. This study suggests that random leaf removal by mechanical leaf removers may have greater advantages than manual leaf removal [38].

Kemp et al. (2011) in New Zealand investigated the effects of mechanical leaf removal on flavan-3-ol composition and concentration in cv. Pinot noir wine. The timing of the intervention was also explored. Flavan-3-ol is one of the phenolics that play a role in forming various tannins and influence the perception of bitterness in the wine. The study indicated that the naturally shaded fruit produced wines with lower monomer flavan-3-ol and tannin concentrations compared with the wines produced from defoliated grapevines. Leaf removal that occurred on the fruiting zone 7 days and 30 days after flowering, led to the highest tannin concentration. The second-year had higher tannin and monomer flavan-3-ols concentrations, which were attributed to either higher alcohol levels or more aggressive leaf removal in the second year or a combination of both factors [39].

Bubola et al. (2019) compared mechanical and manual defoliation with an untreated control group of Croatian cv. Istrian Malvasia grape varieties to understand their effect on wine quality. The study applied treatments at the pea-size stage of berry development. The experiment was carried out in a season characterized by abundant rainfall. Istrian Malvasia is a white *Vitis vinifera* variety that resembles cv. Sauvignon blanc. The study showed that mechanical leaf removal treatment significantly increased the concentration of some aroma compounds such as varietal thiol 3-sulfanylhexan-1-ol, monoterpenes, β-damascenone, and esters. Phenolics such as hydroxycinnamates were lower in wine made from mechanical leaf removal than wine from hand leaf removal. Sensory evaluation indicated wine made from mechanical leaf removal berries contained more enhanced floral, fruity and tropical attributes, which can be attributed to the improvement of aroma compounds [32].

4. Mechanical Shoot and Cluster Thinning

Shoot and cluster thinning are other canopy management practices that reduce crop load to the desired level for optimizing grape and wine quality [40]. Shoot thinning is a highly labor-intensive operation. Dean (2016) reports that the cost of thinning is about $650/ha. Mechanical shoot thinning is a cost-effective practice that reduces vineyard labor by 25 times compared to manual operations [41]. However, it is difficult to adjust the position and orientation of thinning end-effector to the shape of the cordons. The efficiency of mechanical thinning varies from 10% to 85%, depending on the forward speed of the tractor and the rotational speed of the thinning end effector [42]. Majeed et al. (2000) conducted research showing the performance and efficiency of mechanical shoot thinning machines can be greatly improved if the position of thinning end-effector is automatically controlled [43].

Yield management through the mechanical shoot and cluster thinning could induce berry chemical compositional changes, thus affecting wine aroma, taste, and mouthfeel. Diago et al. (2010) conducted an experiment applying mechanical cluster thinning at different intensities and different timings on cv. Grenache and cv. Tempranillo vines in Spain's Rioja region. The results indicated that mechanical cluster thinning was effective in yield reduction and resulted in more ripened fruit and wines with higher alcohol and pH values, more intense color, and increased phenolic content. The extent of the sensory implications seems dependent on several factors, such as the variety and timing of thinning applications. Regardless of the intensity of refinement, Grenache wines had less sensory impact than Tempranillo wines, which had improved aromas, acidity, and astringency from the vines subjected to mechanical cluster thinning [44].

Mechanical thinning could be combined with other vineyard management practices to improve grape and wine quality. Brillante et al. (2018) investigated the interactive effects of mechanical shoot thinning and irrigation management on the accumulation of phenolic and aroma compounds of cv. Syrah grapes and wines under the warm and semi-arid growing conditions of the San Joaquin Valley of California. The results showed that the interaction of two treatments could improve berry skin and wine phenolics and reduce herbaceous aroma, methoxypyrazines, and C_6-alcohol/aldehydes, in wine while achieving high yield if there is no precipitation from fruit set to veraison [45].

Petrie et al. (2006) studied the effects of mechanical crop removal after fruit set (when berries were pea-sized) on cv. Cabernet Sauvignon, at two vineyard sites in Australia (Riverland of South Australia and Sunraysia District of Victoria). The study indicated a significant increase in color density from the Riverland site over several vintages, while phenolic and anthocyanin concentrations in the wines showed similar trends to the grapes. Mechanical thinning successfully reduced crop level to the target yield. Mechanical thinning distorts the distribution for Brix and berry weight but can increase color and is best suited to small pruning situations. The results are consistent with other studies where either hand thinning or mechanical thinning had been used to reduce the crop level [46].

5. Mechanical Pruning

Pruning is a vineyard management practice that removes lignified growth from previous years of vines to promote new growth and fruiting while also controlling yield and growing time. Without pruning, grapevines would stimulate excessive vegetation and the proliferation of small clusters of fruit not suitable for winemaking. Leaving too many buds during pruning can lead to an imbalance in the vine's nutritional and reproductive growth, resulting in poor herbal aromas in the wine [47]. Mechanized vineyard management can reduce labor costs by 45–90%, depending on the region and trellis system [48,49]. Therefore, understanding the impact of mechanical pruning on wine phenolics, aroma, and sensory profile is essential to developing an accurate cost-benefit analysis. The effects of fully mechanized pruning systems on vine physiology, crop load, grapes, and wine quality have been studied since the early 1970s [50]. The studies showed inconsistent results about the application of mechanical pruning on grape and wine quality.

Reynolds (1988) evaluated the response of cv. Riesling vines in the Okanagan Valley, British Columbia, to different training systems and mechanical pruning. The study aimed to determine whether mechanical pruning was a viable option for maintaining profitability in an environment of increasing labor costs without negatively impacting wine quality. Vine training systems, mid-wire bilateral cordon (MBC), Hudson River Umbrella (HBU), and Lenz Moser (LM) were used for mechanical and hand pruning. The ethanol and pH levels of wines made from mechanically trimmed grapes were overall lower than those of hand-pruned wines. The sensory panelists found that mechanical pruning tended to reduce wine quality compared to manual pruning [51]. Santos et al. (2015) conducted an experiment in Brazil using cv. Cabernet Franc, IAC-Máximo, and Merlot to determine the effects of the initial adoption of mechanical pruning on the grape composition, quality, and sensory characteristics of wines. The results showed that applying mechanical pruning in traditional vineyards caused small fluctuations in grape quality [52]. Holt et al. (2008) conducted a study on the relationships between wine phenolic composition and wine sensory properties for cv. Cabernet Sauvignon between 2003–2005 at one vineyard site in the Clare Valley region of South Australia. Machine pruning was compared to hand and spur pruned vines. The results of the study showed that machine-pruned vines resulted in wines that were comparable to hand-pruned wines with respect to wine composition. Machine pruned berry and wine were significantly higher in total anthocyanins. However, color density in the finished wine was the same as in the hand-pruned treatments, suggesting that berry anthocyanins from machine-pruned vines were less extractable than their hand-harvested counterparts [53]. These results are supported by other research, which has shown no direct correlation between berry tannins and their corresponding wine tannins in various red *Vitis vinifera* L. species [54]. Higher concentrations of anthocyanins, tannins, and phenolics in berries from machine-pruned vines did not always correspond to higher concentrations in wine.

Kronfli III (2018) evaluated the sensory effects of mechanical pruning on cv. Syrah via a general descriptive analysis. The mechanically pruned Syrah wines were shown to be not significantly different from the wines made from the hand-pruned vines when harvested at equal levels of ripeness [55]. Between 2013–2015, a study was conducted in Fresno, California, to evaluate the impact of converting a non-mechanized cv. Merlot vineyard to mechanized pruning [49]. The control groups were cane pruned (CP) and bilateral cordon spur pruned (HP) vines and the treatment was a mechanically box-pruned single high-wire sprawling system (SHMP). The effects of mechanization on phenolics profile were determined by evaluating the concentration of gallic acid, total flavan-3-ol, total flavonols, total anthocyanins, and individual anthocyanins. The results of the study showed no adverse effects of mechanical pruning on berry flavonoid concentration throughout the experiment.

The interaction between mechanical pruning and plant nutrition was comprehensively investigated by Botelho et al. (2020, 2021, 2022) [56–58]. The study assessed the interaction between mechanical pruning and soil organic amendments in two trial fields in Portugal. The objective was to evaluate whether a deleterious effect on cv. Syrah berry composition could occur between mechanical pruning and organic soil amendments because their interaction has been known to negatively affect the balance between grapevine vegetative and reproductive growth and berry composition. In particular, the effects of nitrogen (N), added by sources such as organic soil amendments, have been shown to decrease polyphenol content [59]. Treatment groups included mechanical pruning and organic soil amendments of biochar, municipal solid waste compost, cattle manure, and sewage sludge.

These treatments were compared to control groups of spur pruning by hand and no organic soil amendments. The study found that the pruning method had little influence on color intensity and the color hue of wines. However, total anthocyanins were significantly lower in blocks pruned mechanically. Total phenols and tannin power showed little difference between pruning methods. The results of the sensory analysis showed that mechanical pruning reduced aroma balance, body, and astringency. Based on this study, mechanical pruning had significant effects on wine quality when the yield was above a certain level (6 kg/vine in the cooler climate and 8 kg/vine in the warmer climate. Thus, with this pruning system, the choice of the organic amendment and its amount may be critical. The combination of these two treatments is more suitable for the warm-climate region. In warm regions, harvest can be delayed with no threat of *B. cinerea* infections.

6. Conclusions and Future Directions

Wine-grape producers in many viticultural regions face many challenges. The major challenges are the increasing shortage and cost of skilled manual labor for vineyard management tasks that could suppress economic margins. This has driven more research investigating the potential of vineyard mechanization to manage vineyards cost-effectively.

Viticultural practices are known to influence secondary metabolites, phenolics, and aroma compounds. The berries' chemical composition is very sensitive to the microclimate. Vineyards are subjected to a large number of management practices, including row orientation and spacing, density, pruning, clipping, tilling, soil surface management, or manipulation of the canopy structure, among others, which leads to changes in the microclimate of the cluster. Most of them can be converted to mechanization. Some studies have demonstrated that phenolics and aroma compounds are affected by different types of vineyard mechanization (Tables 1 and 2). Some other studies have shown that mechanically treated grapes could produce wine of equal or better quality than wine made from grapes that had been operated manually. Vineyard mechanization could be a promising strategy for grape growers, especially in the regions focusing on yield. However, more research needs to be conducted to understand additional cultivars in different wine regions since this factor could be a variable.

Table 1. The impact of vineyard mechanization on grape and wine phenolics.

Phenolics	Functions	Vineyard Mechanization	Varieties	References	Impact
Hydroxycinnamate	Major phenolic compounds in white wine and are important in white wine color [60]	Mechanical harvesting	Merlot	Kaltbach et al. (2022) [16]	Reduced caffeic and coumaric acids in wines
		Mechanical leaf removal	Istrian Malvasia	Bubola et al. (2019) [32]	Reduced hydroxycinnamatein wines
Anthocyanin	Responsible for red wine color [60]	Mechanical harvesting with optical sorting	Pinot noir	Hendrickson et al. (2016) [12]	Increased total anthocyanins in berries
		Mechanical leaf removal	Barbera	Guidoni et al. (2008) [38]	Comparable to hand leaf removal
		Mechanical crop thinning	Cabernet Sauvignon	Petrie et al. (2006) [46]	Increased total anthocyanins in berries
		Mechanical pruning	Merlot	Kurtural et al. (2019) [49]	Comparable to hand pruning
			Cabernet Sauvignon	Holt et al. (2008) [53]	Increased anthocyanins in both berries and wines

Table 1. *Cont.*

Phenolics	Functions	Vineyard Mechanization	Varieties	References	Impact
		Mechanical pruning and soil amendment	Syrah	Botelho et al. (2020) [58]	Combination of two practices reduced anthocyanins
Flavan-3-ol monomers	Responsible for bitterness in wine and may also have some associated astringency [60]	Mechanical leaf removal	Pinot noir	Kemp et al. (2011) [39]	Increased flavan-3-ols in wines
		Mechanical pruning	Merlot	Kurtural et al. (2019) [49]	Comparable to hand pruning
Proanthocyanidins (condensed tannin)	Impart astringency to red wines [60]	Mechanical leaf removal	Barbera	Guidoni et al. (2008) [38]	Comparable to hand leaf removal
			Pinot noir	Kemp et al. (2011) [39]	Increased tannin in wines
		Mechanical pruning and soil amendment	Syrah	Botelho et al. (2020) [58]	Combination of two practices reduced tannin in wines

Table 2. The impact of vineyard mechanization on grape and wine aroma compounds.

Aroma Compounds		Odor Descriptor	Precursors	Vineyard Mechanization	Varieties	References	Impact
Varietal thiols	3-Mercaptohexanol (3MH or 3SH); 3-mercaptohexyl acetate (3MHA or 3SHA)	Passion fruit, Grapefruit	S-3-(hexan-1-ol)-L-cysteine (Cys-3MH); S-3-(hexan-1-ol)-glutathione (Glut-3MH)	Mechanical harvesting	Sauvignon blanc	Jouanneau (2011) [19] Allen et al. (2011) [20] Herbst-Johnstone et al. (2013) [21] Olejar et al. (2015) [22]	Increased 3MH and 3MHA in both berries and wines
				Mechanical leaf removal	Sauvignon blanc	Bubola et al. (2019) [32]	Increased 3MH in wines
Methoxypyrazine	IBMP	Bell pepper	Leucine	Mechanical shoot thinning and deficit irrigation	Syrah	Brillante et al. (2018) [45]	Reduced IBMP in wines
C13-norisoprenoid	β-Damascenone	Cooked apple, quince, floral	Glycosylated aroma compound	Mechanical leaf removal	Sauvignon blanc	Bubola et al. (2019) [32]	Increased β-Damascenone in wines
				Mechanical harvesting with optical sorting	Pinot noir	Hendrickson et al. (2016) [12]	Increased β-Damascenone in wines
Alcohols	C_6 Alcohol	Grass, green	Linoleic acid and linolenic acid	Mechanical shoot thinning and deficit irrigation	Syrah	Brillante et al. (2018) [45]	Reduced C_6 alcohol in wines
Monoterpenes	Citronellol Nerol Geraniol α-terpinene	Rose, citrus Floral	Glycosylated aroma compound	Mechanical leaf removal	Sauvignon blanc	Bubola et al. (2019) [32]	Increased monoterpenes in wines
				Mechanical harvesting with optical sorting	Pinot noir	Hendrickson et al. (2016) [12]	Increased α-terpinene in berries

Based on the previous studies, vineyard mechanization has a significant impact on aroma compounds, especially grape-derived aroma compounds such as volatile thiols, terpene, C13-norisoprenoid, and methoxypyrazine (Table 2). Grape-derived aroma compounds are one area of particular importance for wine quality due to their distinctiveness and ability to impart 'varietal aromas' to wines. Despite the noticeability of these odorants in the finished wines, these compounds are exclusively produced during fermentation and are not found in distinctive levels in the grape berry or pressed juice. It is the amounts of precursors (glycosides) that provide these aroma potentials. Although their proportion is not directly related to the organoleptic properties of the grapes, the concentration of precursor compounds in grapes could be an indicator of their aromatic potential. Moreover, these aroma precursors change throughout berry development, which is highly dependent on enzyme activity. To date, there are many studies on the impact of manual vineyard operation on grape and wine quality, but few have focused on the effect of vineyard mechanization on aroma precursors. Only Sauvignon blanc wine has been studied. Several volatile thiol precursors have been identified (3S-glutathionylhexan-1-ol (glut-3MH) and 3S-cysteinylhexan-1-ol (cys-3MH). Some studies showed enhanced enzymatic activity that follows mechanical harvesting might be very important in the formation of 3MH precursors in many grape lots and the subsequent release of the free thiols during fermentation. Additional research is needed to identify new aroma precursors, the effects of mechanical practice on enzymatic activity and aroma precursors' synthesis. A scientific understanding of precursors' change during berry development from fruit set to harvest is also critical.

The key to vineyard mechanization is to optimize aroma precursors and phenolics (hydroxycinnamate, anthocyanin, flavan-3-ol, condensed tannin, etc.) using different mechanical treatments during the vine growing and berry development period, which will give growers good guidance on the timing of operation application. More integrated work needs to be explored, such as mechanical harvesting with advanced sorting techniques, mechanical defoliation, mechanical crop thinning, mechanical pruning with vineyard irrigation, soil, or cover crop management.

Author Contributions: Conceptualization, Q.S.; methodology, Q.S.; formal analysis, Q.S.; investigation, Q.S., C.E., D.P.W. and K.C.; data curation, Q.S.; writing—original draft preparation, Q.S., C.E., D.P.W. and K.C.; writing—review and editing, Q.S., C.E., D.P.W. and K.C.; visualization, Q.S.; supervision, Q.S.; project administration, Q.S. All authors have read and agreed to the published version of the manuscript.

Funding: This research received no external funding.

Institutional Review Board Statement: Not applicable.

Informed Consent Statement: Not applicable.

Data Availability Statement: Choose to exclude this statement.

Conflicts of Interest: The authors declare no conflict of interest.

References

1. California Grape Acreage Report, 2021 Summary. Available online: https://www.nass.usda.gov/Statistics_by_State/California/Publications/Specialty_and_Other_Releases/Grapes/Acreage/2022/grpacSUMMARY2021Crop.pdf (accessed on 15 May 2022).
2. U.S. Wine Market Size, Share & Trends Analysis Report By Product (Table Wine, Dessert Wine, Sparkling Wine), By Distribution Channel (On-trade, Off-trade), And Segment Forecasts, 2022–2030. Available online: https://www.grandviewresearch.com/industry-analysis/us-wine-market/request/rs1 (accessed on 30 June 2022).
3. Developing a Viable Vineyard Workforce for the Future. Available online: https://grapesandwine.cals.cornell.edu/newsletters/appellation-cornell/2019-newsletters/issue-36-february-2019/viewpoints/ (accessed on 15 May 2022).
4. Brady, M.P.; Gallardo, R.K.; Badruddozza, S.; Jiang, X. Regional equilibrium wage rate for hired farm workers in the tree fruit industry. *Western Econ. Forum* **2016**, *15*, 20–31.
5. Brat, I. On U.S. Farms, Fewer Hands for the Harvest. Available online: https://www.newamericaneconomy.org/news/u-s-farms-fewer-hands-harvest/ (accessed on 12 May 2022).
6. Winkler, A.J.; Lamouria, L.H.; Abernathy, G.H. Mechanical grape harvest problems and progress. *Am. J. Enol. Vitic.* **1957**, *8*, 182–187.
7. Shaulis, N.J.; Shepardson, E.S.; Moyer, J.C. Grape harvesting research at Cornell NY State. *Hort. Soc.* **1960**, *105*, 250–254.
8. Kurtural, S.K.; Fidelibus, M.W. Mechanization of Pruning, Canopy Management, and Harvest in Winegrape Vineyards. *Catal. Discov. Pract.* **2021**, *5*, 29–44. [CrossRef]
9. Dominici, A.; Boncinelli, F.; Gerini, F.; Marone, E. Consumer Preference for Wine from Hand-Harvested Grapes. *Br. Food J.* **2019**, *122*, 2551–2567. [CrossRef]
10. Johnson, S.S. Mechanical harvesting of wine grapes. USDA, Economic Research Service. *Agric. Econ. Rep.* **1977**, *385*, 1–22.
11. Intrieri, C.; Filippetti, I. Innovations and outlook in grapevine training systems and mechanization in north-Central Italy. In Proceedings of the ASEV 50th Anniversary Annual Meeting, Seattle, WA, USA, 19–23 June 2000.

12. Hendrickson, D.A.; Lerno, L.A.; Hjelmeland, A.K.; Ebeler, S.E.; HHeymann, H.; Hopfer, H.; Block, K.L.; Brenneman, C.A.; Oberholster, A. Impact of Mechanical Harvesting and Optical Berry Sorting on Grape and Wine Composition. *Am. J. Enol. Vitic.* **2016**, *67*, 385–397. [CrossRef]
13. Clary, C.D.; Steinhauer, R.E.; Frisinger, J.E.; Peffer, T.E. Evaluation of machine- vs. hand-harvested Chardonnay. *Am. J. Enol. Vitic.* **1990**, *41*, 176–181.
14. Noble, A.C.; Ough, C.S.; Kasimatis, A.N. Effect of leaf content and mechanical harvest on wine "quality". *Am. J. Enol. Vitic.* **1975**, *26*, 158–163.
15. Kilmartin, P.A.; Oberholster, A. Grape Harvesting and Effects on Wine Composition. In *Managing Wine Quality*, 2nd ed.; Reynold, A., Ed.; Woodhead Publishing: Sawston, UK, 2022; Volume 1, pp. 705–726.
16. Kaltbach, S.B.A.; Kaltbach, P.; Santos, C.G.; Cunha, W.; Giacomini, M.; Domingues, F.; Malgarim, M.; Herter, F.G.; Costa, V.B.; Couto, J.A. Influence of manual and mechanical grape harvest on Merlot wine composition. *J. Food Compos. Anal.* **2022**, *110*, 1045–1048. [CrossRef]
17. Jeffery, D.W. Spotlight on Varietal Thiols and Precursors in Grapes and Wines. *Aust. J. Chem.* **2017**, *69*, 1323–1330. [CrossRef]
18. Bonnaffoux, H.; Delpech, S.; Rémond, E.; Schneider, R.; Roland, A.; Cavelier, F. Revisiting the evaluation strategy of varietal thiol biogenesis. *Food Chem.* **2018**, *268*, 126–133. [CrossRef]
19. Jouanneau, S. Survey of Aroma Compounds in Marlborough Sauvignon Blanc Wines—Regionality and Small Scale Winemaking. Ph.D. Thesis, The University of Auckland, Auckland, New Zealand, 2011.
20. Allen, T.; Herbst-Johnstone, M.; Girault, M.; Butler, P.; Logan, G.; Jouanneau, S. Influence of grape-harvesting steps on varietal thiol aromas in Sauvignon blanc wines. *J. Agric. Food Chem.* **2011**, *59*, 10641–10650. [CrossRef]
21. Herbst-Johnstone, M.; Araujo, L.D.; Allen, T.A.; Logan, G.; Nicolau, L.; Kilmartin, P.A. Effects of Mechanical Harvesting on 'Sauvignon Blanc' Aroma. *Acta Hortic.* **2013**, *978*, 179–186. [CrossRef]
22. Olejar, K.J.; Fedrizzi, B.; Kilmartin, P.A. Influence of Harvesting Technique and Maceration Process on Aroma and Phenolic Attributes of Sauvignon Blanc Wine. *Food Chem.* **2015**, *183*, 181–189. [CrossRef]
23. Downey, M.O.; Dokoozlian, N.K.; Krstic, M.P. Cultural Practice and Environmental Impacts on the Flavonoid Composition of Grapes and Wine: A Review of Recent Research. *Am. J. Enol. Vitic.* **2006**, *57*, 257–268.
24. Caspari, H.V.; Lang, A. Carbohydrate supply limits fruit-set in commercial Sauvignon blanc grapevines. In Proceedings of the 4th International Symposium on Cool Climate Enology and Viticulture, Rochester, NY, USA, 16–20 July 1996.
25. Feng, H.; Qian, M.C.; Skinkis, P.A. Pinot Noir Wine Volatile and Anthocyanin Composition under Different Levels of Vine Fruit Zone Leaf Removal. *Food Chem.* **2017**, *214*, 736–744. [CrossRef]
26. Bubola, M.; Sivilotti, P.; Janjanin, D.; Poni, S. Early Leaf Removal Has a Larger Effect than Cluster Thinning on Grape Phenolic Composition in Cv. Teran. *Am. J. Enol. Vitic.* **2017**, *68*, 234–242. [CrossRef]
27. Osrečak, M.; Karoglan, M.; Kozina, B. Influence of Leaf Removal and Reflective Mulch on Phenolic Composition and Antioxidant Activity of Merlot, Teran and Plavac Mali Wines (*Vitis Vinifera* L.). *Sci. Hortic.* **2016**, *209*, 261–269. [CrossRef]
28. Palliotti, A.; Gardi, T.; Berrios, J.G.; Civardi, S.; Poni, S. Early Source Limitation as a Tool for Yield Control and Wine Quality Improvement in a High-Yielding Red *Vitis Vinifera* L. Cultivar. *Sci. Hortic.* **2012**, *145*, 10–16. [CrossRef]
29. Friedel, M.; Frotscher, J.; Nitsch, M.; Hofmann, M.; Bogs, J.; Stoll, M.; Dietrich, H. Light Promotes Expression of Monoterpene and Flavonol Metabolic Genes and Enhances Flavour of Winegrape Berries (*Vitis Vinifera* L. Cv. Riesling). *Aust. J. Grape Wine Res.* **2016**, *22*, 409–421. [CrossRef]
30. Gregan, S.M.; Wargent, J.J.; Liu, L.; Shinkle, J.; Hofmann, R.; Winefield, C.; Trought, M.; Jordan, B. Effects of Solar Ultraviolet Radiation and Canopy Manipulation on the Biochemical Composition of Sauvignon Blanc Grapes. *Aust. J. Grape Wine Res.* **2012**, *18*, 227–238. [CrossRef]
31. Martin, D.; Albright, A.; McLachlan, A.; Fedrizzi, B.; Grose, C.; Stuart, L. Grape Cluster Microclimate Influences the Aroma Composition of Sauvignon Blanc Wine. *Food Chem.* **2016**, *210*, 640–647. [CrossRef] [PubMed]
32. Bubola, M.; Vanzo, A.; Bavčar, D.; Lukić, I.; Lisjak, K.; Grozić, K.; Sivilotti, P.; Radeka, S. Enhancement of Istrian Malvasia Wine Aroma and Hydroxycinnamate Composition by Hand and Mechanical Leaf Removal. *J. Sci. Food Agric.* **2019**, *99*, 904–914. [CrossRef]
33. Lemut, M.S.; Sivilotti, P.; Franceschi, P.; Wehrens, R.; Vrhovsek, U. Use of Metabolic Profiling to Study Grape Skin Polyphenol Behavior as a Result of Canopy Microclimate Manipulation in a 'Pinot Noir' Vineyard. *J. Agric. Food Chem.* **2013**, *61*, 8976–8986. [CrossRef]
34. Coniberti, A.; Ferrari, V.; Dellacassa, E.; Boido, E.; Carrau, F.; Gepp, V.; Disegna, E. Kaolin over Sun-Exposed Fruit Affects Berry Temperature, Must Composition and Wine Sensory Attributes of Sauvignon Blanc. *Eur. J. Agron.* **2013**, *50*, 75–81. [CrossRef]
35. Šuklje, K.; Antalick, G.; Coetzee, Z.; Schmidtke, L.M.; Baša Česnik, H.; Brandt, J.; Toit, W.J.; Lisjak, K.; Deloire, A. Effect of Leaf Removal and Ultraviolet Radiation on the Composition and Sensory Perception of *Vitis Vinifera* L. Cv. Sauvignon Blanc Wine. *Aust. J. Grape Wine Res.* **2014**, *20*, 223–233. [CrossRef]
36. Šuklje, K.; Antalick, G.; Buica, A.; Langlois, J.; Coetzee, Z.A.; Gouot, J.; Schmidtke, L.M.; Deloire, A. Clonal Differences and Impact of Defoliation on Sauvignon Blanc (*Vitis Vinifera* L.) Wines: A Chemical and Sensory Investigation. *J. Sci. Food Agric.* **2016**, *96*, 915–926. [CrossRef]
37. Diago, M.P.; Vilanova, M.; Blanco, J.A.; Tardaguila, J. Effects of Timing of Manual and Mechanical Early Defoliation on the Aroma of *Vitis vinifera* L. Tempranillo Wine. *Am. J. Enol. Vitic.* **2010**, *61*, 382–391.

38. Guidoni, S.; Oggero, G.; Cravero, S.; Rabino, M.; Cravero, M.C.; Balsari, P. Manual and Mechanical Leaf Removal in the Bunch Zone (*Vitis Vinifera* L., Cv Barbera): Effects on Berry Composition, Health, Yield and Wine Quality, in a Warm Temperate Area. *OENO One* **2008**, *42*, 49–58. [CrossRef]
39. Kemp, B.S.; Harrison, R.; Creasy, G.L. Effect of Mechanical Leaf Removal and Its Timing on Flavan-3-ol Composition and Concentrations in *Vitis Vinifera* L. Cv. Pinot Noir Wine. *Aust. J. Grape Wine Res.* **2011**, *17*, 270–279. [CrossRef]
40. Smart, R.E. Shoot spacing and canopy light microclimate. *Am. J. Enol. Vitic.* **1988**, *39*, 325–333.
41. Dean, S. Mechanical Shoot & Leaf Removal Practices. 2016. Available online: https://www.awri.com.au/wp-content/uploads/2016/06/4-Mechanical-Shoot-Leaf-Removal_Sean_Dean.pdf (accessed on 15 May 2022).
42. Dokoozlian, N. The evolution of mechanized vineyard production systems in California. *Acta Hortic.* **2013**, *978*, 265–278. [CrossRef]
43. Majeed, Y.; Karkee, M.; Zhang, Q.; Fu, L.; Whiting, M.D. Determining grapevine cordon shape for automated green shoot thinning using semantic segmentation-based deep learning networks. *Comput. Electron. Agric.* **2000**, *171*, 105308. [CrossRef]
44. Diago, M.P.; Vilanova, M.; Blanco, J.A.; Tardaguila, J. Effects of Mechanical Thinning on Fruit and Wine Composition and Sensory Attributes of Grenache and Tempranillo Varieties (*Vitis Vinifera* L.): Mechanical Thinning on Fruit and Wine Composition. *Aust. J. Grape Wine Res.* **2010**, *16*, 314–326. [CrossRef]
45. Brillante, L.; Martínez-Lüscher, J.; Kurtural, S.K. Applied Water and Mechanical Canopy Management Affect Berry and Wine Phenolic and Aroma Composition of Grapevine (*Vitis Vinifera* L., Cv. Syrah) in Central California. *Sci. Hortic.* **2018**, *227*, 261–271. [CrossRef]
46. Petrie, P.R.; Clingeleffer, P.R. Crop Thinning (Hand versus Mechanical), Grape Maturity and Anthocyanin Concentration: Outcomes from Irrigated Cabernet Sauvignon (*Vitis Vinifera* L.) in a Warm Climate. *Aust. J. Grape Wine Res.* **2006**, *12*, 21–29. [CrossRef]
47. Terry, D.B.; Kurtural, S.K. Achieving Vine Balance of Syrah with Mechanical Canopy Management and Regulated Deficit Irrigation. *Am. J. Enol. Vitic.* **2011**, *62*, 426–437. [CrossRef]
48. Morris, J.R. Development and Commercialization of a Complete Vineyard Mechanization System. *Horttechnology* **2007**, *17*, 411–420. [CrossRef]
49. Kurtural, S.K.; Beebe, A.E.; Martínez-Lüscher, J.; Zhuang, S.J.; Lund, K.T.; McGourty, G.; Bettiga, L.J. Conversion to Mechanical Pruning in Vineyards Maintains Fruit Composition While Reducing Labor Costs in "Merlot" Grape Production. *Horttechnology* **2019**, *29*, 128–139. [CrossRef]
50. Poni, S.; Tombesi, S.; Palliotti, A.; Ughinia, V.; Gatti, M. Mechanical winter pruning of grapevine: Physiological bases and applications. *Sci. Hortic.* **2016**, *204*, 88–98. [CrossRef]
51. Reynolds, A.G. Response of Okanagan Riesling Vines to Training System and Simulated Mechanical Pruning. *Am. J. Enol. Vitic.* **1988**, *39*, 205–212.
52. Santos, A.O.; Pereira, S.E.; Moreira, C.A. Physical-Chemical Quality of Grapes and wine sensory profile for different varieties of vine subjected to mechanical pruning. *Rev. Bras. Frutic.* **2015**, *37*, 432–441. [CrossRef]
53. Holt, H.E.; Francis, I.L.; Field, J.; Herderich, M.J.; Iland, P.G. Relationships between Wine Phenolic Composition and Wine Sensory Properties for Cabernet Sauvignon (*Vitis Vinifera* L.). *Aust. J. Grape Wine Res.* **2008**, *14*, 162–176.
54. Harbertson, J.F.; Kennedy, J.A.; Adams, D.O. Tannin in Skins and Seeds of Cabernet Sauvignon, Syrah, and Pinot Noir Berries during Ripening. *Am. J. Enol. Vitic.* **2002**, *53*, 54–59.
55. Kronfli, E.J., III. Sensory Effect of Regulated Deficit Irrigation and Mechanical Pruning on Washington State Wines. Master's Thesis, University of California, Davis, CA, USA, 2018.
56. Botelho, M.; Cruz, A.; Ricardo-da-Silva, J.; de Castro, R.; Ribeiro, H. Mechanical Pruning and Soil Fertilization with Distinct Organic Amendments in Vineyards of Syrah: Effects on Vegetative and Reproductive Growth. *Agronomy* **2020**, *10*, 1090. [CrossRef]
57. Botelho, M.; Ribeiro, H.; Cruz, A.; Duarte, D.F.; Faria, D.L.; de Castro, R.; Ricardo-Da-Silva, J. Mechanical Pruning and Soil Organic Amendments in Vineyards of Syrah: Effects on Grape Composition. *OENO One* **2021**, *55*, 267–277. [CrossRef]
58. Botelho, M.; Ribeiro, H.; Cruz, A.; Duarte, D.F.; Faria, D.L.; Khairnar, K.S.; Pardal, R.; Susini, M.; Correia, C.; Catarino, S.; et al. Mechanical Pruning and Soil Organic Amending in Two Terroirs. Effects on Wine Chemical Composition and Sensory Profile. *Am. J. Enol. Vitic.* **2022**, *73*, 26–38. [CrossRef]
59. Delgado, R.; Martin, P.; del Alamo, M.; Gonzalez, M.R. Changes in the Phenolic Composition of Grape Berries during Ripening in Relation to Vineyard Nitrogen and Potassium Fertilisation Rates. *J. Sci. Food Agric.* **2004**, *84*, 623–630. [CrossRef]
60. Kennedy, J.A.; Saucier, C.; Glories, Y. Grape and Wine Phenolics: History and Perspective. *Am. J. Enol. Vitic.* **2006**, *57*, 239–248.

MDPI
St. Alban-Anlage 66
4052 Basel
Switzerland
www.mdpi.com

Fermentation Editorial Office
E-mail: fermentation@mdpi.com
www.mdpi.com/journal/fermentation

Disclaimer/Publisher's Note: The statements, opinions and data contained in all publications are solely those of the individual author(s) and contributor(s) and not of MDPI and/or the editor(s). MDPI and/or the editor(s) disclaim responsibility for any injury to people or property resulting from any ideas, methods, instructions or products referred to in the content.

www.ingramcontent.com/pod-product-compliance
Lightning Source LLC
LaVergne TN
LVHW070949170726
843515LV00005B/954

* 9 7 8 3 0 3 6 5 9 5 5 9 7 *